P9-AOG-776

Applied Logistic Regression

WILEY SERIES IN PROBABILITY AND STATISTICS
TEXTS AND REFERENCES SECTION

Established by WALTER A. SHEWHART and SAMUEL S. WILKS

A complete list of the titles in this series appears at the end of this volume.

Applied Logistic Regression

Second Edition

DAVID W. HOSMER
University of Massachusetts
Amherst, Massachusetts

STANLEY LEMESHOW
The Ohio State University
Columbus, Ohio

A Wiley-Interscience Publication
JOHN WILEY & SONS, INC.
New York • Chichester • Weinheim • Brisbane • Singapore • Toronto

To Trina, Wylie, Tri,
D. W. H.

To Elaine, Jenny, Adina, Steven,
S. L.

This text is printed on acid-free paper. ∞

To order books or for customer service please, call 1(800)-CALL-WILEY (225-5945).

Library of Congress Cataloging in Publication Data:

Hosmer, David W.
Applied logistic regression / David W. Hosmer, Jr., Stanley Lemeshow.—2nd ed.
p. cm.
Includes bibliographical references and index.
ISBN 0-471-35632-8 (cloth : alk. paper)
1. Regression analysis. I. Lemeshow, Stanley. II. Title.
QA278.2.H67 2000
519.5'36—dc21 00-036843

Printed in the United States of America

10 9 8 7 6 5 4

CONTENTS

Preface To The Second Edition

The use of logistic regression modeling has exploded during the past decade. From its original acceptance in epidemiologic research, the method is now commonly employed in many fields including but not nearly limited to biomedical research, business and finance, criminology, ecology, engineering, health policy, linguistics and wildlife biology. At the same time there has been an equal amount of effort in research on all statistical aspects of the logistic regression model. A literature search that we did in preparing this Second Edition turned up more than 1000 citations that have appeared in the 10 years since the First Edition of this book was published.

When we worked on the First Edition of this book we were very limited by software that could carry out the kinds of analyses we felt were important. Specifically, beyond estimation of regression coefficients, we were interested in such issues as measures of model performance, diagnostic statistics, conditional analyses and multinomial response data. Software is now readily available in numerous easy to use and widely available statistical packages to address these and other extremely important modeling issues. Enhancements to these capabilities are being added to each new version. As is well-recognized in the statistical community, the inherent danger of this easy-to-use software is that investigators are using a very powerful tool about which they may have only limited understanding. It is our hope that this Second Edition will bridge the gap between the outstanding theoretical developments and the need to apply these methods to diverse fields of inquiry.

Numerous texts have sections containing a limited discussion of logistic regression modeling but there are still very few comprehensive texts on this subject. Among the textbooks written at a level similar to this one are: Cox and Snell (1989), Collett (1991) and Kleinbaum (1994).

As was the case in our First Edition, the primary objective of the Second Edition is to provide a focused introduction to the logistic regression model and its use in methods for modeling the relationship between a categorical outcome variable and a set of covariates. Topics that have been added to this edition include: numerous new techniques for model building including determination of scale of continuous covariates; a greatly expanded discussion of assessing model performance; a discussion of logistic regression modeling using complex sample survey data; a comprehensive treatment of the use of logistic regression modeling in matched studies; completely new sections dealing with logistic regression models for multinomial, ordinal and correlated response data, exact methods for logistic regression and sample size issues. An underlying theme throughout this entire book is the focus on providing guidelines for effective model building and interpreting the resulting fitted model within the context of the applied problem.

The materials in the book have evolved considerably over the past ten years as a result of our teaching and consulting experiences. We have used this book to teach parts of graduate level survey courses, quarter- or semester-long courses, and focused short courses to working professionals. We assume that students have a solid foundation in linear regression methodology and contingency table analysis.

The approach we take is to develop the model from a regression analysis point of view. This is accomplished by approaching logistic regression in a manner analogous to what would be considered good statistical practice for linear regression. This differs from the approach used by other authors who have begun their discussion from a contingency table point of view. While the contingency table approach may facilitate the interpretation of the results, we believe that it obscures the regression aspects of the analysis. Thus, discussion of the interpretation of the model is deferred until the regression approach to the analysis is firmly established.

To a large extent there are no major differences in the capabilities of the various software packages. When a particular approach is available in a limited number of packages, it will be noted in this text. In general, analyses in this book have been performed in STATA [Stata Corp. (1999)]. This easy to use package combines excellent graphics and analysis routines, is fast, is compatible across Macintosh, Windows and UNIX platforms and interacts well with Microsoft Word. Other major statistical packages employed at various points during the preparation of this text include SAS [SAS Institute Inc. (1999)], SPSS [SPSS Inc. (1998)], and BMDP [BMDP Statistical Software (1992)]. In general, the results produced were the same regardless of which package was used. Reported numeric results have been rounded from figures obtained from computer output and thus may differ slightly from those that would be obtained in a replication of our analyses or from calculations based on the reported results. When features or capabilities of the programs differ in an important way, we note them by the names given rather than by their bibliographic citation.

This text was prepared in camera ready format using Microsoft Word 98 on a Power Macintosh platform. Mathematical equations and symbols were built using Math Type 3.6a [Math Type: Mathematical Equation Editor (1998)].

Early on in the preparation of the Second Edition we made a decision that data sets used in the text would be made available to readers via the World Wide Web. The ftp site at John Wiley & Sons, Inc. for the data in this text is

ftp://ftp.wiley.com/public/sci_tech_med/logistic.

In addition, the data may also be found, by permission of John Wiley & Sons Inc., in the archive of statistical data sets maintained at the University of Massachusetts at Internet address

http://www-unix.oit.umass.edu/~statdata

in the logistic regression section. Another advantage to having a text web site is that it provides a convenient medium for conveying to readers text changes after publication. In particular, as errata become known to us they will be added to an errata section of the text's web site at John Wiley & Sons, Inc. Another use that we envision for the web is the addition, over time, of additional data sets to the statistical data set archive at the University of Massachusetts.

We are deeply appreciative of the efforts of our students and colleagues who carefully read and contributed to the clarity of this manuscript. In particular we are indebted to Elizabeth Donohoe-Cook, Sunny Kim and Soon-Kwi Kim for their careful and meticulous reading of the drafts of this manuscript. Special thanks also goes to Rita Popat for helping us make the transition between the software we used for the first and second editions. We appreciate Alan Agresti's comments on the section dealing with the analysis of correlated data. Cyrus Mehta was particularly helpful in sharing key papers and for providing us with the LogXact 4 (2000) program used for computations in Section 8.4. Others contributed significantly to the First Edition and their original suggestions made this Second Edition stronger. These include Gordon Fitzgerald, Sander Greenland, Bob Harris and Ed Stanek.

There have been many other contributors to this book. Data sets were made available by our colleagues, Donn Young, Jane McCusker, Carol Bigelow, Anne Stoddard, Harris Pastides, and Jane Zapka, as well as by Doctors Daniel Teres and Laurence E. Lundy at Baystate Medical Center in Springfield, Massachusetts. Cliff Johnson at NCHS was helpful in providing us with a data set from the NHANES III that we used extensively in Section 6.4 as well as for sharing insights with us into analytic strategies used by that agency. We are very grateful to Professor Petter Laake, Section of Medical Statistics at the University of Oslo and Professeur Roger Salamon of the University of Bordeaux, II who provided us with support to work on this manuscript during visits to their universities. Comments by many of our students and colleagues at the University of Massachusetts, The Ohio State University, the New England Epidemiology Summer Program, the Erasmus Summer Program, the Summer Program in Applied Statistical Methods at The Ohio State University, the University of Oslo and the University of Bordeaux as well as at innumerable short courses that we have had the privilege to be invited to teach over the past ten years, were extremely useful.

Finally, we would like to thank Steve Quigley and the production staff at John Wiley & Sons for their help in bringing this project to completion.

DAVID W. HOSMER, JR.
STANLEY LEMESHOW

Amherst, Massachusetts
Columbus Ohio
June, 2000

CHAPTER 1

Introduction to the Logistic Regression Model

1.1 INTRODUCTION

Regression methods have become an integral component of any data analysis concerned with describing the relationship between a response variable and one or more explanatory variables. It is often the case that the outcome variable is discrete, taking on two or more possible values. Over the last decade the logistic regression model has become, in many fields, the standard method of analysis in this situation.

Before beginning a study of logistic regression it is important to understand that the goal of an analysis using this method is the same as that of any model-building technique used in statistics: to find the best fitting and most parsimonious, yet biologically reasonable model to describe the relationship between an outcome (dependent or response) variable and a set of independent (predictor or explanatory) variables. These independent variables are often called *covariates*. The most common example of modeling, and one assumed to be familiar to the readers of this text, is the usual linear regression model where the outcome variable is assumed to be continuous.

What distinguishes a logistic regression model from the linear regression model is that the outcome variable in logistic regression is *binary* or *dichotomous*. This difference between logistic and linear regression is reflected both in the choice of a parametric model and in the assumptions. Once this difference is accounted for, the methods employed in an analysis using logistic regression follow the same general principles used in linear regression. Thus, the techniques used in linear regression analysis will motivate our approach to logistic regression. We illustrate both the similarities and differences between logistic regression and linear regression with an example.

Example

Table 1.1 lists age in years (AGE), and presence or absence of evidence of significant coronary heart disease (CHD) for 100 subjects selected to participate in a study. The table also contains an identifier variable (ID) and an age group variable (AGRP). The outcome variable is CHD, which is coded with a value of zero to indicate CHD is absent, or 1 to indicate that it is present in the individual.

It is of interest to explore the relationship between age and the presence or absence of CHD in this study population. Had our outcome variable been continuous rather than binary, we probably would begin by forming a scatterplot of the outcome versus the independent variable. We would use this scatterplot to provide an impression of the nature and strength of any relationship between the outcome and the independent variable. A scatterplot of the data in Table 1.1 is given in Figure 1.1.

In this scatterplot all points fall on one of two parallel lines representing the absence of CHD ($y=0$) and the presence of CHD ($y=1$). There is some tendency for the individuals with no evidence of CHD to be younger than those with evidence of CHD. While this plot does depict the dichotomous nature of the outcome variable quite clearly, it does not provide a clear picture of the nature of the relationship between CHD and age.

A problem with Figure 1.1 is that the variability in CHD at all ages is large. This makes it difficult to describe the functional relationship between age and CHD. One common method of removing some variation while still maintaining the structure of the relationship between the outcome and the independent variable is to create intervals for the independent variable and compute the mean of the outcome variable within each group. In Table 1.2 this strategy is carried out by using the age group variable, AGRP, which categorizes the age data of Table 1.1. Table 1.2 contains, for each age group, the frequency of occurrence of each outcome as well as the mean (or proportion with CHD present) for each group.

By examining this table, a clearer picture of the relationship begins to emerge. It appears that as age increases, the proportion of individuals with evidence of CHD increases. Figure 1.2 presents a plot of the proportion of individuals with CHD versus the midpoint of each age interval. While this provides considerable insight into the relationship between CHD and age in this study, a functional form for this relationship needs to be described. The plot in this figure is similar to what one

Table 1.1 Age and Coronary Heart Disease (CHD) Status of 100 Subjects

ID	AGE	AGRP	CHD	ID	AGE	AGRP	CHD
1	20	1	0	51	44	4	1
2	23	1	0	52	44	4	1
3	24	1	0	53	45	5	0
4	25	1	0	54	45	5	1
5	25	1	1	55	46	5	0
6	26	1	0	56	46	5	1
7	26	1	0	57	47	5	0
8	28	1	0	58	47	5	0
9	28	1	0	59	47	5	1
10	29	1	0	60	48	5	0
11	30	2	0	61	48	5	1
12	30	2	0	62	48	5	1
13	30	2	0	63	49	5	0
14	30	2	0	64	49	5	0
15	30	2	0	65	49	5	1
16	30	2	1	66	50	6	0
17	32	2	0	67	50	6	1
18	32	2	0	68	51	6	0
19	33	2	0	69	52	6	0
20	33	2	0	70	52	6	1
21	34	2	0	71	53	6	1
22	34	2	0	72	53	6	1
23	34	2	1	73	54	6	1
24	34	2	0	74	55	7	0
25	34	2	0	75	55	7	1
26	35	3	0	76	55	7	1
27	35	3	0	77	56	7	1
28	36	3	0	78	56	7	1
29	36	3	1	79	56	7	1
30	36	3	0	80	57	7	0
31	37	3	0	81	57	7	0
32	37	3	1	82	57	7	1
33	37	3	0	83	57	7	1
34	38	3	0	84	57	7	1
35	38	3	0	85	57	7	1
36	39	3	0	86	58	7	0
37	39	3	1	87	58	7	1
38	40	4	0	88	58	7	1
39	40	4	1	89	59	7	1
40	41	4	0	90	59	7	1
41	41	4	0	91	60	8	0
42	42	4	0	92	60	8	1
43	42	4	0	93	61	8	1
44	42	4	0	94	62	8	1
45	42	4	1	95	62	8	1
46	43	4	0	96	63	8	1
47	43	4	0	97	64	8	0
48	43	4	1	98	64	8	1
49	44	4	0	99	65	8	1
50	44	4	0	100	69	8	1

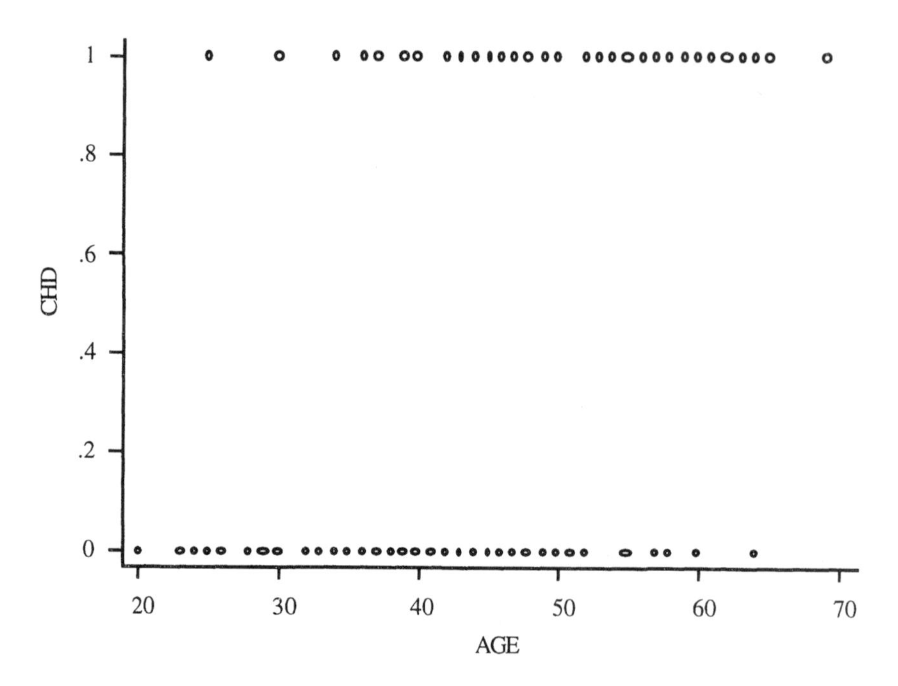

Figure 1.1 Scatterplot of CHD by AGE for 100 subjects.

might obtain if this same process of grouping and averaging were performed in a linear regression. We will note two important differences.

The first difference concerns the nature of the relationship between the outcome and independent variables. In any regression problem the key quantity is the mean value of the outcome variable, given the value of the independent variable. This quantity is called the *conditional mean* and will be expressed as "$E(Y \mid x)$" where Y denotes the outcome

Table 1.2 Frequency Table of Age Group by CHD

		CHD		
Age Group	n	Absent	Present	Mean (Proportion)
20 – 29	10	9	1	0.10
30 – 34	15	13	2	0.13
35 – 39	12	9	3	0.25
40 – 44	15	10	5	0.33
45 – 49	13	7	6	0.46
50 – 54	8	3	5	0.63
55 – 59	17	4	13	0.76
60 – 69	10	2	8	0.80
Total	100	57	43	0.43

conditional distribution of the outcome variable given x will be normal with mean $E(Y \mid x)$, and a variance that is constant. This is not the case with a dichotomous outcome variable. In this situation we may express the value of the outcome variable given x as $y = \pi(x) + \varepsilon$. Here the quantity ε may assume one of two possible values. If $y = 1$ then $\varepsilon = 1 - \pi(x)$ with probability $\pi(x)$, and if $y = 0$ then $\varepsilon = -\pi(x)$ with probability $1 - \pi(x)$. Thus, ε has a distribution with mean zero and variance equal to $\pi(x)[1 - \pi(x)]$. That is, the conditional distribution of the outcome variable follows a binomial distribution with probability given by the conditional mean, $\pi(x)$.

In summary, we have seen that in a regression analysis when the outcome variable is dichotomous:

(1) The conditional mean of the regression equation must be formulated to be bounded between zero and 1. We have stated that the logistic regression model, $\pi(x)$ given in equation (1.1), satisfies this constraint.

(2) The binomial, not the normal, distribution describes the distribution of the errors and will be the statistical distribution upon which the analysis is based.

(3) The principles that guide an analysis using linear regression will also guide us in logistic regression.

1.2 FITTING THE LOGISTIC REGRESSION MODEL

Suppose we have a sample of n independent observations of the pair (x_i, y_i), $i = 1, 2, \ldots, n$, where y_i denotes the value of a dichotomous outcome variable and x_i is the value of the independent variable for the ith subject. Furthermore, assume that the outcome variable has been coded as 0 or 1, representing the absence or the presence of the characteristic, respectively. This coding for a dichotomous outcome is used throughout the text. To fit the logistic regression model in equation (1.1) to a set of data requires that we estimate the values of β_0 and β_1, the unknown parameters.

In linear regression, the method used most often for estimating unknown parameters is *least squares*. In that method we choose those values of β_0 and β_1 which minimize the sum of squared deviations of the observed values of Y from the predicted values based upon the model. Under the usual assumptions for linear regression the method of least squares yields estimators with a number of desirable statistical properties. Unfortunately, when the method of least squares is applied to a model with a dichotomous outcome the estimators no longer have these same properties.

The general method of estimation that leads to the least squares function under the linear regression model (when the error terms are normally distributed) is called *maximum likelihood.* This method will provide the foundation for our approach to estimation with the logistic regression model. In a very general sense the method of maximum likelihood yields values for the unknown parameters which maximize the probability of obtaining the observed set of data. In order to apply this method we must first construct a function, called the *likelihood function.* This function expresses the probability of the observed data as a function of the unknown parameters. The *maximum likelihood estimators* of these parameters are chosen to be those values that maximize this function. Thus, the resulting estimators are those which agree most closely with the observed data. We now describe how to find these values from the logistic regression model.

If Y is coded as 0 or 1 then the expression for $\pi(x)$ given in equation (1.1) provides (for an arbitrary value of $\boldsymbol{\beta}=(\beta_0,\beta_1)$, the vector of parameters) the conditional probability that Y is equal to 1 given x. This will be denoted as $P(Y=1\mid x)$. It follows that the quantity $1-\pi(x)$ gives the conditional probability that Y is equal to zero given x, $P(Y=0\mid x)$. Thus, for those pairs (x_i,y_i), where $y_i=1$, the contribution to the likelihood function is $\pi(x_i)$, and for those pairs where $y_i=0$, the contribution to the likelihood function is $1-\pi(x_i)$, where the quantity $\pi(x_i)$ denotes the value of $\pi(x)$ computed at x_i. A convenient way to express the contribution to the likelihood function for the pair (x_i,y_i) is through the expression

$$\pi(x_i)^{y_i}\left[1-\pi(x_i)\right]^{1-y_i} . \tag{1.2}$$

Since the observations are assumed to be independent, the likelihood function is obtained as the product of the terms given in expression (1.2) as follows:

$$l(\boldsymbol{\beta})=\prod_{i=1}^{n}\pi(x_i)^{y_i}\left[1-\pi(x_i)\right]^{1-y_i} . \tag{1.3}$$

The principle of maximum likelihood states that we use as our estimate of $\boldsymbol{\beta}$ the value which maximizes the expression in equation (1.3). However, it is easier mathematically to work with the log of equation (1.3). This expression, the *log likelihood*, is defined as

$$L(\boldsymbol{\beta}) = \ln[l(\boldsymbol{\beta})] = \sum_{i=1}^{n} \{y_i \ln[\pi(x_i)] + (1 - y_i)\ln[1 - \pi(x_i)]\} . \tag{1.4}$$

To find the value of $\boldsymbol{\beta}$ that maximizes $L(\boldsymbol{\beta})$ we differentiate $L(\boldsymbol{\beta})$ with respect to β_0 and β_1 and set the resulting expressions equal to zero. These equations, known as the *likelihood equations*, are:

$$\sum [y_i - \pi(x_i)] = 0 \tag{1.5}$$

and

$$\sum x_i [y_i - \pi(x_i)] = 0. \tag{1.6}$$

In equations (1.5) and (1.6) it is understood that the summation is over i varying from 1 to n. (The practice of suppressing the index and range of summation, when these are clear, is followed throughout the text.)

In linear regression, the likelihood equations, obtained by differentiating the sum of squared deviations function with respect to $\boldsymbol{\beta}$ are linear in the unknown parameters and thus are easily solved. For logistic regression the expressions in equations (1.5) and (1.6) are nonlinear in β_0 and β_1, and thus require special methods for their solution. These methods are iterative in nature and have been programmed into available logistic regression software. For the moment we need not be concerned about these iterative methods and will view them as a computational detail taken care of for us. The interested reader may see the text by McCullagh and Nelder (1989) for a general discussion of the methods used by most programs. In particular, they show that the solution to equations (1.5) and (1.6) may be obtained using an iterative weighted least squares procedure.

The value of $\boldsymbol{\beta}$ given by the solution to equations (1.5) and (1.6) is called the maximum likelihood estimate and will be denoted as $\hat{\boldsymbol{\beta}}$. In general, the use of the symbol "^" denotes the maximum likelihood estimate of the respective quantity. For example, $\hat{\pi}(x_i)$ is the maximum likelihood estimate of $\pi(x_i)$. This quantity provides an estimate of the conditional probability that Y is equal to 1, given that x is equal to x_i.

Table 1.3 Results of Fitting the Logistic Regression Model to the Data in Table 1.1

Variable	Coeff.	Std. Err.	z	P>|z|
AGE	0.111	0.0241	4.61	<0.001
Constant	–5.309	1.1337	–4.68	<0.001

Log likelihood = –53.67656

As such, it represents the fitted or predicted value for the logistic regression model. An interesting consequence of equation (1.5) is that

$$\sum_{i=1}^{n} y_i = \sum_{i=1}^{n} \hat{\pi}(x_i).$$

That is, the sum of the observed values of y is equal to the sum of the predicted (expected) values. This property will be especially useful in later chapters when we discuss assessing the fit of the model.

As an example, consider the data given in Table 1.1. Use of a logistic regression software package, with continuous variable AGE as the independent variable, produces the output in Table 1.3. The maximum likelihood estimates of β_0 and β_1 are thus seen to be $\hat{\beta}_0 = -5.309$ and $\hat{\beta}_1 = 0.111$. The fitted values are given by the equation

$$\hat{\pi}(x) = \frac{e^{-5.309+0.111\times\text{AGE}}}{1+e^{-5.309+0.111\times\text{AGE}}} \tag{1.7}$$

and the estimated logit, $\hat{g}(x)$, is given by the equation

$$\hat{g}(x) = -5.309 + 0.111\times\text{AGE}. \tag{1.8}$$

The log likelihood given in Table 1.3 is the value of equation (1.4) computed using $\hat{\beta}_0$ and $\hat{\beta}_1$.

Three additional columns are present in Table 1.3. One contains estimates of the standard errors of the estimated coefficients, the next column displays the ratios of the estimated coefficients to their estimated standard errors and the last column displays a p-value. These quantities are discussed in the next section.

Following the fitting of the model we begin to evaluate its adequacy.

1.3 TESTING FOR THE SIGNIFICANCE OF THE COEFFICIENTS

In practice, the modeling of a set of data, as we show in Chapters 4, 7, and 8, is a much more complex process than one of fitting and testing. The methods we present in this section, while simplistic, do provide essential building blocks for the more complex process.

After estimating the coefficients, our first look at the fitted model commonly concerns an assessment of the significance of the variables in the model. This usually involves formulation and testing of a statistical hypothesis to determine whether the independent variables in the model are "significantly" related to the outcome variable. The method for performing this test is quite general and differs from one type of model to the next only in the specific details. We begin by discussing the general approach for a single independent variable. The multivariate case is discussed in Chapter 2.

One approach to testing for the significance of the coefficient of a variable in any model relates to the following question. *Does the model that includes the variable in question tell us more about the outcome (or response) variable than a model that does not include that variable?* This question is answered by comparing the observed values of the response variable to those predicted by each of two models; the first with and the second without the variable in question. The mathematical function used to compare the observed and predicted values depends on the particular problem. If the predicted values with the variable in the model are better, or more accurate in some sense, than when the variable is not in the model, then we feel that the variable in question is "significant." It is important to note that we are not considering the question of whether the predicted values are an accurate representation of the observed values in an absolute sense (this would be called *goodness-of-fit*). Instead, our question is posed in a relative sense. The assessment of goodness-of-fit is a more complex question which is discussed in detail in Chapter 5.

The general method for assessing significance of variables is easily illustrated in the linear regression model, and its use there will motivate the approach used for logistic regression. A comparison of the two approaches will highlight the differences between modeling continuous and dichotomous response variables.

In linear regression, the assessment of the significance of the slope coefficient is approached by forming what is referred to as an *analysis of variance table*. This table partitions the total sum of squared deviations of observations about their mean into two parts: (1) the sum of squared deviations of observations about the regression line SSE, (or *residual sum-of-squares*), and (2) the sum of squares of predicted values, based on the regression model, about the mean of the dependent variable SSR, (or *due regression sum-of-squares*). This is just a convenient way of displaying the comparison of observed to predicted values under two models. In linear regression, the comparison of observed and predicted values is based on the square of the distance between the two. If y_i denotes the observed value and $\hat{y}_i$ denotes the predicted value for the ith individual under the model, then the statistic used to evaluate this comparison is

$$\text{SSE} = \sum_{i=1}^{n} (y_i - \hat{y}_i)^2 .$$

Under the model not containing the independent variable in question the only parameter is β_0, and $\hat{\beta}_0 = \bar{y}$, the mean of the response variable. In this case, $\hat{y}_i = \bar{y}$ and SSE is equal to the total variance. When we include the independent variable in the model any decrease in SSE will be due to the fact that the slope coefficient for the independent variable is not zero. The change in the value of SSE is the due to the regression source of variability, denoted SSR. That is,

$$\text{SSR} = \left[\sum_{i=1}^{n} (y_i - \bar{y}_i)^2\right] - \left[\sum_{i=1}^{n} (y_i - \hat{y}_i)^2\right] .$$

In linear regression, interest focuses on the size of SSR. A large value suggests that the independent variable is important, whereas a small value suggests that the independent variable is not helpful in predicting the response.

The guiding principle with logistic regression is the same: *Compare observed values of the response variable to predicted values obtained from models with and without the variable in question*. In logistic regression, comparison of observed to predicted values is based on the log likelihood function defined in equation (1.4). To better understand this comparison, it is helpful conceptually to think of an observed value of the response variable as also being a predicted value resulting from a *saturated model*. A saturated model is one that contains as many parameters as there are data points. (A simple example of a saturated model is fitting a linear regression model when there are only two data points, $n = 2$.)

The comparison of observed to predicted values using the likelihood function is based on the following expression:

$$D = -2\ln\left[\frac{(\text{likelihood of the fitted model})}{(\text{likelihood of the saturated model})}\right]. \tag{1.9}$$

The quantity inside the large brackets in the expression above is called the *likelihood ratio.* Using minus twice its log is necessary to obtain a quantity whose distribution is known and can therefore be used for hypothesis testing purposes. Such a test is called the *likelihood ratio test.* Using equation (1.4), equation (1.9) becomes

$$D = -2\sum_{i=1}^{n}\left[y_i \ln\left(\frac{\hat{\pi}_i}{y_i}\right) + (1-y_i)\ln\left(\frac{1-\hat{\pi}_i}{1-y_i}\right)\right], \tag{1.10}$$

where $\hat{\pi}_i = \hat{\pi}(x_i)$.

The statistic, D, in equation (1.10) is called the *deviance* by some authors [see, for example, McCullagh and Nelder (1983)], and plays a central role in some approaches to assessing goodness-of-fit. The deviance for logistic regression plays the same role that the residual sum of squares plays in linear regression. In fact, the deviance as shown in equation (1.10), when computed for linear regression, is identically equal to the SSE.

Furthermore, in a setting such as the one shown in Table 1.1, where the values of the outcome variable are either 0 or 1, the likelihood of the saturated model is 1. Specifically, it follows from the definition of a saturated model that $\hat{\pi}_i = y_i$ and the likelihood is

$$l(\text{saturated model}) = \prod_{i=1}^{n} y_i^{y_i} \times (1-y_i)^{(1-y_i)} = 1.$$

Thus it follows from equation (1.9) that the deviance is

$$D = -2\ln(\text{likelihood of the fitted model}). \tag{1.11}$$

Some software packages, such as SAS, report the value of the deviance in (1.11) rather than the log likelihood for the fitted model. We discuss the deviance in more detail in Chapter 5 in the context of evaluating model goodness-of-fit. At this stage we want to emphasize that we think of the deviance in the same terms that we think of the residual sum of squares in linear regression in the context of testing for the significance of a fitted model.

For purposes of assessing the significance of an independent variable we compare the value of D with and without the independent variable in the equation. The change in D due to the inclusion of the independent variable in the model is obtained as:

$$G = D(\text{model without the variable}) - D(\text{model with the variable}).$$

This statistic plays the same role in logistic regression as the numerator of the partial F test does in linear regression. Because the likelihood of the saturated model is common to both values of D being differenced to compute G, it can be expressed as

$$G = -2\ln\left[\frac{(\text{likelihood without the variable})}{(\text{likelihood with the variable})}\right]. \tag{1.12}$$

For the specific case of a single independent variable, it is easy to show that when the variable is not in the model, the maximum likelihood estimate of β_0 is $\ln(n_1/n_0)$ where $n_1 = \Sigma y_i$ and $n_0 = \Sigma(1-y_i)$ and the predicted value is constant, n_1/n. In this case, the value of G is:

$$G = -2\ln\left[\frac{\left(\frac{n_1}{n}\right)^{n_1}\left(\frac{n_0}{n}\right)^{n_0}}{\prod_{i=1}^{n} \hat{\pi}_i^{y_i}(1-\hat{\pi}_i)^{(1-y_i)}}\right] \tag{1.13}$$

or

$$G = 2\left\{\sum_{i=1}^{n}\left[y_i \ln(\hat{\pi}_i) + (1-y_i)\ln(1-\hat{\pi}_i)\right] - \left[n_1\ln(n_1) + n_0\ln(n_0) - n\ln(n)\right]\right\}. \tag{1.14}$$

Under the hypothesis that β_1 is equal to zero, the statistic G follows a chi-square distribution with 1 degree of freedom. Additional mathematical assumptions are also needed; however, for the above case they are rather nonrestrictive and involve having a sufficiently large sample size, n.

As an example, we consider the model fit to the data in Table 1.1, whose estimated coefficients and log likelihood are given in Table 1.3. For these data, $n_1 = 43$ and $n_0 = 57$; thus, evaluating G as shown in equation (1.14) yields

$$G = 2\{-53.677 - [43\ \ln(43) + 57\ \ln(57) - 100\ \ln(100)]\}$$
$$= 2[-53.677 - (-68.331)] = 29.31.$$

The first term in this expression is the log likelihood from the model containing AGE (see Table 1.3), and the remainder of the expression simply substitutes n_1 and n_0 into the second part of equation (1.14). We use the symbol $\chi^2(v)$ to denote a chi-square random variable with v degrees-of-freedom. Using this notation, the p-value associated with this test is $P[\chi^2(1) > 29.31] < 0.001$; thus, we have convincing evidence that AGE is a significant variable in predicting CHD. This is merely a statement of the statistical evidence for this variable. Other important factors to consider before concluding that the variable is clinically important would include the appropriateness of the fitted model, as well as inclusion of other potentially important variables.

The calculation of the log likelihood and the likelihood ratio test are standard features of all logistic regression software. This makes it easy to check for the significance of the addition of new terms to the model. In the simple case of a single independent variable, we first fit a model containing only the constant term. We then fit a model containing the independent variable along with the constant. This gives rise to a new log likelihood. The likelihood ratio test is obtained by multiplying the difference between these two values by −2.

In the current example, the log likelihood for the model containing only a constant term is −68.331. Fitting a model containing the independent variable (AGE) along with the constant term results in the log likelihood shown in Table 1.3 of −53.677. Multiplying the difference in these log likelihoods by −2 gives

$$-2 \times [-68.331 - (-53.677)] = -2 \times (-14.655) = 29.31.$$

This result, along with the associated p-value for the chi-square distribution, may be obtained from most software packages.

Two other similar, statistically equivalent tests have been suggested. These are the Wald test and the Score test. The assumptions needed for these tests are the same as those of the likelihood ratio test in equation (1.13). A more complete discussion of these tests and their assumptions may be found in Rao (1973).

The Wald test is obtained by comparing the maximum likelihood estimate of the slope parameter, $\hat{\beta}_1$, to an estimate of its standard error. The resulting ratio, under the hypothesis that $\beta_1 = 0$, will follow a standard normal distribution. While we have not yet formally discussed how the estimates of the standard errors of the estimated parameters are obtained, they are routinely printed out by computer software. For example, the Wald test for the logistic regression model in Table 1.3 is provided in the column headed z and is

$$W = \frac{\hat{\beta}_1}{\widehat{\text{SE}}(\hat{\beta}_1)} = \frac{0.111}{0.024} = 4.61$$

and the two tailed p-value, provided in the last column of Table 1.3, is $P(|z| > 4.61)$, where z denotes a random variable following the standard normal distribution. Hauck and Donner (1977) examined the performance of the Wald test and found that it behaved in an aberrant manner, often failing to reject the null hypothesis when the coefficient was significant. They recommended that the likelihood ratio test be used.

Jennings (1986a) has also looked at the adequacy of inferences in logistic regression based on Wald statistics. His conclusions are similar to those of Hauck and Donner. Both the likelihood ratio test, G, and the Wald test, W, require the computation of the maximum likelihood estimate for β_1.

A test for the significance of a variable which does not require these computations is the Score test. Proponents of the Score test cite this reduced computational effort as its major advantage. Use of the test is limited by the fact that it cannot be obtained from some software packages. The Score test is based on the distribution theory of the derivatives of the log likelihood. In general, this is a multivariate test requiring matrix calculations which are discussed in Chapter 2.

In the univariate case, this test is based on the conditional distribution of the derivative in equation (1.6), given the derivative in equation

(1.5). In this case, we can write down an expression for the Score test. The test uses the value of equation (1.6), computed using $\beta_0 = \ln(n_1 / n_0)$ and $\beta_1 = 0$. As noted earlier, under these parameter values, $\hat{\pi} = n_1/n = \bar{y}$. Thus, the left-hand side of equation (1.6) becomes $\Sigma x_i(y_i - \bar{y})$. It may be shown that the estimated variance is $\bar{y}(1-\bar{y})\Sigma(x_i - \bar{x})^2$. The test statistic for the Score test (ST) is

$$\text{ST} = \frac{\sum_{i=1}^{n} x_i(y_i - \bar{y})}{\sqrt{\bar{y}(1-\bar{y})\sum_{i=1}^{n}(x_i - \bar{x})^2}}.$$

As an example of the Score test, consider the model fit to the data in Table 1.1. The value of the test statistic for this example is

$$\text{ST} = \frac{296.66}{\sqrt{3333.742}} = 5.14$$

and the two tailed p-value is $P(|z| > 5.14) < 0.001$. We note that, for this example, the values of the three test statistics are nearly the same (*note*: $\sqrt{G} = 5.41$).

In summary, the method for testing the significance of the coefficient of a variable in logistic regression is similar to the approach used in linear regression; however, it uses the likelihood function for a dichotomous outcome variable.

1.4 CONFIDENCE INTERVAL ESTIMATION

An important adjunct to testing for significance of the model, discussed in Section 1.3, is calculation and interpretation of confidence intervals for parameters of interest. As is the case in linear regression we can obtain these for the slope, intercept and the "line", (i.e., the logit). In some settings it may be of interest to provide interval estimates for the fitted values (i.e., the predicted probabilities).

The basis for construction of the interval estimators is the same statistical theory we used to formulate the tests for significance of the model. In particular, the confidence interval estimators for the slope and intercept are based on their respective Wald tests. The endpoints of a $100(1-\alpha)\%$ confidence interval for the slope coefficient are

$$\hat{\beta}_1 \pm z_{1-\alpha/2}\widehat{\mathrm{SE}}\left(\hat{\beta}_1\right), \tag{1.15}$$

and for the intercept they are

$$\hat{\beta}_0 \pm z_{1-\alpha/2}\widehat{\mathrm{SE}}\left(\hat{\beta}_0\right), \tag{1.16}$$

where $z_{1-\alpha/2}$ is the upper $100(1-\alpha/2)\%$ point from the standard normal distribution and $\widehat{\mathrm{SE}}(\cdot)$ denotes a model-based estimator of the standard error of the respective parameter estimator. We defer discussion of the actual formula used for calculating the estimators of the standard errors to Chapter 2. For the moment we use the fact that estimated values are provided in the output following the fit of a model and, in addition, many packages also provide the endpoints of the interval estimates.

As an example, consider the model fit to the data in Table 1.1 regressing age on the presence or absence of CHD. The results are presented in Table 1.3. The endpoints of a 95 percent confidence interval for the slope coefficient from (1.15) are $0.111 \pm 1.96 \times 0.0241$, yielding the interval $(0.064, 0.158)$. We defer a detailed discussion of the interpretation of these results to Chapter 3. Briefly, the results suggest that the change in the log-odds of CHD per one year increase in age is 0.111 and the change could be as little as 0.064 or as much as 0.158 with 95 percent confidence.

As is the case with any regression model, the constant term provides an estimate of the response in the absence of x unless the independent variable has been centered at some clinically meaningful value. In our example, the constant provides an estimate of the log-odds ratio of CHD at zero years of age. As a result, the constant term, by itself, has no useful clinical interpretation. In any event, from expression (1.16), the endpoints of a 95 percent confidence interval for the constant are $-5.309 \pm 1.96 \times 1.1337$, yielding the interval $(-7.531, -3.087)$. The constant is important when considering point and interval estimators of the logit.

The logit is the linear part of the logistic regression model and, as such, is most like the fitted line in a linear regression model. The estimator of the logit is

$$\hat{g}(x) = \hat{\beta}_0 + \hat{\beta}_1 x. \tag{1.17}$$

The estimator of the variance of the estimator of the logit requires obtaining the variance of a sum. In this case it is

$$\widehat{\text{Var}}[\hat{g}(x)] = \widehat{\text{Var}}(\hat{\beta}_0) + x^2\widehat{\text{Var}}(\hat{\beta}_1) + 2x\widehat{\text{Cov}}(\hat{\beta}_0, \hat{\beta}_1). \tag{1.18}$$

In general the variance of a sum is equal to the sum of the variance of each term and twice the covariance of each possible pair of terms formed from the components of sum. The endpoints of a $100(1-\alpha)\%$ Wald-based confidence interval for the logit are

$$\hat{g}(x) \pm z_{1-\alpha/2}\widehat{\text{SE}}[\hat{g}(x)], \tag{1.19}$$

where $\widehat{\text{SE}}[\hat{g}(x)]$ is the positive square root of the variance estimator in (1.18).

The estimated logit for the fitted model in Table 1.3 is shown in (1.8). In order to evaluate (1.18) for a specific age we need the estimated covariance matrix. This matrix can be obtained from the output from all logistic regression software packages. How it is displayed varies from package to package, but the triangular form shown in Table 1.4 is a common one.

The estimated logit from (1.8) for a subject of age 50 is

$$\hat{g}(50) = -5.31 + 0.111 \times 50 = 0.240.$$

The estimated variance, using (1.18) and the results in Table 1.4, is

$$\widehat{\text{Var}}[\hat{g}(50)] = 1.28517 + (50)^2 \times 0.000579 + 2 \times 50 \times (-0.026677) = 0.0650$$

and the estimated standard error is $\widehat{\text{SE}}[\hat{g}(50)] = 0.2549$. Thus the end points of a 95 percent confidence interval for the logit at age 50 are $0.240 \pm 1.96 \times 0.2550 = (-0.260,\ 0.740)$. We discuss the interpretation and use of the estimated logit in providing estimates of odds ratios in Chapter 3.

The estimator of the logit and its confidence interval provide the basis for the estimator of the fitted value, in this case the logistic probability, and its associated confidence interval. In particular, using (1.7) at age 50 the estimated logistic probability is

Table 1.4 Estimated Covariance Matrix of the Estimated Coefficients in Table 1.3

	AGE	Constant
AGE	0.000579	
Constant	-0.026677	1.28517

$$\hat{\pi}(50) = \frac{e^{\hat{g}(50)}}{1+e^{\hat{g}(50)}} = \frac{e^{-5.31+0.111\times 50}}{1+e^{-5.31+0.111\times 50}} = 0.560 \quad (1.20)$$

and the endpoints of a 95 percent confidence interval are obtained from the respective endpoints of the confidence interval for the logit. The endpoints of the $100(1-\alpha)\%$ Wald-based confidence interval for the fitted value are

$$\frac{e^{\hat{g}(x) \pm z_{1-\alpha/2}\hat{\text{SE}}[\hat{g}(x)]}}{1+e^{\hat{g}(x) \pm z_{1-\alpha/2}\hat{\text{SE}}[\hat{g}(x)]}}. \quad (1.21)$$

Using the example at age 50 to demonstrate the calculations, the lower limit is

$$\frac{e^{-0.260}}{1+e^{-0.260}} = 0.435,$$

and the upper limit is

$$\frac{e^{0.740}}{1+e^{0.740}} = 0.677.$$

We have found that a major mistake often made by persons new to logistic regression modeling is to try and apply estimates on the probability scale to individual subjects. The fitted value computed in (1.20) is analogous to a particular point on the line obtained from a linear regression. In linear regression each point on the fitted line provides an estimate of the mean of the dependent variable in a population of subjects with covariate value "x". Thus the value of 0.56 in (1.20) is an estimate of the mean (i.e., proportion) of 50 year old subjects in the population sampled that have evidence of CHD. Each individual 50 year old subject either does or does not have evidence of CHD. The confidence interval suggests that this mean could be between 0.435 and 0.677 with 95 percent confidence. We discuss the use and interpretation of fitted values in greater detail in Chapter 3.

One application of fitted logistic regression models that has received a lot of attention in the subject matter literature is the use of model-based fitted values like the one in (1.20) to predict the value of a binary dependent value in individual subjects. This process is called *classification* and has a long history in statistics where it is referred to as *discriminant analysis*. We discuss the classification problem in detail in Chapter 4. We discuss discriminant analysis within the context of a method for obtaining estimators of the coefficients in the next section.

1.5 OTHER METHODS OF ESTIMATION

The method of maximum likelihood described in Section 1.2 is the estimation method used in the logistic regression routines of the major software packages. However, two other methods have been and may still be used for estimating the coefficients. These methods are: (1) noniterative weighted least squares, and (2) discriminant function analysis.

A linear models approach to the analysis of categorical data was proposed by Grizzle, Starmer, and Koch (1969), which uses estimators based on noniterative weighted least squares. They demonstrate that the logistic regression model is an example of a general class of models that can be handled with their methods. We should add that the maximum likelihood estimators are usually calculated using an iterative reweighted least squares algorithm, and thus are also "least squares" estimators. The approach suggested by Grizzle et al. uses only one iteration in the process.

A major limitation of this method is that we must have an estimate of $\pi(x)$ which is not zero or 1 for most values of x. An example where we could use both maximum likelihood and noniterative weighted least squares is the data in Table 1.2. In cases such as this, the two methods are *asymptotically equivalent*, meaning that as n gets large, the distributional properties of the estimators become identical.

The discriminant function approach to estimation of the coefficients is of historical importance as it was popularized by Cornfield (1962) in some of the earliest work on logistic regression. These estimators take their name from the fact that the posterior probability in the usual discriminant function model is the logistic regression function given in equation (1.1). More precisely, if the independent variable, X, follows a normal distribution within each of two groups (subpopulations) defined by the two values of y having different means and the same variance, then the conditional distribution of Y given $X = x$ is the logistic regression model. That is, if

$$X \mid Y \sim \mathrm{N}\left(\mu_j, \sigma^2\right),\ j = 0,1$$

then $P(Y=1 \mid x) = \pi(x)$. The symbol "$\sim$" is read "is distributed" and the "$\mathrm{N}\left(\mu, \sigma^2\right)$" denotes the normal distribution with mean equal to μ and variance equal to σ^2. Under these assumptions it is easy to show [Lachenbruch (1975)] that the logistic coefficients are

$$\beta_0 = \ln\left(\frac{\theta_1}{\theta_0}\right) - 0.5\left(\mu_1^2 - \mu_0^2\right)/\sigma^2 \tag{1.22}$$

and

$$\beta_1 = \left(\mu_1 - \mu_0\right)/\sigma^2, \tag{1.23}$$

where $\theta_j = P(Y = j),\ j = 0,\ 1$. The discriminant function estimators of β_0 and β_1 are found by substituting estimators for μ_j, θ_j, $j =$ 0, 1 and σ^2 into the above equations. The estimators usually used are $\hat{\mu}_j = \bar{x}_j$, the mean of x in the subgroup defined by $y = j,\ j = 0,\ 1$, $\theta_1 = n_1 / n$ the mean of y with $\hat{\theta}_0 = 1 - \hat{\theta}_1$ and

$$\hat{\sigma}^2 = \left[\left(n_0 - 1\right)s_0^2 + \left(n_1 - 1\right)s_1^2\right]/\left(n_0 + n_1 - 2\right),$$

where s_j^2 is the unbiased estimator of σ^2 computed within the subgroup of the data defined by $y = j,\ j = 0,1$. The above expressions are for a single variable x; the multivariable case is presented in Chapter 2.

It is natural to ask why, if the discriminant function estimators are so easy to compute, are they not used in place of the maximum likelihood estimators? Halpern, Blackwelder, and Verter (1971) and Hosmer, Hosmer, and Fisher (1983) have compared the two methods when the model contains a mixture of continuous and discrete variables, with the general conclusion that the discriminant function estimators are sensitive to the assumption of normality. In particular, the estimators of the coefficients for nonnormally distributed variables are biased away from zero when the coefficient is, in fact, different from zero. The practical implication of this is that for dichotomous independent variables (which occur in many situations), the discriminant function estimators will overestimate the magnitude of the coefficient.

At this point it may be helpful to delineate more carefully the various uses of the term "maximum likelihood," as it applies to the estimation of the logistic regression coefficients. Under the assumptions of the discriminant function model stated above, the estimators obtained from equations (1.22) and (1.15) are maximum likelihood estimators. Those obtained from equations (1.5) and (1.6) are based on the conditional distribution of Y given X and, as such, are actually "conditional maximum likelihood estimators." Because discriminant function estimators are rarely used anymore, the word conditional has been dropped when describing the estimators given in equations (1.5) and (1.6). We use the word *conditional* to describe estimators in logistic regression with matched data as discussed in Chapter 7.

In summary there are alternative methods of estimation for some data configurations that are computationally quicker; however, we use the method of maximum likelihood described in Section 1.2 throughout the rest of this text.

1.6 DATA SETS

A number of different data sets are used in the examples as well as the exercises for the purpose of demonstrating various aspects of logistic regression modeling. Four data sets used throughout the text are described below. Other data sets will be introduced as needed in later chapters. All data sets used in this text may be obtained from the text web sites at John Wiley & Sons Inc. and the University of Massachusetts as described in the Preface.

1.6.1 The ICU Study

The ICU study data set consists of a sample of 200 subjects who were part of a much larger study on survival of patients following admission to an adult intensive care unit (ICU). The major goal of this study was to develop a logistic regression model to predict the probability of sur-

Table 1.5 Code Sheet for the ICU Data

Variable	Description	Codes/Values	Name
1	Identification Code	ID Number	ID
2	Vital Status	0 = Lived 1 = Died	STA
3	Age	Years	AGE
4	Sex	0 = Male 1 = Female	SEX
5	Race	1 = White 2 = Black 3 = Other	RACE
6	Service at ICU Admission	0 = Medical 1 = Surgical	SER
7	Cancer Part of Present Problem	0 = No 1 = Yes	CAN
8	History of Chronic Renal Failure	0 = No 1 = Yes	CRN
9	Infection Probable at ICU Admission	0 = No 1 = Yes	INF
10	CPR Prior to ICU Admission	0 = No 1 = Yes	CPR
11	Systolic Blood Pressure at ICU Admission	mm Hg	SYS
12	Heart Rate at ICU Admission	Beats/min	HRA
13	Previous Admission to an ICU Within 6 Months	0 = No 1 = Yes	PRE
14	Type of Admission	0 = Elective 1 = Emergency	TYP
15	Long Bone, Multiple, Neck, Single Area, or Hip Fracture	0 = No 1 = Yes	FRA
16	PO2 from Initial Blood Gases	$0 = > 60$ $1 = \leq 60$	PO2
17	PH from Initial Blood Gases	$0 = \geq 7.25$ $1 = < 7.25$	PH
18	PCO2 from Initial Blood Gases	$0 = \leq 45$ $1 = > 45$	PCO
19	Bicarbonate from Initial Blood Gases	$0 = \geq 18$ $1 = < 18$	BIC
20	Creatinine from Initial Blood Gases	$0 = \leq 2.0$ $1 = > 2.0$	CRE
21	Level of Consciousness at ICU Admission	0 = No Coma or Deep Stupor 1 = Deep Stupor 2 = Coma	LOC

vival to hospital discharge of these patients. A number of publications have appeared which have focused on various facets of this problem. The reader wishing to learn more about the clinical aspects of this study should start with Lemeshow, Teres, Avrunin, and Pastides (1988). For a more up-to-date discussion of modeling the outcome of ICU patients the reader is referred to Lemeshow and Le Gall (1994) and to Lemeshow, Teres, Klar, Avrunin, Gehlbach and Rapoport (1993). Actual observed variable values have been modified to protect subject confidentiality.

A code sheet for the variables to be considered in this text is given in Table 1.5.

1.6.2 The Low Birth Weight Study

Low birth weight, defined as birth weight less than 2500 grams, is an outcome that has been of concern to physicians for years. This is due to the fact that infant mortality rates and birth defect rates are very high for low birth weight babies. A woman's behavior during pregnancy (including diet, smoking habits, and receiving prenatal care) can greatly alter the chances of carrying the baby to term and, consequently, of delivering a baby of normal birth weight.

Data were collected as part of a larger study at Baystate Medical Center in Springfield, Massachusetts. This data set contains information on 189 births to women seen in the obstetrics clinic. Fifty-nine of these births were low birth weight. The variables identified in the code sheet given in Table 1.6 have been shown to be associated with low birth weight in the obstetrical literature. The goal of the current study was to determine whether these variables were risk factors in the clinic population being served by Baystate Medical Center. Actual observed variable values have been modified to protect subject confidentiality.

1.6.3 The Prostate Cancer Study

A third data set involves a study of patients with cancer of the prostate. These data have been provided to us by Dr. Donn Young at The Ohio State University Comprehensive Cancer Center. The goal of the analysis is to determine whether variables measured at a baseline exam can be used to predict whether the tumor has penetrated the prostatic capsule. The data presented are a subset of variables from the main study. Of

Table 1.6 Code Sheet for the Variables in the Low Birth Weight Data

Variable	Description	Codes/Values	Name
1	Identification Code	ID Number	ID
2	Low Birth Weight	0 = ≥ 2500 g 1 = < 2500 g	LOW
3	Age of Mother	Years	AGE
4	Weight of Mother at Last Menstrual Period	Pounds	LWT
5	Race	1 = White 2 = Black 3 = Other	RACE
6	Smoking Status During Pregnancy	0 = No 1 = Yes	SMOKE
7	History of Premature Labor	0 = None 1 = One 2 = Two, etc.	PTL
8	History of Hypertension	0 = No 1 = Yes	HT
9	Presence of Uterine Irritability	0 = No 1 = Yes	UI
10	Number of Physician Visits During the First Trimester	0 = None 1 = One 2 = Two, etc.	FTV
11	Birth Weight	Grams	BWT

the 380 subjects considered here, 153 had a cancer that penetrated the prostatic capsule. Actual observed variable values have been modified to protect subject confidentiality. These data will be used primarily for exercises. A code sheet for the variables to be considered in this text is shown in Table 1.7.

1.6.4 The UMARU IMPACT Study

Our colleagues, Drs. Jane McCusker, Carol Bigelow, and Anne Stoddard, have provided us with a subset of data from the University of Massachusetts Aids Research Unit (UMARU) IMPACT Study (UIS). This was a 5-year (1989–1994) collaborative research project (Benjamin F. Lewis, P.I., National Institute on Drug Abuse Grant #R18-DA06151) com-

Table 1.7 Code Sheet for the Prostate Cancer Study

Variable	Description	Codes/Values	Name
1	Identification Code	1 - 380	ID
2	Tumor Penetration of Prostatic Capsule	0 = No Penetration 1 = Penetration	CAPSULE
3	Age	Years	AGE
4	Race	1= White 2 = Black	RACE
5	Results of the Digital Rectal Exam	1 = No Nodule 2 = Unilobar Nodule (Left) 3 = Unilobar Nodule (Right) 4 = Bilobar Nodule	DPROS
6	Detection of Capsular Involvement in Rectal Exam	1 = No 2 = Yes	DCAPS
7	Prostatic Specific Antigen Value	mg/ml	PSA
8	Tumor Volume Obtained from Ultrasound	cm^3	VOL
9	Total Gleason Score	0 – 10	GLEASON

prised of two concurrent randomized trials of residential treatment for drug abuse. The purpose of the study was to compare treatment programs of different planned durations designed to reduce drug abuse and to prevent high-risk HIV behavior. The UIS sought to determine whether alternative residential treatment approaches are variable in effectiveness and whether efficacy depends on planned program duration.

We refer to the two treatment program sites as A and B in this text. The trial at site A randomized 444 participants and was a comparison of 3- and 6-month modified therapeutic communities which incorporated elements of health education and relapse prevention. Clients in the relapse prevention/health education program (site A) were taught to recognize "high-risk" situations that are triggers to relapse and were taught the skills to enable them to cope with these situations without using drugs. In the trial at site B, 184 clients were randomized to receive either a 6- or 12-month therapeutic community program involving a highly structured life-style in a communal living setting. Our colleagues have published a number of papers reporting the results of this study, see McCusker et. al. (1995, 1997a, 1997b).

Table 1.8 Description of Variables in the UMARU IMPACT Study

Variable	Description	Codes/Values	Name
1	Identification Code	1–575	ID
2	Age at Enrollment	Years	AGE
3	Beck Depression Score at Admission	0.000–54.000	BECK
4	IV Drug Use History at Admission	1 = Never 2 = Previous 3 = Recent	IVHX
5	Number of Prior Drug Treatments	0–40	NDRUGTX
6	Subject's Race	0 = White 1 = Other	RACE
7	Treatment Randomization Assignment	0 = Short 1 = Long	TREAT
8	Treatment Site	0 = A 1 = B	SITE
9	Returned to Drug Use Prior to the Scheduled End of the Treatment Program	1 = Remained Drug Free 0 = Otherwise	DFREE

As is shown in the coming chapters, the data from the UIS provide a rich setting for illustrating methods for logistic regression modeling. The data presented here are a subset of both variables and subjects of the data used to demonstrate methods for survival analysis in Hosmer and Lemeshow (1999). The small subset of variables from the main study we use in this text is described in Table 1.8. Since the analyses we report in this text are based on this small subset of variables and subjects, the results reported here should not be thought of as being in any way comparable to results of the main study. In addition we have taken the liberty in this text of simplifying the study design by representing the planned duration as short versus long. Thus, short versus long represents 3 months versus 6 months planned duration at site A, and 6 months versus 12 months planned duration at site B. The dichotomous outcome variable considered in this text is defined as having returned to drug use prior to the scheduled completion of the treatment program. The original data have been modified in such a way as to preserve subject confidentiality.

EXERCISES

1. In the ICU data described in Section 1.6.1 the primary outcome variable is vital status at hospital discharge, STA. Clinicians associated with the study felt that a key determinant of survival was the patient's age at admission, AGE.
   (a) Write down the equation for the logistic regression model of STA on AGE. Write down the equation for the logit transformation of this logistic regression model. What characteristic of the outcome variable, STA, leads us to consider the logistic regression model as opposed to the usual linear regression model to describe the relationship between STA and AGE?
   (b) Form a scatterplot of STA versus AGE.
   (c) Using the intervals [15, 24], [25, 34], [35, 44], [45, 54], [55, 64], [65, 74], [75, 84], [85, 94] for AGE, compute the STA mean over subjects within each AGE interval. Plot these values of mean STA versus the midpoint of the AGE interval using the same set of axes as was used in Exercise 1(b).
   (d) Write down an expression for the likelihood and log likelihood for the logistic regression model in Exercise 1(a) using the ungrouped, $n = 200$, data. Obtain expressions for the two likelihood equations.
   (e) Using a logistic regression package of your choice obtain the maximum likelihood estimates of the parameters of the logistic regression model in Exercise 1(a). These estimates should be based on the ungrouped, $n = 200$, data. Using these estimates, write down the equation for the fitted values, that is, the estimated logistic probabilities. Plot the equation for the fitted values on the axes used in the scatterplots in Exercises 1(b) and 1(c).
   (f) Summarize (describe in words) the results presented in the plot obtained from Exercises 1(b), 1(c), and 1(e).
   (g) Using the results of the output from the logistic regression package used for Exercise 1(e), assess the significance of the slope coefficient for AGE using the likelihood ratio test, the Wald test, and, if possible, the Score test. What assumptions are needed for the p-values computed for each of these tests to be valid? Are the results of these tests consistent with one another? What is the value of the deviance for the fitted model?

(h) Using the results from Exercise 1(e) compute 95 percent confidence intervals for the slope and constant term. Write a sentence interpreting the confidence interval for the slope.

(i) Obtain the estimated covariance matrix for the model fit in Exercise 1(e). Compute the logit and estimated logistic probability for a 60-year old subject. Compute a 95 percent confidence intervals for the logit and estimated logistic probability. Write a sentence or two interpreting the estimated probability and its confidence interval.

(j) Use the logistic regression package to obtain the estimated logit and its standard error for each subject in the ICU study. Graph the estimated logit and the pointwise 95 percent confidence limits versus AGE for each subject. Explain (in words) the similarities and differences between the appearance of this graph and a graph of a fitted linear regression model and its pointwise 95 percent confidence bands.

2. Use the ICU Study and repeat Exercises 1(a), 1(b), 1(d), 1(e) and 1(g) using the variable "type of admission," TYP, as the covariate.

3. In the Low Birth Weight Study described in Section 1.6.2, one variable that physicians felt was important to control for was the weight of the mother at the last menstrual period, LWT. Repeat steps (a) – (g) of Exercise 1, but for Exercise 3(c) use intervals [80, 99], [100, 109], [110, 114], [115, 119], [120, 124], [125, 129], [130, 250].

   (h) The graph in Exercises 3(c) does not look "S-Shaped". The primary reason is that the range of plotted values is from approximately 0.2 to 0.56. Explain why a model for the probability of low birth weight as a function of LWT could still be the logistic regression model.

4. In the Prostate Cancer Study described in Section 1.6.3, one variable thought to be particularly predictive of capsule penetration is the prostate specific antigen level, PSA. Repeat steps (a) – (g) and (j) of Exercise 1 using CAPSULE as the outcome variable and PSA as the covariate. For Exercises 4(c) use intervals for PSA of [0, 2.4], [2.5, 4.4], [4.5, 6.4], [6.5, 8.4], [8.5, 10.4], [10.5, 12.4], [12.5, 20.4], [20.5, 140].

CHAPTER 2

Multiple Logistic Regression

2.1 INTRODUCTION

In the previous chapter we introduced the logistic regression model in the univariate context. As in the case of linear regression, the strength of a modeling technique lies in its ability to model many variables, some of which may be on different measurement scales. In this chapter we will generalize the logistic model to the case of more than one independent variable. This will be referred to as the "multivariable case." Central to the consideration of multiple logistic models will be estimation of the coefficients in the model and testing for their significance. This will follow along the same lines as the univariate model. An additional modeling consideration which will be introduced in this chapter is the use of design variables for modeling discrete, nominal scale independent variables. In all cases it will be assumed that there is a predetermined collection of variables to be examined. The question of variable selection is dealt with in Chapter 4.

2.2 THE MULTIPLE LOGISTIC REGRESSION MODEL

Consider a collection of p independent variables denoted by the vector $\mathbf{x}' = (x_1, x_2, \ldots, x_p)$. For the moment we will assume that each of these variables is at least interval scale. Let the conditional probability that the outcome is present be denoted by $P(Y=1 \mid \mathbf{x}) = \pi(\mathbf{x})$. The logit of the multiple logistic regression model is given by the equation

$$g(\mathbf{x}) = \beta_0 + \beta_1 x_1 + \beta_2 x_2 + \ldots + \beta_p x_p, \tag{2.1}$$

in which case the logistic regression model is

$$\pi(\mathbf{x}) = \frac{e^{g(\mathbf{x})}}{1+e^{g(\mathbf{x})}} \ . \tag{2.2}$$

If some of the independent variables are discrete, nominal scale variables such as race, sex, treatment group, and so forth, it is inappropriate to include them in the model as if they were interval scale variables. The numbers used to represent the various levels of these nominal scale variables are merely identifiers, and have no numeric significance. In this situation the method of choice is to use a collection of *design variables* (or *dummy variables*). Suppose, for example, that one of the independent variables is race, which has been coded as "white," "black" and "other." In this case, two design variables are necessary. One possible coding strategy is that when the respondent is "white," the two design variables, D_1 and D_2, would both be set equal to zero; when the respondent is "black," D_1 would be set equal to 1 while D_2 would still equal 0; when the race of the respondent is "other," we would use $D_1 = 0$ and $D_2 = 1$. Table 2.1 illustrates this coding of the design variables.

Most logistic regression software will generate design variables, and some programs have a choice of several different methods. The different strategies for creation and interpretation of design variables are discussed in detail in Chapter 3.

In general, if a nominal scaled variable has k possible values, then $k-1$ design variables will be needed. This is true since, unless stated otherwise, all of our models have a constant term. To illustrate the notation used for design variables in this text, suppose that the jth independent variable x_j has k_j levels. The $k_j - 1$ design variables will be denoted as D_{jl} and the coefficients for these design variables will be denoted as $\beta_{jl}, l = 1,2,\ldots,k_j - 1$. Thus, the logit for a model with p vari-

Table 2.1 An Example of the Coding of the Design Variables for Race, Coded at Three Levels

	Design Variable	
RACE	D_1	D_2
White	0	0
Black	1	0
Other	0	1

ables and the jth variable being discrete would be

$$g(\mathbf{x}) = \beta_0 + \beta_1 x_1 + \cdots + \sum_{l=1}^{k_j - 1} \beta_{jl} D_{jl} + \beta_p x_p .$$

When discussing the multiple logistic regression model we will, in general, suppress the summation and double subscripting needed to indicate when design variables are being used. The exception to this will be the discussion of modeling strategies when we need to use the specific value of the coefficients for any design variables in the model.

2.3 FITTING THE MULTIPLE LOGISTIC REGRESSION MODEL

Assume that we have a sample of n independent observations $(\mathbf{x}_i, y_i)$, $i = 1, 2, \ldots, n$. As in the univariate case, fitting the model requires that we obtain estimates of the vector $\boldsymbol{\beta}' = (\beta_0, \beta_1, \ldots, \beta_p)$. The method of estimation used in the multivariable case will be the same as in the univariate situation – maximum likelihood. The likelihood function is nearly identical to that given in equation (1.3) with the only change being that $\pi(\mathbf{x})$ is now defined as in equation (2.2). There will be $p+1$ likelihood equations that are obtained by differentiating the log likelihood function with respect to the $p+1$ coefficients. The likelihood equations that result may be expressed as follows:

$$\sum_{i=1}^{n} \left[y_i - \pi(\mathbf{x}_i) \right] = 0$$

and

$$\sum_{i=1}^{n} x_{ij} \left[y_i - \pi(\mathbf{x}_i) \right] = 0$$

for $j = 1, 2, \ldots, p$.

As in the univariate model, the solution of the likelihood equations requires special software that is available in most, if not all, statistical packages. Let $\hat{\boldsymbol{\beta}}$ denote the solution to these equations. Thus, the fitted values for the multiple logistic regression model are $\hat{\pi}(\mathbf{x}_i)$, the value of the expression in equation (2.2) computed using $\hat{\boldsymbol{\beta}}$, and $\mathbf{x}_i$.

In the previous chapter only a brief mention was made of the method for estimating the standard errors of the estimated coefficients. Now that the logistic regression model has been generalized both in concept and notation to the multivariable case, we consider estimation of standard errors in more detail.

The method of estimating the variances and covariances of the estimated coefficients follows from well-developed theory of maximum likelihood estimation [see, for example, Rao (1973)]. This theory states that the estimators are obtained from the matrix of second partial derivatives of the log likelihood function. These partial derivatives have the following general form

$$\frac{\partial^2 L(\beta)}{\partial \beta_j^2} = -\sum_{i=1}^{n} x_{ij}^2 \pi_i (1-\pi_i) \tag{2.3}$$

and

$$\frac{\partial^2 L(\beta)}{\partial \beta_j \partial \beta_l} = -\sum_{i=1}^{n} x_{ij} x_{il} \pi_i (1-\pi_i) \tag{2.4}$$

for $j,\ l = 0,1,2,...,p$ where π_i denotes $\pi(\mathbf{x}_i)$. Let the $(p+1)\times(p+1)$ matrix containing the negative of the terms given in equations (2.3) and (2.4) be denoted as $\mathbf{I}(\boldsymbol{\beta})$. This matrix is called the *observed information matrix*. The variances and covariances of the estimated coefficients are obtained from the inverse of this matrix which we denote as $\text{Var}(\boldsymbol{\beta}) = \mathbf{I}^{-1}(\boldsymbol{\beta})$. Except in very special cases it is not possible to write down an explicit expression for the elements in this matrix. Hence, we will use the notation $\text{Var}(\beta_j)$ to denote the jth diagonal element of this matrix, which is the variance of $\hat{\beta}_j$, and $\text{Cov}(\beta_j, \beta_l)$ to denote an arbitrary off-diagonal element, which is the covariance of $\hat{\beta}_j$ and $\hat{\beta}_l$. The estimators of the variances and covariances, which will be denoted by $\widehat{\text{Var}}(\hat{\boldsymbol{\beta}})$, are obtained by evaluating $\text{Var}(\boldsymbol{\beta})$ at $\hat{\boldsymbol{\beta}}$. We will use $\widehat{\text{Var}}(\hat{\beta}_j)$ and $\widehat{\text{Cov}}(\hat{\beta}_j, \hat{\beta}_l)$, $j,l = 0,\ 1,\ 2,...,\ p$ to denote the values in this matrix.

For the most part, we will have occasion to use only the estimated standard errors of the estimated coefficients, which we will denote as

$$\hat{\text{SE}}(\hat{\beta}_j) = \left[\hat{\text{Var}}(\hat{\beta}_j)\right]^{1/2} \tag{2.5}$$

for $j = 0, 1, 2, \ldots, p$. We will use this notation in developing methods for coefficient testing and confidence interval estimation.

A formulation of the information matrix which will be useful when discussing model fitting and assessment of fit is $\hat{\mathbf{I}}(\hat{\boldsymbol{\beta}}) = \mathbf{X}'\mathbf{V}\mathbf{X}$ where $\mathbf{X}$ is an n by $p+1$ matrix containing the data for each subject, and $\mathbf{V}$ is an n by n diagonal matrix with general element $\hat{\pi}_i(1-\hat{\pi}_i)$. That is, the matrix $\mathbf{X}$ is

$$\mathbf{X} = \begin{bmatrix} 1 & x_{11} & x_{12} & \cdots & x_{1p} \\ 1 & x_{21} & x_{22} & \cdots & x_{2p} \\ \vdots & \vdots & \vdots & \cdots & \vdots \\ 1 & x_{n1} & x_{n2} & \cdots & x_{np} \end{bmatrix}$$

and the matrix $\mathbf{V}$ is

$$\mathbf{V} = \begin{bmatrix} \hat{\pi}_1(1-\hat{\pi}_1) & 0 & \cdots & 0 \\ 0 & \hat{\pi}_2(1-\hat{\pi}_2) & \cdots & 0 \\ \vdots & 0 & \ddots & \vdots \\ 0 & \cdots & 0 & \hat{\pi}_n(1-\hat{\pi}_n) \end{bmatrix}.$$

Before proceeding further we present an example that illustrates the formulation of a multiple logistic regression model and the estimation of its coefficients using a subset of the variables from the data for the low birth weight study described in Section 1.6.2. The code sheet for the full data set is given in Table 1.6. As discussed in Section 1.6.2, the goal of this study was to identify risk factors associated with giving birth to a low birth weight baby (weighing less than 2500 grams). Data were collected on 189 women, $n_1 = 59$ of whom had low birth weight babies and $n_0 = 130$ of whom had normal birth weight babies. Four variables thought to be of importance were age, weight of the mother at her last menstrual period, race, and number of physician visits during the first trimester of the pregnancy. In this example, the variable race has been

Table 2.2 Estimated Coefficients for a Multiple Logistic Regression Model Using the Variables AGE, Weight at Last Menstrual Period (LWT), RACE, and Number of First Trimester Physician Visits (FTV) from the Low Birth Weight Study

Variable	Coeff.	Std. Err.	z	P>|z|
AGE	–0.024	0.0337	–0.71	0.480
LWT	–0.014	0.0065	–2.18	0.029
RACE_2	1.004	0.4979	2.02	0.044
RACE_3	0.433	0.3622	1.20	0.232
FTV	–0.049	0.1672	–0.30	0.768
Constant	1.295	1.0714	1.21	0.227

Log likelihood = –111.286

recoded using the two design variables in Table 2.1. The results of fitting the logistic regression model to these data are shown in Table 2.2.

In Table 2.2 the estimated coefficients for the two design variables for race are indicated by RACE_2 and RACE_3. The estimated logit is given by the following expression:

$$\hat{g}(\mathbf{x}) = 1.295 - 0.024 \times AGE - 0.014 \times LWT + 1.004 \times RACE\_2$$
$$+ 0.433 \times RACE\_3 - 0.049 \times FTV.$$

The fitted values are obtained using the estimated logit, $\hat{g}(\mathbf{x})$.

2.4 TESTING FOR THE SIGNIFICANCE OF THE MODEL

Once we have fit a particular multiple (multivariable) logistic regression model, we begin the process of model assessment. As in the univariate case presented in Chapter 1, the first step in this process is usually to assess the significance of the variables in the model. The likelihood ratio test for overall significance of the p coefficients for the independent variables in the model is performed in exactly the same manner as in the univariate case. The test is based on the statistic G given in equation (1.12). The only difference is that the fitted values, $\hat{\pi}$, under the model are based on the vector containing $p+1$ parameters, $\hat{\boldsymbol{\beta}}$. Under the null hypothesis that the p "slope" coefficients for the covariates in the model are equal to zero, the distribution of G will be chi-square with p degrees-of-freedom.

Consider the fitted model whose estimated coefficients are given in Table 2.2. For that model, the value of the log likelihood, shown at the bottom of the table, is $L = -111.286$. The log likelihood for the constant only model may be obtained by evaluating the numerator of equation (1.13) or by fitting the constant only model. Either method yields the log likelihood $L = -117.336$. Thus the value of the likelihood ratio test is, from equation (1.12),

$$G = -2[(-117.336) - (-111.286))] = 12.099$$

and the p-value for the test is $P[\chi^2(5) > 12.099] = 0.034$ which is significant at the $\alpha = 0.05$ level. We reject the null hypothesis in this case and conclude that at least one and perhaps all p coefficients are different from zero, an interpretation analogous to that in multiple linear regression.

Before concluding that any or all of the coefficients are nonzero, we may wish to look at the univariate Wald test statistics,

$$W_j = \hat{\beta}_j / \widehat{\mathrm{SE}}(\hat{\beta}_j).$$

These are given in the fourth column in Table 2.2. Under the hypothesis that an individual coefficient is zero, these statistics will follow the standard normal distribution. The p-values are given in the fifth column of Table 2.2. If we use a level of significance of 0.05, then we would conclude that the variables LWT and possibly RACE are significant, while AGE and FTV are not significant.

If our goal is to obtain the best fitting model while minimizing the number of parameters, the next logical step is to fit a reduced model containing only those variables thought to be significant, and compare it to the full model containing all the variables. The results of fitting the reduced model are given in Table 2.3.

The difference between the two models is the exclusion of the variables AGE and FTV from the full model. The likelihood ratio test comparing these two models is obtained using the definition of G given in equation (1.12). It will have a distribution that is chi-square with 2 degrees-of-freedom under the hypothesis that the coefficients for the

Table 2.3 Estimated Coefficients for a Multiple Logistic Regression Model Using the Variables LWT and RACE from the Low Birth Weight Study

Variable	Coeff.	Std. Err.	z	P>|z|
LWT	−0.015	0.0064	−2.36	0.018
RACE_2	1.081	0.4881	2.22	0.027
RACE_3	0.481	0.3567	1.35	0.178
Constant	0.806	0.8452	0.95	0.340

Log likelihood = −111.630

variables excluded are equal to zero. The value of the test statistic comparing the models in Tables 2.2 and 2.3 is

$$G = -2[(-111.630) - (-111.286)] = 0.688,$$

which, with 2 degrees-of-freedom, has a p-value of $P[\chi^2(2) > 0.688] =$ 0.709. Since the p-value is large, exceeding 0.05, we conclude that the reduced model is as good as the full model. Thus there is no advantage to including AGE and FTV in the model. However, we must not base our models entirely on tests of statistical significance. As we will see in Chapter 5, there are numerous other considerations that will influence our decision to include or exclude variables from a model.

Whenever a categorical independent variable is included (or excluded) from a model, all of its design variables should be included (or excluded); to do otherwise implies that we have recoded the variable. For example, if we only include design variable D_1 as defined in Table 2.1, then race is entered into the model as a dichotomous variable coded as black or not black. If k is the number of levels of a categorical variable, then the contribution to the degrees-of-freedom for the likelihood ratio test for the exclusion of this variable will be $k-1$. For example, if we exclude race from the model, and race is coded at three levels using the design variables shown in Table 2.1, then there would be 2 degrees-of-freedom for the test, one for each design variable.

Because of the multiple degrees-of-freedom we must be careful in our use of the Wald (W) statistics to assess the significance of the coefficients. For example, if the W statistics for both coefficients exceed 2, then we could conclude that the design variables are significant. Alternatively, if one coefficient has a W statistic of 3.0 and the other a value of 0.1, then we cannot be sure about the contribution of the variable to the model. The estimated coefficients for the variable RACE in Table 2.3 provide a good example. The Wald statistic for the coefficient for the first design variable is 2.22, and 1.35 for the second. The likelihood ratio test comparing the model containing LWT and RACE to the one containing only LWT yields

$$G = -2[-(114.345)-(-111.630)] = 5.43,$$

which, with 2 degrees-of-freedom, yields a p-value of 0.066. Strict adherence to the $\alpha = 0.05$ level of significance would justify excluding RACE from the model. However, RACE is known to be a "clinically important" variable. In this case the decision to include or exclude RACE should be made in conjunction with subject matter experts.

In the previous chapter we described, for the univariate model, two other tests equivalent to the likelihood ratio test for assessing the significance of the model, the Wald and Score tests. We will briefly discuss the multivariable versions of these tests, as their use appears occasionally in the literature. These tests are available in some software packages. SAS computes both the likelihood ratio and score tests for a fitted model and STATA has the capability to perform the Wald test easily. For the most part we will use likelihood ratio tests in this text. As noted earlier, we favor the likelihood ratio test as the quantities needed to carry it out may be obtained from all computer packages.

The multivariable analog of the Wald test is obtained from the following vector–matrix calculation:

$$\begin{aligned} W &= \hat{\boldsymbol{\beta}}'\left[\widehat{\text{Var}}\left(\hat{\boldsymbol{\beta}}\right)\right]^{-1}\hat{\boldsymbol{\beta}} \\ &= \hat{\boldsymbol{\beta}}'(\mathbf{X}'\mathbf{V}\mathbf{X})\hat{\boldsymbol{\beta}}, \end{aligned}$$

which will be distributed as chi-square with $p+1$ degrees-of-freedom under the hypothesis that each of the $p+1$ coefficients is equal to zero. Tests for just the p slope coefficients are obtained by eliminating $\hat{\beta}_0$ from $\hat{\boldsymbol{\beta}}$ and the relevant row (first or last) and column (first or last) from $(\mathbf{X}'\mathbf{V}\mathbf{X})$. Since evaluation of this test requires the capability to perform vector-matrix operations and to obtain $\hat{\boldsymbol{\beta}}$, there is no gain over the likelihood ratio test of the significance of the model. Extensions of the Wald test which can be used to examine functions of the coefficients are quite useful and are illustrated in subsequent chapters. In addition, the modeling approach of Grizzle, Starmer, and Koch (1969), noted earlier, contains many such examples.

The multivariable analog of the Score test for the significance of the model is based on the distribution of the p derivatives of $L(\boldsymbol{\beta})$ with respect to $\boldsymbol{\beta}$. The computation of this test is of the same order of complication as the Wald test. To define it in detail would require introduction of additional notation which would find little use in the remainder of this text. Thus, we refer the interested reader to Cox and Hinkley (1974) or Dobson (1990).

2.5 CONFIDENCE INTERVAL ESTIMATION

We discussed confidence interval estimators for the coefficients, logit and logistic probabilities for the simple logistic regression model in Section 1.4. The methods used for confidence interval estimators for a multiple variable model are essentially the same.

The endpoints for a $100(1-\alpha)\%$ confidence interval for the coefficients are obtained from (1.4.1) for slope coefficients and from (1.4.2) for the constant term. For example, using the fitted model presented in Table 2.3, the 95 percent confidence interval for LWT is

$$-0.015 \pm 1.96 \times 0.0064 = (-0.028,\ -0.002).$$

The interpretation of this interval is that we are 95 percent confident that the decrease in the log-odds per one pound increase in weight of the mother is between –0.028 and –0.002. As we noted in Section 1.4 many software packages automatically provide confidence intervals for all model coefficients in the output.

The confidence interval estimator for the logit is a bit more complicated for the multiple variable model than the result presented in (1.19). The basic idea is the same, only there are now more terms involved in the summation. It follows from (2.1) that a general expression for the estimator of the logit for a model containing p covariates is

$$\hat{g}(\mathbf{x}) = \hat{\beta}_0 + \hat{\beta}_1 x_1 + \hat{\beta}_2 x_2 + \cdots + \hat{\beta}_p x_p. \tag{2.6}$$

An alternative way to express the estimator of the logit in (2.6) is through the use of vector notation as $\hat{g}(\mathbf{x}) = \mathbf{x}'\hat{\boldsymbol{\beta}}$, where the vector

$\hat{\boldsymbol{\beta}}' = \left(\hat{\beta}_0, \hat{\beta}_1, \hat{\beta}_2, \ldots, \hat{\beta}_p\right)$ denotes the estimator of the $p+1$ coefficients and the vector $\mathbf{x}' = \left(x_0, x_1, x_2, \ldots, x_p\right)$ represents the constant and a set of values of the p-covariates in the model, where $x_0 = 1$.

It follows from (1.18) that an expression for the estimator of the variance of the estimator of the logit in (2.6) is

$$\hat{\text{Var}}[\hat{g}(\mathbf{x})] = \sum_{j=0}^{p} x_j^2 \hat{\text{Var}}\left(\hat{\beta}_j\right) + \sum_{j=0}^{p} \sum_{k=j+1}^{p} 2x_j x_k \hat{\text{Cov}}\left(\hat{\beta}_j, \hat{\beta}_k\right). \tag{2.7}$$

We can express this result much more concisely by using the matrix expression for the estimator of the variance of the estimator of the coefficients. From the expression for the observed information matrix, we have that

$$\hat{\text{Var}}\left(\hat{\boldsymbol{\beta}}\right) = (\mathbf{X}'\mathbf{V}\mathbf{X})^{-1}. \tag{2.8}$$

It follows from (2.8) that an equivalent expression for the estimator in (2.7) is

$$\begin{aligned}\hat{\text{Var}}[(\hat{g}(\mathbf{x}))] &= \mathbf{x}'\hat{\text{Var}}\left(\hat{\boldsymbol{\beta}}\right)\mathbf{x} \\ &= \mathbf{x}'(\mathbf{X}'\mathbf{V}\mathbf{X})^{-1}\mathbf{x}\,. \end{aligned} \tag{2.9}$$

Fortunately, all good logistic regression software packages provide the option for the user to create a new variable containing the estimated values of (2.9) or the standard error for all subjects in the data set. This feature eliminates the computational burden associated with the matrix calculations in (2.9) and allows the user to routinely calculate fitted values and confidence interval estimates. However it is useful to illustrate the details of the calculations.

Using the model in Table 2.3, the estimated logit for a 150 pound white woman is

$$\begin{aligned}\hat{g}(LWT = 150, RACE = White) &= 0.806 - 0.015 \times 150 + 1.081 \times 0 + 0.481 \times 0 \\ &= -1.444\end{aligned}$$

and the estimated logistic probability is

$$\hat{\pi}(LWT=150, RACE=White)=\frac{e^{-1.444}}{1+e^{-1.444}}=0.191.$$

The interpretation of the fitted value is that the estimated proportion of low birthweight babies among 150 pound white women is 0.191.

In order to use (2.7) to estimate the variance of this estimated logit we need to obtain the estimated covariance matrix shown in Table 2.4. Thus the estimated variance of the logit is

$$\begin{aligned}\widehat{\text{Var}}[\hat{g}(LWT=150, RACE=White)] &= \widehat{\text{Var}}(\hat{\beta}_0)+(150)^2\times\widehat{\text{Var}}(\hat{\beta}_1)+\\ &(0)^2\times\widehat{\text{Var}}(\hat{\beta}_2)+(0)^2\times\widehat{\text{Var}}(\hat{\beta}_3)+2\times150\times\widehat{\text{Cov}}(\hat{\beta}_0,\hat{\beta}_1)\\ &+2\times0\times\widehat{\text{Cov}}(\hat{\beta}_0,\hat{\beta}_2)+2\times0\times\widehat{\text{Cov}}(\hat{\beta}_0,\hat{\beta}_3)+2\times150\times0\times\widehat{\text{Cov}}(\hat{\beta}_1,\hat{\beta}_2)\\ &+2\times150\times0\times\widehat{\text{Cov}}(\hat{\beta}_1,\hat{\beta}_3)+2\times0\times0\times\widehat{\text{Cov}}(\hat{\beta}_2,\hat{\beta}_3)\end{aligned}$$

$$\begin{aligned}&=0.7143+(150)^2\times0.000041+0\times0.2382+0\times0.1272\\ &\quad+2\times150\times(-0.0052)+2\times0\times0.0226+2\times0\times(-0.1035)\\ &\quad+2\times150\times0\times(-0.000647)+2\times150\times0\times0.000036\\ &\quad+2\times0\times0\times0.0532=0.0768\end{aligned}$$

and the standard error is $\widehat{\text{SE}}[\hat{g}(LWT=150, RACE=White)]=0.2771$. The 95 percent confidence interval for the estimated logit is

$$-1.444\pm1.96\times0.2771=(-1.988,\ -0.901).$$

The associated confidence interval for the fitted value is (0.120, 0.289). We defer further discussion and interpretation of the estimated logit, fitted values and their respective confidence intervals until Chapter 3.

Table 2.4 Estimated Covariance Matrix of the Estimated Coefficients in Table 2.3

	LWT	RACE_2	RACE_3	Constant
LWT	0.000041			
RACE_2	−0.000647	0.2382		
RACE_3	0.000036	0.0532	0.1272	
Constant	−0.005211	0.0226	−0.1035	0.7143

2.6 OTHER METHODS OF ESTIMATION

In Section 1.5, two alternative methods of estimating the parameters of the logistic regression model were discussed. These were the methods of noniteratively weighted least squares and discriminant function. Each may also be employed in the multivariable case, though application of the noniteratively weighted least squares estimators is limited by the need for nonzero estimates of $\pi(\mathbf{x})$ for most values of $\mathbf{x}$ in the data set. With a large number of independent variables, or even a few continuous variables, this condition is not likely to hold. The discriminant function estimators do not have this limitation and may be easily extended to the multivariable case.

The discriminant function approach to estimation of the logistic coefficients is based on the assumption that the distribution of the independent variables, given the value of the outcome variable, is multivariate normal. Two points should be kept in mind: (1) the assumption of multivariate normality will rarely if ever be satisfied because of the frequent occurrence of dichotomous independent variables, and (2) the discriminant function estimators of the coefficients for nonnormally distributed independent variables, especially dichotomous variables, will be biased away from zero when the true coefficient is nonzero. For these reasons we, in general, do not recommend its use. However, these estimators are of some historical importance as a number of the classic papers in the applied literature, such as Truett, Cornfield, and Kannel (1967), have used them. These estimators are easily computed and, in the absence of a logistic regression program, should be adequate for a preliminary examination of your data. Thus, it seems worthwhile to include the relevant formulae for their computation.

The assumptions necessary to employ the discriminant function approach to estimating the logistic regression coefficients state that the conditional distribution of X (the vector of p covariate random variables) given the outcome variable, $Y = y$, is multivariate normal with a mean vector that depends on y, but a covariance matrix that does not. Using notation defined in Section 1.5 we say $\mathbf{X} \mid y = j \sim N(\mu_j, \Sigma_j)$ where μ_j contains the means of the p independent variables for the subpopulation defined by $y = j$ and $\mathbf{\Sigma}$ is the $p \times p$ covariance matrix of these variables. Under these assumptions, $P(Y = 1 \mid \mathbf{x}) = \pi(\mathbf{x})$, where the coefficients are given by:

$$\beta_0 = \ln\left(\frac{\theta_1}{\theta_0}\right) - 0.5(\mu_1 - \mu_0)' \, \Sigma^{-1}(\mu_1 + \mu_0) \qquad (2.10)$$

and

$$\boldsymbol{\beta} = (\mu_1 - \mu_0)' \Sigma^{-1}, \tag{2.11}$$

where $\theta_1 = P(Y=1)$ and $\theta_0 = 1 - \theta_1$ denote the proportion of the population with y equal to 1 or 0, respectively. Equations (2.10) and (2.11) are the multivariable analogs of equations (1.22) and (1.23).

The discriminant function estimators of β_0 and $\boldsymbol{\beta}$ are found by substituting estimators for μ_j, j = 0, 1, Σ, and θ_1 into equations (2.10) and (2.11). The estimators most often used are the maximum likelihood estimators under the multivariate normal model. That is, we let

$$\hat{\mu}_j = \bar{\mathbf{x}}_j$$

the mean of $\mathbf{x}$ in the subgroup of the sample with $y = j$, $j = 0,1$.

The estimator of the covariance matrix, Σ, is the multivariable extension of the pooled sample variance given in Section 1.5. This may be represented as

$$\mathbf{S} = \frac{(n_0 - 1)\mathbf{S}_0 + (n_1 - 1)\mathbf{S}_1}{(n + n - 2)},$$

where $\mathbf{S}_j$, $j = 0,1$ is the $p \times p$ matrix of the usual unbiased estimators of the variances and covariances computed within the subgroup defined by $y = j$, $j = 0,1$.

Because of the bias in the discriminant function estimators when normality does not hold, they should be used only when logistic regression software is not available, and then only in preliminary analyses. Any final analyses should be based on the maximum likelihood estimators of the coefficients.

EXERCISES

1. Use the ICU data described in Section 1.6.1 and consider the multiple logistic regression model of vital status, STA, on age (AGE), cancer part of the present problem (CAN), CPR prior to ICU admission (CPR), infection probable at ICU admission (INF), and race (RACE).

(a) The variable RACE is coded at three levels. Prepare a table showing the coding of the two design variables necessary for including this variable in a logistic regression model.

(b) Write down the equation for the logistic regression model of STA on AGE, CAN, CPR, INF, and RACE. Write down the equation for the logit transformation of this logistic regression model. How many parameters does this model contain?

(c) Write down an expression for the likelihood and log likelihood for the logistic regression model in Exercise 1(b). How many likelihood equations are there? Write down an expression for a typical likelihood equation for this problem.

(d) Using a logistic regression package, obtain the maximum likelihood estimates of the parameters of the logistic regression model in Exercise 1(b). Using these estimates write down the equation for the fitted values, that is, the estimated logistic probabilities.

(e) Using the results of the output from the logistic regression package used in Exercise 1(d), assess the significance of the slope coefficients for the variables in the model using the likelihood ratio test. What assumptions are needed for the p-values computed for this test to be valid? What is the value of the deviance for the fitted model?

(f) Use the Wald statistics to obtain an approximation to the significance of the individual slope coefficients for the variables in the model. Fit a reduced model that eliminates those variables with nonsignificant Wald statistics. Assess the joint (conditional) significance of the variables excluded from the model. Present the results of fitting the reduced model in a table.

(g) Using the results from Exercise 1(f), compute 95 percent confidence intervals for all coefficients in the model. Write a sentence interpreting the confidence intervals for the non-constant covariates.

(h) Obtain the estimated covariance matrix for the final model fit in Exercise 1(f). Choose a set of values for the covariates in that model and estimate the logit and logistic probability for a subject with these characteristics. Compute 95 percent confidence intervals for the logit and estimated logistic probability. Write a sentence or two interpreting the estimated probability and its confidence interval.

2. Use the Prostate Cancer data described in Section 1.6.3 and consider the multiple logistic regression model of capsule penetration (CAPSULE),

on AGE, RACE, results of the digital rectal exam (DPROS and DCAPS), prostate specific antigen (PSA), Gleason score (GLEASON) and tumor volume (VOL).

(a) The variable DPROS is coded at four levels. Prepare a table showing the coding of the three design variables necessary for including this variable in a logistic regression model.

(b) The variable DCAPS is coded 1 and 2. Can this variable be used in its original coding or must a design variable be created? Explore this question by comparing the estimated coefficients obtained from fitting a model containing DCAPS as originally coded with those obtained from one using a 0–1 coded design variable, $DCAPSnew = DCAPS - 1$.

(c) Repeat parts 1(b) – 1(h) of Exercise 1.

CHAPTER 3

Interpretation of the Fitted Logistic Regression Model

3.1 INTRODUCTION

In Chapters 1 and 2 we discussed the methods for fitting and testing for the significance of the logistic regression model. After fitting a model the emphasis shifts from the computation and assessment of significance of the estimated coefficients to the interpretation of their values. Strictly speaking, an assessment of the adequacy of the fitted model should precede any attempt at interpreting it. In the case of logistic regression the methods for assessment of fit are rather technical in nature and thus are deferred until Chapter 5, at which time the reader should have a good working knowledge of the logistic regression model. Thus, we begin this chapter assuming that a logistic regression model has been fit, that the variables in the model are significant in either a clinical or statistical sense, and that the model fits according to some statistical measure of fit.

The interpretation of any fitted model requires that we be able to draw practical inferences from the estimated coefficients in the model. The question being addressed is: *What do the estimated coefficients in the model tell us about the research questions that motivated the study?* For most models this involves the estimated coefficients for the independent variables in the model. On occasion, the intercept coefficient is of interest; but this is the exception, not the rule. The estimated coefficients for the independent variables represent the slope (i.e., rate of change) of a function of the dependent variable per unit of change in the independent variable. Thus, interpretation involves two issues: determining the functional relationship between the dependent variable and the independent variable, and appropriately defining the unit of change for the independent variable.

The first step is to determine what function of the dependent variable yields a linear function of the independent variables. This is called the *link function* [see McCullagh and Nelder (1983) or Dobson (1990)]. In the case of a linear regression model, it is the identity function since the dependent variable, by definition, is linear in the parameters. (For those unfamiliar with the term "identity function," it is the function $y = y$.) In the logistic regression model the link function is the logit transformation $g(x) = \ln\{\pi(x)/[1-\pi(x)]\} = \beta_0 + \beta_1 x$.

For a linear regression model recall that the slope coefficient, β_1, is equal to the difference between the value of the dependent variable at $x+1$ and the value of the dependent variable at x, for any value of x. For example, if $y(x) = \beta_0 + \beta_1 x$, it follows that $\beta_1 = y(x+1) - y(x)$. In this case, the interpretation of the coefficient is relatively straightforward as it expresses the resulting change in the measurement scale of the dependent variable for a unit change in the independent variable. For example, if in a regression of weight on height of male adolescents the slope is 5, then we would conclude that an increase of 1 inch in height is associated with an increase of 5 pounds in weight.

In the logistic regression model, the slope coefficient represents the change in the logit corresponding to a change of one unit in the independent variable (i.e., $\beta_1 = g(x+1) - g(x)$). Proper interpretation of the coefficient in a logistic regression model depends on being able to place meaning on the difference between two logits. Interpretation of this difference is discussed in detail on a case-by-case basis as it relates directly to the definition and meaning of a one-unit change in the independent variable. In the following sections of this chapter we consider the interpretation of the coefficients for a univariate logistic regression model for each of the possible measurement scales of the independent variable. In addition we discuss interpretation of the coefficients in multivariable models.

3.2 DICHOTOMOUS INDEPENDENT VARIABLE

We begin our consideration of the interpretation of logistic regression coefficients with the situation where the independent variable is nominal scale and dichotomous (i.e., measured at two levels). This case provides the conceptual foundation for all the other situations.

We assume that the independent variable, x, is coded as either zero or one. The difference in the logit for a subject with $x = 1$ and $x = 0$ is

$$g(1)-g(0)=[\beta_0+\beta_1]-[\beta_0]=\beta_1.$$

The algebra shown in this equation is rather straightforward. We present it in this level of detail to emphasize that the first step in interpreting the effect of a covariate in a model is to express the desired logit difference in terms of the model. In this case the logit difference is equal to β_1. In order to interpret this result we need to introduce and discuss a measure of association termed the *odds ratio*.

The possible values of the logistic probabilities may be conveniently displayed in a 2 × 2 table as shown in Table 3.1. The *odds* of the outcome being present among individuals with $x=1$ is defined as $\pi(1)/[1-\pi(1)]$. Similarly, the odds of the outcome being present among individuals with $x=0$ is defined as $\pi(0)/[1-\pi(0)]$. The *odds ratio*, denoted OR, is defined as the ratio of the odds for $x=1$ to the odds for $x=0$, and is given by the equation

$$\text{OR}=\frac{\pi(1)/[1-\pi(1)]}{\pi(0)/[1-\pi(0)]}. \tag{3.1}$$

Substituting the expressions for the logistic regression model shown in Table 3.1 into (3.1) we obtain

Table 3.1 Values of the Logistic Regression Model When the Independent Variable Is Dichotomous

Outcome Variable (Y)	Independent Variable (X) $x=1$	$x=0$
$y=1$	$\pi(1)=\dfrac{e^{\beta_0+\beta_1}}{1+e^{\beta_0+\beta_1}}$	$\pi(0)=\dfrac{e^{\beta_0}}{1+e^{\beta_0}}$
$y=0$	$1-\pi(1)=\dfrac{1}{1+e^{\beta_0+\beta_1}}$	$1-\pi(0)=\dfrac{1}{1+e^{\beta_0}}$
Total	1.0	1.0

$$\text{OR} = \frac{\left(\dfrac{e^{\beta_0+\beta_1}}{1+e^{\beta_0+\beta_1}}\right) \Big/ \left(\dfrac{1}{1+e^{\beta_0+\beta_1}}\right)}{\left(\dfrac{e^{\beta_0}}{1+e^{\beta_0}}\right) \Big/ \left(\dfrac{1}{1+e^{\beta_0}}\right)}$$

$$= \frac{e^{\beta_0+\beta_1}}{e^{\beta_0}}$$

$$= e^{(\beta_0+\beta_1)-\beta_0}$$

$$= e^{\beta_1}.$$

Hence, for logistic regression with a dichotomous independent variable coded 1 and 0, the relationship between the odds ratio and the regression coefficient is

$$\text{OR} = e^{\beta_1} \quad . \tag{3.2}$$

This simple relationship between the coefficient and the odds ratio is the fundamental reason why logistic regression has proven to be such a powerful analytic research tool.

The odds ratio is a measure of association which has found wide use, especially in epidemiology, as it approximates how much more likely (or unlikely) it is for the outcome to be present among those with $x = 1$ than among those with $x = 0$. For example, if y denotes the presence or absence of lung cancer and if x denotes whether the person is a smoker, then $\widehat{\text{OR}} = 2$ estimates that lung cancer is twice as likely to occur among smokers than among nonsmokers in the study population. As another example, suppose y denotes the presence or absence of heart disease and x denotes whether or not the person engages in regular strenuous physical exercise. If the estimated odds ratio is $\widehat{\text{OR}} = 0.5$, then occurrence of heart disease is one half as likely to occur among those who exercise than among those who do not in the study population.

The interpretation given for the odds ratio is based on the fact that in many instances it approximates a quantity called the relative risk. This parameter is equal to the ratio $\pi(1)/\pi(0)$. It follows from (3.1) that the odds ratio approximates the relative risk if $[1-\pi(0)]/[1-\pi(1)] \approx 1$. This holds when $\pi(x)$ is small for both $x = 1$ and 0.

Readers who have not had experience with the odds ratio as a measure of association would be advised to spend some time reading about this measure in one of the following texts: Breslow and Day (1980), Kelsey, Thompson, and Evans (1986), Rothman and Greenland (1998) and Schlesselman (1982).

An example may help to clarify what the odds ratio is and how it is computed from the results of a logistic regression program or from a 2×2 table. In many examples of logistic regression encountered in the literature we find that a continuous variable has been dichotomized at some biologically meaningful cutpoint. A more detailed discussion of the rationale and implications for the modeling of such a decision is presented in Chapter 4. With this in mind we use the data displayed in Table 1.1 and create a new variable, AGED, which takes on the value 1 if the age of the subject is greater than or equal to 55 and zero otherwise. The result of cross classifying the dichotomized age variable with the outcome variable CHD is presented in Table 3.2.

The data in Table 3.2 tell us that there were 21 subjects with values ($x = 1, y = 1$), 22 with ($x = 0, y = 1$), 6 with ($x = 1, y = 0$), and 51 with ($x = 0, y = 0$). Hence, for these data, the likelihood function shown in (1.3) simplifies to

$$l(\boldsymbol{\beta}) = \pi(1)^{21} \times \left[1 - \pi(1)\right]^{6} \times \pi(0)^{22} \times \left[1 - \pi(0)\right]^{51}.$$

Use of a logistic regression program to obtain the estimates of β_0 and β_1 yields the results shown in Table 3.3.

The estimate of the odds ratio from (3.2) is $\widehat{\text{OR}} = e^{2.094} = 8.1$. Readers who have had some previous experience with the odds ratio undoubtedly wonder why a logistic regression package was used to obtain the maximum likelihood estimate of the odds ratio, when it could have been obtained directly from the cross-product ratio from Table 3.2, namely,

Table 3.2 Cross-Classification of AGE Dichotomized at 55 Years and CHD for 100 Subjects

	AGED(x)		
CHD(y)	≥ 55 (1)	< 55 (0)	Total
Present (1)	21	22	43
Absent (0)	6	51	57
Total	27	73	100

$$\widehat{\text{OR}} = \frac{21/6}{22/51} = 8.11 .$$

Thus $\hat{\beta}_1 = \ln[(21/6)/(22/51)] = 2.094$. We emphasize here that logistic regression is, in fact, regression even in the simplest case possible. The fact that the data may be formulated in terms of a contingency table provides the basis for interpretation of estimated coefficients as the log of odds ratios.

Along with the point estimate of a parameter, it is a good idea to use a confidence interval estimate to provide additional information about the parameter value. In the case of the odds ratio, OR, for a 2×2 table there is an extensive literature dealing with this problem, much of which is focused on methods when the sample size is small. The reader who wishes to learn more about the available exact and approximate methods should see the papers by Fleiss (1979) and Gart and Thomas (1972). A good summary may be found in the texts by Breslow and Day (1980), Kleinbaum, Kupper, and Morgenstern (1982), and Rothman and Greenland (1998).

The odds ratio, OR, is usually the parameter of interest in a logistic regression due to its ease of interpretation. However, its estimate, $\widehat{\text{OR}}$, tends to have a distribution that is skewed. The skewness of the sampling distribution of $\widehat{\text{OR}}$ is due to the fact that possible values range between 0 and ∞, with the null value equaling 1. In theory, for large enough sample sizes, the distribution of $\widehat{\text{OR}}$ is normal. Unfortunately, this sample size requirement typically exceeds that of most studies. Hence, inferences are usually based on the sampling distribution of $\ln\left(\widehat{\text{OR}}\right) = \hat{\beta}_1$, which tends to follow a normal distribution for much smaller sample sizes. A $100 \times (1-\alpha)\%$ confidence interval (CI) estimate for the odds ratio is obtained by first calculating the endpoints of a con-

Table 3.3 Results of Fitting the Logistic Regression Model to the Data in Table 3.2

Variable	Coeff.	Std. Err.	z	P>|z|
AGED	2.094	0.5285	3.96	<0.001
Constant	−0.841	0.2551	−3.30	0.001

Log likelihood = −58.9795

fidence interval for the coefficient, β_1, and then exponentiating these values. In general, the endpoints are given by the expression

$$\exp\left[\hat{\beta}_1 \pm z_{1-\alpha/2} \times \widehat{\mathrm{SE}}\left(\hat{\beta}_1\right)\right].$$

As an example, consider the estimation of the odds ratio for the dichotomized variable AGED. The point estimate is $\widehat{\mathrm{OR}} = 8.1$ and the endpoints of a 95% CI are

$$\exp(2.094 \pm 1.96 \times 0.529) = (2.9, 22.9).$$

This interval is typical of the confidence intervals seen for odds ratios when the point estimate exceeds 1. The confidence interval is skewed to the right. This confidence interval suggests that CHD among those 55 and older in the study population could be as little as 2.9 times or much as 22.9 times more likely than those under 55, at the 95 percent level of confidence.

Because of the importance of the odds ratio as a measure of association, many software packages automatically provide point and confidence interval estimates based on the exponentiation of each coefficient in a fitted logistic regression model. These quantities provide estimates of odds ratios of interest in only a few special cases (e.g., a dichotomous variable coded zero or one that is not involved in any interactions with other variables). The major goal of this chapter is to provide the methods for using the results of fitted models to provide point and confidence interval estimates of odds ratios that are of interest, regardless of how complex the fitted model may be.

Before concluding the dichotomous variable case, it is important to consider the effect that the coding of the variable has on the computation of the estimated odds ratio. In the previous discussion we noted that the estimate of the odds ratio was $\widehat{\mathrm{OR}} = \exp\left(\hat{\beta}_1\right)$. This is correct when the independent variable is coded as 0 or 1. Other coding may require that we calculate the value of the logit difference for the specific coding used, and then exponentiate this difference to estimate the odds ratio.

We illustrate these computations in detail, as they demonstrate the general method for computing estimates of odds ratios in logistic regression. The estimate of the log of the odds ratio for any independent variable at two different levels, say $x=a$ versus $x=b$, is the difference between the estimated logits computed at these two values,

$$\begin{aligned}\ln\left[\hat{\text{OR}}(a,b)\right] &= \hat{g}(x=a)-\hat{g}(x=b)\\ &= \left(\hat{\beta}_0+\hat{\beta}_1\times a\right)-\left(\hat{\beta}_0+\hat{\beta}_1\times b\right)\\ &= \hat{\beta}_1\times(a-b).\end{aligned} \tag{3.3}$$

The estimate of the odds ratio is obtained by exponentiating the logit difference,

$$\hat{\text{OR}}(a,b)=\exp\left[\hat{\beta}_1\times(a-b)\right]. \tag{3.4}$$

Note that this expression is equal to $\exp\left(\hat{\beta}_1\right)$ only when $(a-b)=1$. In (3.3) and (3.4) the notation $\hat{\text{OR}}(a,b)$ is used to represent the odds ratio

$$\hat{\text{OR}}(a,b)=\frac{\hat{\pi}(x=a)/\left[1-\hat{\pi}(x=a)\right]}{\hat{\pi}(x=b)/\left[1-\hat{\pi}(x=b)\right]} \tag{3.5}$$

and when $a=1$ and $b=0$ we let $\hat{\text{OR}}=\hat{\text{OR}}(1,0)$.

Some software packages offer a choice of methods for coding design variables. The "zero-one" coding used so far in this section is frequently referred to as *reference cell* coding. The reference cell method typically assigns the value of zero to the lower code for x and one to the higher code. For example, if SEX was coded as 1 = male and 2 = female, then the resulting design variable under this method, D, would be coded 0 = male and 1 = female. Exponentiation of the estimated coefficient for D would estimate the odds ratio of female relative to male. This same result would have been obtained had sex been coded originally as 0 = male and 1 = female, and then treating the variable SEX as if it were interval scaled.

Another coding method is frequently referred to as *deviation from means* coding. This method assigns the value of -1 to the lower code, and a value of 1 to the higher code. The coding for the variable SEX discussed above is shown in Table 3.4.

Table 3.4 Illustration of the Coding of the Design Variable Using the Deviation from Means Method

SEX (Code)	Design Variable D
Male (1)	-1
Female (2)	1

Suppose we wish to estimate the odds ratio of female versus male when deviation from means coding is used. We do this by using the general method shown in (3.3) and (3.4),

$$\begin{aligned}\ln\left[\widehat{\mathrm{OR}}(\text{female},\text{male})\right] &= \hat{g}(\text{female}) - \hat{g}(\text{male}) \\ &= g(D=1) - g(D=-1) \\ &= \left[\hat{\beta}_0 + \hat{\beta}_1 \times (D=1)\right] - \left[\hat{\beta}_0 + \hat{\beta}_1 \times (D=-1)\right] \\ &= 2\hat{\beta}_1\end{aligned}$$

and the estimated odds ratio is $\widehat{\mathrm{OR}}(\text{female},\text{male}) = \exp\left(2\hat{\beta}_1\right)$. Thus, if we had exponentiated the coefficient from the computer output we would have obtained the wrong estimate of the odds ratio. This points out quite clearly that we must pay close attention to the method used to code the design variables.

The method of coding also influences the calculation of the endpoints of the confidence interval. For the above example, using the deviation from means coding, the estimated standard error needed for confidence interval estimation is $\widehat{\mathrm{SE}}\left(2\hat{\beta}_1\right)$ which is $2 \times \widehat{\mathrm{SE}}\left(\hat{\beta}_1\right)$. Thus the endpoints of the confidence interval are

$$\exp\left[2\hat{\beta}_1 \pm z_{1-\alpha/2} 2\widehat{\mathrm{SE}}\left(\hat{\beta}_1\right)\right].$$

In general, the endpoints of the confidence interval for the odds ratio given in (3.5) are

$$\exp\left[\hat{\beta}_1 (a-b) \pm z_{1-\alpha/2} |a-b| \times \widehat{\mathrm{SE}}\left(\hat{\beta}_1\right)\right],$$

where $|a-b|$ is the absolute value of $(a-b)$. Since we can control how we code our dichotomous variables, we recommend that, in most situations, they be coded as 0 or 1 for analysis purposes. Each dichotomous variable is then treated as an interval scale variable.

In summary, for a dichotomous variable the parameter of interest is the odds ratio. An estimate of this parameter may be obtained from the estimated logistic regression coefficient, regardless of how the variable is coded. This relationship between the logistic regression coefficient and the odds ratio provides the foundation for our interpretation of all logistic regression results.

3.3 POLYCHOTOMOUS INDEPENDENT VARIABLE

Suppose that instead of two categories the independent variable has $k > 2$ distinct values. For example, we may have variables that denote the county of residence within a state, the clinic used for primary health care within a city, or race. Each of these variables has a fixed number of discrete values and the scale of measurement is nominal. We saw in Chapter 2 that it is inappropriate to model a nominal scale variable as if it were an interval scale variable. Therefore, we must form a set of design variables to represent the categories of the variable. In this section we present methods for creating design variables for polychotomous independent variables. The choice of a particular method depends to some extent on the goals of the analysis and the stage of model development.

We begin by extending the method presented in Table 2.1 for a dichotomous variable. For example, suppose that in a study of CHD the variable RACE is coded at four levels, and that the cross-classification of

Table 3.5 Cross-Classification of Hypothetical Data on RACE and CHD Status for 100 Subjects

CHD Status	White	Black	Hispanic	Other	Total
Present	5	20	15	10	50
Absent	20	10	10	10	50
Total	25	30	25	20	100
Odds Ratio	1	8	6	4	
95 % CI		(2.3, 27.6)	(1.7, 21.3)	(1.1, 14.9)	
$\ln(\widehat{OR})$	0.0	2.08	1.79	1.39	

Table 3.6 Specification of the Design Variables for RACE Using Reference Cell Coding with White as the Reference Group

	Design Variables		
RACE(Code)	RACE_2	RACE_3	RACE_4
White (1)	0	0	0
Black (2)	1	0	0
Hispanic (3)	0	1	0
Other (4)	0	0	1

RACE by CHD status yields the data in Table 3.5. These data are hypothetical and have been formulated for ease of computation. The extension to a situation where the variable has more than four levels is not conceptually different, so all the examples in this section use $k = 4$.

At the bottom of Table 3.5, the odds ratio is given for each race, using White as the reference group. For example, for Hispanic the estimated odds ratio is $15 \times 20/5 \times 10$. The log of each odds ratio is given in the last row of Table 3.5. This table is typical of what is found in the literature. The reference group is indicated by a value of 1 for the odds ratio. These same estimates of the odds ratio may be obtained from a logistic regression program with an appropriate choice of design variables. The method for specifying the design variables involves setting all of them equal to zero for the reference group, and then setting a single design variable equal to 1 for each of the other groups. This is illustrated in Table 3.6. As noted in Section 3.2 this method is usually referred to as *reference cell* coding and is the default method in many packages.

Use of any logistic regression program with design variables coded as shown in Table 3.6 yields the estimated logistic regression coefficients given in Table 3.7.

A comparison of the estimated coefficients in Table 3.7 to the log odds ratios in Table 3.5 shows that

$$\ln\left[\widehat{\text{OR}}(\text{Black, White})\right] = \hat{\beta}_1 = 2.079,$$

$$\ln\left[\widehat{\text{OR}}(\text{Hispanic, White})\right] = \hat{\beta}_2 = 1.792,$$

and

Table 3.7 Results of Fitting the Logistic Regression Model to the Data in Table 3.5 Using the Design Variables in Table 3.6

Variable	Coeff.	Std. Err.	z	P>|z|
RACE_2	2.079	0.6325	3.29	0.001
RACE_3	1.792	0.6466	2.78	0.006
RACE_4	1.386	0.6708	2.07	0.039
Constant	−1.386	0.5000	−2.77	0.006

Log likelihood = −62.2937

$$\ln\left[\widehat{\text{OR}}(\text{Other, White})\right] = \hat{\beta}_3 = 1.386.$$

Did this happen by chance? Calculation of the logit difference shows that it is by design. The comparison of Black to White is as follows:

$$\begin{aligned}\ln\left[\widehat{\text{OR}}(\text{Black, White})\right] &= \hat{g}(\text{Black}) - \hat{g}(\text{White}) \\ &= \begin{bmatrix} \hat{\beta}_0 + \hat{\beta}_1 \times (RACE\_2 = 1) + \hat{\beta}_2 \times (RACE\_3 = 0) \\ + \hat{\beta}_3 \times (RACE\_4 = 0) \end{bmatrix} \\ &\quad - \begin{bmatrix} \hat{\beta}_0 + \hat{\beta}_1 \times (\text{RACE\_2} = 0) + \hat{\beta}_2 \times (RACE\_3 = 0) \\ + \hat{\beta}_3 \times (RACE\_4 = 0) \end{bmatrix} \\ &= \hat{\beta}_1.\end{aligned}$$

Similar calculations would demonstrate that the other coefficients estimated using logistic regression are also equal to the log of odds ratios computed from the data in Table 3.5.

A comment about the estimated standard errors may be helpful at this point. In the univariate case the estimates of the standard errors found in the logistic regression output are identical to the estimates obtained using the cell frequencies from the contingency table. For example, the estimated standard error of the estimated coefficient for the design variable RACE_2 is

$$\widehat{\text{SE}}\left(\hat{\beta}_1\right) = \left[\frac{1}{5} + \frac{1}{20} + \frac{1}{20} + \frac{1}{10}\right]^{0.5} = 0.6325.$$

A derivation of this result may be found in Bishop, Feinberg, and Holland (1975).

Confidence limits for odds ratios are obtained using the same approach used in Section 3.2 for a dichotomous variable. We begin by computing the confidence limits for the log odds ratio (the logistic regression coefficient) and then exponentiate these limits to obtain limits for the odds ratio. In general, the limits for a $100(1-\alpha)\%$ CIE for the coefficient are of the form

$$\hat{\beta}_j \pm z_{1-\alpha/2} \times \widehat{\mathrm{SE}}\left(\hat{\beta}_j\right).$$

The corresponding limits for the odds ratio, obtained by exponentiating these limits, are as follows:

$$\exp\left[\hat{\beta}_j \pm z_{1-\alpha/2} \times \widehat{\mathrm{SE}}\left(\hat{\beta}_j\right)\right]. \quad (3.6)$$

The confidence limits given in Table 3.5 in the row beneath the estimated odds ratios were obtained using the estimated coefficients and standard errors in Table 3.7 with (3.6) for $j=1,2,3$ with $\alpha=0.05$.

Reference cell coding is the most commonly employed coding method appearing in the literature. The primary reason for the widespread use of this method is the interest in estimating the risk of an "exposed" group relative to that of a "control" or "unexposed" group.

As discussed in Section 3.2 a second method of coding design variables is called *deviation from means* coding. This coding expresses effect as the deviation of the "group mean" from the "overall mean." In the case of logistic regression, the "group mean" is the logit for the

Table 3.8 Specification of the Design Variables for RACE Using Deviation from Means Coding

	Design Variables		
RACE(Code)	RACE_2	RACE_3	RACE_4
White (1)	−1	−1	−1
Black (2)	1	0	0
Hispanic (3)	0	1	0
Other (4)	0	0	1

Table 3.9 Results of Fitting the Logistic Regression Model to the Data in Table 3.5 Using the Design Variables in Table 3.8

Variable	Coeff.	Std. Err.	z	P>|z|
RACE_2	0.765	0.3506	2.18	0.029
RACE_3	0.477	0.3623	1.32	0.188
RACE_4	0.072	0.3846	0.19	0.852
Constant	−0.072	0.2189	−0.33	0.742

Log likelihood = −62.2937

group and the "overall mean" is the average logit over all groups. This method of coding is obtained by setting the value of all the design variables equal to −1 for one of the categories, and then using the 0, 1 coding for the remainder of the categories. Use of the deviation from means coding for race shown in Table 3.8 yields the estimated logistic regression coefficients in Table 3.9.

In order to interpret the estimated coefficients in Table 3.9 we need to refer to Table 3.5 and calculate the logit for each of the four categories of RACE. These are

$$\hat{g}_1 = \ln\left(\frac{5/25}{20/25}\right) = \ln\left(\frac{5}{20}\right) = -1.386$$

$\hat{g}_2 = \ln(20/10) = 0.693$, $\hat{g}_3 = \ln(15/10) = 0.405$, $\hat{g}_4 = \ln(10/10) = 0$, and their average is $\bar{g} = \sum \hat{g}_i/4 - 0.072$. The estimated coefficient for design variable RACE_2 in Table 3.9 is $\hat{g}_2 - \bar{g} = 0.693 - (-0.072) = 0.765$. The general relationship for the estimated coefficient for design variable RACE_j is $\hat{g}_j - \bar{g}$, for $j = 2,3,4$.

The interpretation of the estimated coefficients is not as easy or clear as in the situation when a reference group is used. Exponentiation of the estimated coefficients yields the ratio of the odds for the particular group to the geometric mean of the odds. Specifically, for RACE_2 in Table 3.9 we have

$$\begin{aligned}\exp(0.765) &= \exp(\hat{g}_2 - \bar{g}) \\ &= \exp(\hat{g}_2)\Big/\exp\left(\sum \hat{g}_j/4\right) \\ &= (20/10)\Big/[(5/20)\times(20/10)\times(15/10)\times(10/10)]^{0.25} \\ &= 2.15\ .\end{aligned}$$

This number, 2.15, is not a true odds ratio because the quantities in the numerator and denominator do not represent the odds for two distinct categories. The exponentiation of the estimated coefficient expresses the odds relative to an "average" odds, the geometric mean. The interpretation of this value depends on whether the "average" odds is in fact meaningful.

The estimated coefficients obtained using deviation from means coding may be used to estimate the odds ratio for one category relative to a reference category. The equation for the estimate is more complicated than the one obtained using the reference cell coding. However, it provides an excellent example of the basic principle of using the logit difference to compute an odds ratio.

To illustrate this we calculate the log odds ratio of Black versus White using the coding for design variables given in Table 3.8. The logit difference is as follows:

$$\begin{aligned}\ln\left[\widehat{\text{OR}}(\text{Black, White})\right] &= \hat{g}(\text{Black}) - \hat{g}(\text{White}) \\ &= \begin{bmatrix}\hat{\beta}_0 + \hat{\beta}_1 \times (RACE\_2 = 1) + \hat{\beta}_2 \times (RACE\_3 = 0) \\ + \hat{\beta}_3 \times (RACE\_4 = 0)\end{bmatrix} \\ &\quad - \begin{bmatrix}\hat{\beta}_0 + \hat{\beta}_1 \times (RACE\_2 = -1) + \hat{\beta}_2 \times (RACE\_3 = -1) \\ + \hat{\beta}_3 \times (RACE\_4 = -1)\end{bmatrix} \\ &= 2\hat{\beta}_1 + \hat{\beta}_2 + \hat{\beta}_3. \end{aligned} \tag{3.7}$$

To obtain a confidence interval we must estimate the variance of the sum of the coefficients in (3.7). In this example, the estimator is

$$\begin{aligned}\widehat{\text{Var}}\left\{\ln\left[\widehat{\text{OR}}(\text{Black, White})\right]\right\} &= 4 \times \widehat{\text{Var}}(\hat{\beta}_1) + \widehat{\text{Var}}(\hat{\beta}_2) \\ &\quad + \widehat{\text{Var}}(\hat{\beta}_3) + 4 \times \widehat{\text{Cov}}(\hat{\beta}_1, \hat{\beta}_2) \\ &\quad + 4 \times \widehat{\text{Cov}}(\hat{\beta}_1, \hat{\beta}_3) + 2 \times \widehat{\text{Cov}}(\hat{\beta}_2, \hat{\beta}_3). \end{aligned} \tag{3.8}$$

Values for each of the estimators in (3.8) may be obtained from output that is available from logistic regression software. Confidence intervals for the odds ratio are obtained by exponentiating the endpoints of the confidence limits for the sum of the coefficients in (3.7). Evaluation of (3.7) for the current example gives

$$\ln\left[\widehat{OR}(\text{Black},\text{White})\right]=2(0.765)+0.477+0.072=2.079.$$

The estimate of the variance is obtained by evaluating (3.8) which, for the current example, yields

$$\widehat{\text{Var}}\left\{\ln\left[\widehat{OR}(\text{Black},\text{White})\right]\right\}=4(0.351)^2+(0.362)^2+(0.385)^2$$
$$+4(-0.031)+4(-0.040)+2(-0.044)=0.400$$

and the estimated standard error is

$$\widehat{\text{SE}}\left\{\ln\left[\widehat{OR}(\text{Black},\text{White})\right]\right\}=0.6325.$$

We note that the values of the estimated log odds ratio, 2.079, and the estimated standard error, 0.6325, are identical to the values of the estimated coefficient and standard error for the first design variable in Table 3.7. This is expected, since the design variables used to obtain the estimated coefficients in Table 3.7 were formulated specifically to yield the log odds ratio relative to the White race category.

It should be apparent that, if the objective is to obtain odds ratios, use of deviation from means coding for design variables is computationally much more complex than reference cell coding.

In summary, we have shown that discrete nominal scale variables are included properly into the analysis only when they have been recoded into design variables. The particular choice of design variables depends on the application, though the reference cell coding is the easiest to interpret, and thus is the one we use in the remainder of this text.

3.4 CONTINUOUS INDEPENDENT VARIABLE

When a logistic regression model contains a continuous independent variable, interpretation of the estimated coefficient depends on how it is entered into the model and the particular units of the variable. For purposes of developing the method to interpret the coefficient for a continuous variable, we assume that the logit is linear in the variable. Other modeling strategies that examine this assumption are presented in Chapter 4.

Under the assumption that the logit is linear in the continuous covariate, x, the equation for the logit is $g(x)=\beta_0+\beta_1 x$. It follows that the slope coefficient, β_1, gives the change in the log odds for an increase of "1" unit in x, that is, $\beta_1=g(x+1)-g(x)$ for any value of x. Most often the value of "1" is not clinically interesting. For example, a 1 year increase in age or a 1 mm Hg increase in systolic blood pressure may be too small to be considered important. A change of 10 years or 10 mm Hg might be considered more useful. On the other hand, if the range of x is from zero to 1, then a change of 1 is too large and a change of 0.01 may be more realistic. Hence, to provide a useful interpretation for continuous scale covariates we need to develop a method for point and interval estimation for an arbitrary change of "c" units in the covariate.

The log odds ratio for a change of c units in x is obtained from the logit difference $g(x+c)-g(x)=c\beta_1$ and the associated odds ratio is obtained by exponentiating this logit difference, $\text{OR}(c)=\text{OR}(x+c,x)$ $=\exp(c\beta_1)$. An estimate may be obtained by replacing β_1 with its maximum likelihood estimate $\hat{\beta}_1$. An estimate of the standard error needed for confidence interval estimation is obtained by multiplying the estimated standard error of $\hat{\beta}_1$ by c. Hence the endpoints of the $100(1-\alpha)\%$ CI estimate of $\text{OR}(c)$ are

$$\exp\left[c\hat{\beta}_1 \pm z_{1-\alpha/2}c\ \widehat{\text{SE}}\left(\hat{\beta}_1\right)\right].$$

Since both the point estimate and endpoints of the confidence interval depend on the choice of c, the particular value of c should be clearly specified in all tables and calculations. The rather arbitrary nature of the choice of c may be troublesome to some. For example, why use a change of 10 years when 5 or 15 or even 20 years may be equally good? We, of course, could use any reasonable value; but the goal must be kept in mind: to provide the reader of your analysis with a clear indication of how the risk of the outcome being present changes with the variable in question. Changes in multiples of 5 or 10 may be most meaningful and easily understood.

As an example, consider the univariate model in Table 1.3. In that example a logistic regression of AGE on CHD status using the data of Table 1.1 was reported. The resulting estimated logit was $\hat{g}(\text{AGE}) = -5.310 + 0.111 \times \text{AGE}$. The estimated odds ratio for an increase of 10 years in age is $\widehat{\text{OR}}(10) = \exp(10 \times 0.111) = 3.03$. This indicates that for every increase of 10 years in age, the risk of CHD increases 3.03 times. The validity of such a statement is questionable in this example, since the additional risk of CHD for a 40 year-old compared to a 30 year-old may be quite different from the additional risk of CHD for a 60 year-old compared to a 50 year-old. This is an unavoidable dilemma when continuous covariates are modeled linearly in the logit. If it is believed that the logit is not linear in the covariate, then grouping and use of dummy variables should be considered. Alternatively, use of higher order terms (e.g., $x^2, x^3, \ldots$) or other nonlinear scaling in the covariate (e.g., $\log(x)$) could be considered. Thus, we see that an important modeling consideration for continuous covariates is their scale in the logit. We consider this in considerable detail in Chapter 4. The endpoints of a 95% confidence interval for this odds ratio are

$$\exp(10 \times 0.111 \pm 1.96 \times 10 \times 0.024) = (1.90, 4.86).$$

Results similar to these may be placed in tables displaying the results of a fitted logistic regression model.

In summary, the interpretation of the estimated coefficient for a continuous variable is similar to that of nominal scale variables: an estimated log odds ratio. The primary difference is that a meaningful change must be defined for the continuous variable.

3.5 THE MULTIVARIABLE MODEL

In the previous sections in this chapter we discussed the interpretation of an estimated logistic regression coefficient in the case when there is a single variable in the fitted model. Fitting a series of univariate models rarely provides an adequate analysis of the data in a study since the independent variables are usually associated with one another and may have different distributions within levels of the outcome variable. Thus, one generally considers a multivariable analysis for a more comprehensive modeling of the data. One goal of such an analysis is to *statistically adjust* the estimated effect of each variable in the model for differences in the distributions of and associations among the other independent variables. Applying this concept to a multivariable logistic regression model, we may surmise that each estimated coefficient provides an estimate of the log odds adjusting for all other variables included in the model.

A full understanding of the estimates of the coefficients from a multivariable logistic regression model requires that we have a clear understanding of what is actually meant by the term *adjusting, statistically, for other variables*. We begin by examining adjustment in the context of a linear regression model, and then extend the concept to logistic regression.

The multivariable situation we examine is one in which the model contains two independent variables — one dichotomous and one continuous — but primary interest is focused on the effect of the dichotomous variable. This situation is frequently encountered in epidemiologic research when an exposure to a risk factor is recorded as being either present or absent, and we wish to adjust for a variable such as age. The analogous situation in linear regression is called *analysis of covariance*.

Suppose we wish to compare the mean weight of two groups of boys. It is known that weight is associated with many characteristics, one of which is age. Assume that on all characteristics except age the two groups have nearly identical distributions. If the age distribution is also the same for the two groups, then a univariate analysis would suffice and we could compare the mean weight of the two groups. This comparison would provide us with a correct estimate of the difference in weight between the two groups. However, if one group was much younger than the other group, then a comparison of the two groups would be meaningless, since at least a portion of any difference observed would likely be due to the difference in age. It would not be possible to determine the effect of group without first eliminating the discrepancy in ages between the groups.

This situation is described graphically in Figure 3.1. In this figure it is assumed that the relationship between age and weight is linear, with the same significant nonzero slope in each group. Both of these assumptions would usually be tested in an analysis of covariance before making any inferences about group differences. We defer a discussion

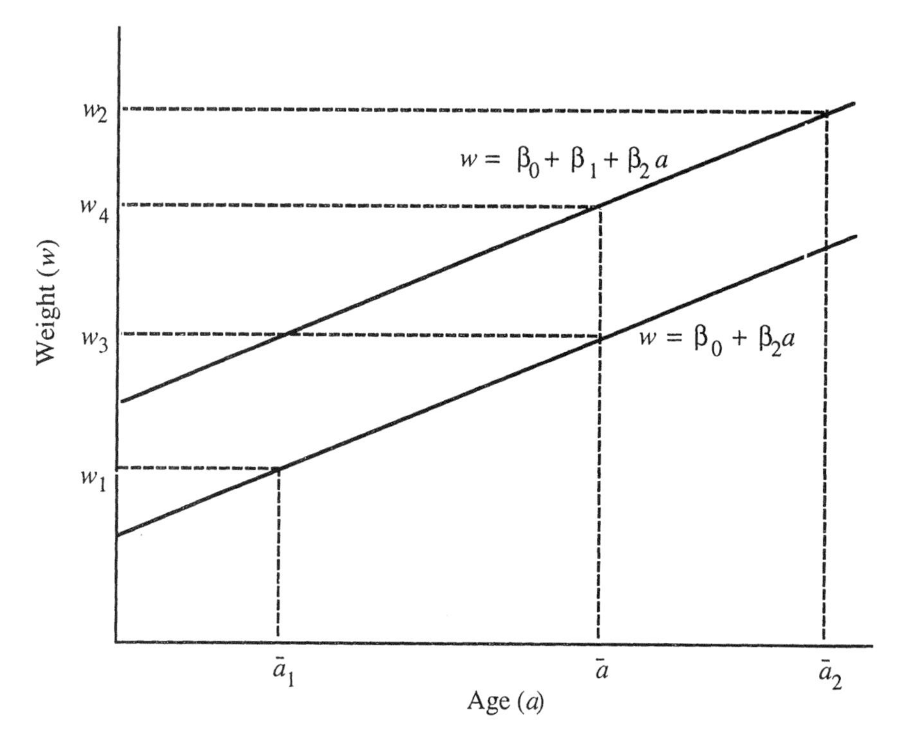

Figure 3.1 Comparison of the weight of two groups of boys with different distributions of age.

of this until Chapter 4, as it gets to the heart of modeling with logistic regression. We proceed as if these assumptions have been checked and are supported by the data.

The statistical model that describes the situation in Figure 3.1 states that the value of weight, w, may be expressed as $w = \beta_0 + \beta_1 x + \beta_2 a$, where $x = 0$ for group 1 and $x = 1$ for group 2 and "a" denotes age. In this model the parameter β_1 represents the true difference in weight between the two groups and β_2 is the rate of change in weight per year of age. Suppose that the mean age of group 1 is $\bar{a}_1$ and the mean age of group 2 is $\bar{a}_2$. These values are indicated in Figure 3.1. Comparison of the mean weight of group 1 to the mean weight of group 2 amounts to a comparison of w_1 to w_2. In terms of the model this difference is $(w_2 - w_1) = \beta_1 + \beta_2(\bar{a}_2 - \bar{a}_1)$. Thus the comparison involves not only the

Table 3.10 Descriptive Statistics for Two Groups of 50 Men on AGE and Whether They Had Seen a Physician (PHY) (1 = Yes, 0 = No) Within the Last Six Months

	Group 1		Group 2	
Variable	Mean	Std. Dev.	Mean	Std. Dev
PHY	0.36	0.485	0.80	0.404
AGE	39.60	5.272	47.34	5.259

true difference between the groups, β_1, but a component, $\beta_2(\bar{a}_2 - \bar{a}_1)$, which reflects the difference between the ages of the groups.

The process of statistically adjusting for age involves comparing the two groups at some common value of age. The value usually used is the mean of the two groups which, for the example, is denoted by $\bar{a}$ in Figure 3.1. In terms of the model this yields a comparison of w_4 to w_3, $(w_4 - w_3) = \beta_1 + \beta_2(\bar{a} - \bar{a}) = \beta_1$, the true difference between the two groups. In theory any common value of age could be used, as it would yield the same difference between the two lines. The choice of the overall mean makes sense for two reasons: it is biologically reasonable and lies within the range for which we believe that the association between age and weight is linear and constant within each group.

Consider the same situation shown in Figure 3.1, but instead of weight being the dependent variable, assume it is a dichotomous variable and that the vertical axis denotes the logit. That is, under the model the logit is given by the equation $g(x,a) = \beta_0 + \beta_1 x + \beta_2 a$. A univariate comparison obtained from the 2 × 2 table cross-classifying outcome and group would yield a log odds ratio approximately equal to $\beta_1 + \beta_2(\bar{a}_2 - \bar{a}_1)$. This would incorrectly estimate the effect of group due to the difference in the distribution of age. To account or adjust for this difference, we include age in the model and calculate the logit difference at a common value of age, such as the combined mean, $\bar{a}$. This logit difference is $g(x=1,\bar{a}) - g(x=0,\bar{a}) = \beta_1$. Thus, the coefficient β_1 is the log odds ratio that we would expect to obtain from a univariate comparison if the two groups had the same distribution of age.

The data summarized in Table 3.10 provide the basis for an example of interpreting the estimated logistic regression coefficient for a dichotomous variable when the coefficient is adjusted for a continuous variable.

It follows from the descriptive statistics in Table 3.10 that the univariate log odds ratio for group 2 versus group 1 is

$$\ln\left(\widehat{OR}\right) = \ln(0.8/0.2) - \ln(0.36/0.64) = 1.962,$$

and the unadjusted estimated odds ratio is $\widehat{OR} = 7.11$. We can also see that there is a considerable difference in the age distribution of the two groups, the men in group 2 being on average more than 7 years older than those in group 1. We would guess that much of the apparent difference in the proportion of men seeing a physician might be due to age. Analyzing the data with a bivariate model using a coding of GROUP = 0 for group 1, and GROUP = 1 for group 2, yields the estimated logistic regression coefficients shown in Table 3.11. The age-adjusted log odds ratio is given by the estimated coefficient for group in Table 3.11 and is $\hat{\beta}_1 = 1.263$. The age adjusted odds ratio is $\widehat{OR} = \exp(1.263) = 3.54$. Thus, much of the apparent difference between the two groups is, in fact, due to differences in age.

Let us examine this adjustment in more detail using Figure 3.1. An approximation to the unadjusted odds ratio is obtained by exponentiating the difference $w_2 - w_1$. In terms of the fitted logistic regression model shown in Table 3.11 this difference is

$$[-4.866 + 1.263 + 0.107(47.34)] - [-4.866 + 0.107(39.60)] =$$
$$1.263 + 0.107(47.34 - 39.60).$$

The value of this odds ratio is

$$e^{[1.263+0.107(47.34-39.60)]} = 8.09.$$

The discrepancy between 8.09 and the actual unadjusted odds ratio, 7.11, is due to the fact that the above comparison is based on the difference in the average logit, while the crude odds ratio is approximately equal to a calculation based on the average estimated logistic probability for the two groups. The age adjusted odds ratio is obtained by exponentiating the difference $w_4 - w_3$, which is equal to the estimated coefficient for GROUP. In the example the difference is

$$[-4.866 + 1.263 + 0.107(43.47)] - [-4.866 + 0.107(43.47)] = 1.263.$$

Table 3.11 Results of Fitting the Logistic Regression Model to the Data Summarized in Table 3.10

Variable	Coeff.	Std. Err.	z	P>|z|
GROUP	1.263	0.5361	2.36	0.018
AGE	0.107	0.0465	2.31	0.021
Constant	−4.866	1.9020	−2.56	0.011

Log likelihood = −54.8292

Bachand and Hosmer (1999) compare two different sets of criteria for defining a covariate to be a confounder. They show that the numeric approach used in this Section, examining the change in the magnitude of the coefficient for the risk factor from logistic regression models fit with and without the potential confounder, is appropriate when the logistic regression model containing both risk factor and confounder is not fully S-shaped. A more detailed evaluation is needed when the fitted model yields fitted values producing a full S-shaped function within the levels of the risk factor. This is discussed in greater detail in Chapter 4.

The method of adjustment when the variables are all dichotomous, polychotomous, continuous or a mixture of these is identical to that just described for the dichotomous-continuous variable case. For example, suppose that instead of treating age as continuous it was dichotomized using a cutpoint of 45 years. To obtain the age-adjusted effect of group we fit the bivariate model containing the two dichotomous variables and calculate a logit difference at the two levels of group and a common value of the dichotomous variable for age. The procedure is similar for any number and mix of variables. Adjusted odds ratios are obtained by comparing individuals who differ only in the characteristic of interest and have the values of all other variables constant. The adjustment is statistical as it only estimates what might be expected to be observed had the subjects indeed differed only on the particular characteristic being examined, with all other variables having identical distributions within the two levels of outcome.

One point should be kept clearly in mind when interpreting statistically adjusted log odds ratios and odds ratios. The effectiveness of the adjustment is entirely dependent on the adequacy of the assumptions of the model: linearity and constant slope. Departures from these may render the adjustment useless. One such departure, where the relationship is linear but the slopes differ, is called *interaction*. Modeling interactions is discussed in Section 3.6 and again in Chapter 4.

3.6 INTERACTION AND CONFOUNDING

In the last section we saw how the inclusion of additional variables in a model provides a way of statistically adjusting for potential differences in their distributions. The term confounder is used by epidemiologists to describe a covariate that is associated with both the outcome variable of interest and a primary independent variable or risk factor. When both associations are present then the relationship between the risk factor and the outcome variable is said to be *confounded*. The procedure for adjusting for confounding, described in Section 3.5, is appropriate when there is no interaction. In this section we introduce the concept of interaction and show how we can control for its effect in the logistic regression model. In addition, we illustrate with an example how confounding and interaction may affect the estimated coefficients in the model.

Interaction can take many different forms, so we begin by describing the situation when it is absent. Consider a model containing a dichotomous risk factor variable and a continuous covariate, such as in the example discussed in Section 3.5. If the association between the covariate (i.e., age) and the outcome variable is the same within each level of the risk factor (i.e., group), then there is no interaction between the covariate and the risk factor. Graphically, the absence of interaction yields a model with two parallel lines, one for each level of the risk factor variable. In general, the absence of interaction is characterized by a model that contains no second or higher order terms involving two or more variables.

When interaction is present, the association between the risk factor and the outcome variable differs, or depends in some way on the level of the covariate. That is, the covariate modifies the effect of the risk factor. Epidemiologists use the term *effect modifier* to describe a variable that interacts with a risk factor. In the previous example, if the logit is linear in age for the men in group 1, then interaction implies that the logit does not follow a line with the same slope for the second group. In theory, the association in group 2 could be described by almost any model except one with the same slope as the logit for group 1.

The simplest and most commonly used model for including interaction is one in which the logit is also linear in the confounder for the second group, but with a different slope. Alternative models can be formulated which would allow for a relationship that is non-linear between the logit and the variables in the model within each group. In any

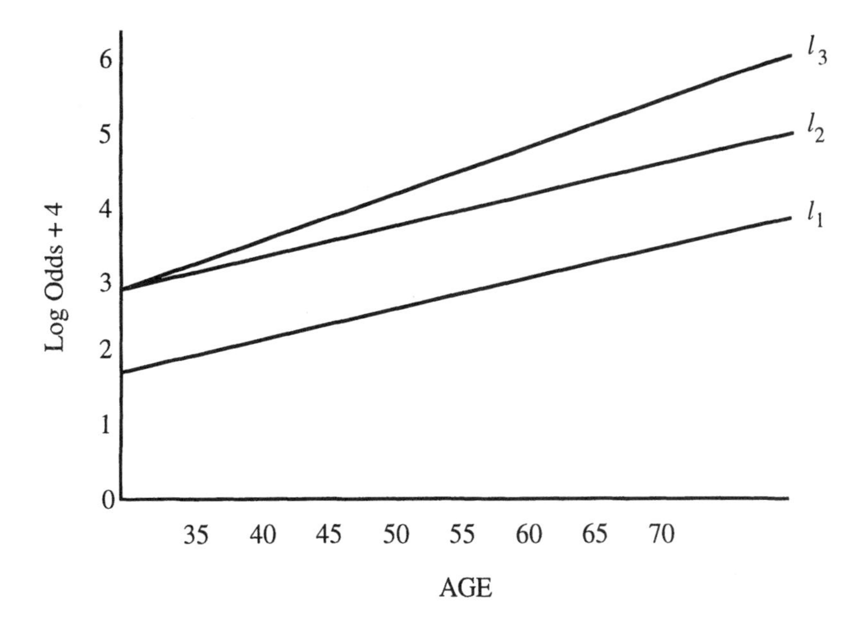

Figure 3.2 Plot of the logits under three different models showing the presence and absence of interaction.

model, interaction is incorporated by the inclusion of appropriate higher order terms.

An important step in the process of modeling a set of data is determining whether there is evidence of interaction in the data. This aspect of modeling is discussed in Chapter 4. In this section we assume that when interaction is present it can be modeled by nonparallel straight lines.

Figure 3.2 presents the graphs of three different logits. In this graph, 4 has been added to each of the logits to make plotting more convenient. The graphs of these logits are used to explain what is meant by interaction. Consider an example where the outcome variable is the presence or absence of CHD, the risk factor is sex, and the covariate is age. Suppose that the line labeled l_1 corresponds to the logit for females as a function of age. Line l_2 represents the logit for males. These two lines are parallel to each other, indicating that the relationship between age and CHD is the same for males and females. In this situation there is no interaction and the log odds ratios for sex (male versus female), controlling for age, is given by the difference between line l_2

Table 3.12 Estimated Logistic Regression Coefficients, Deviance, and the Likelihood Ratio Test Statistic (G) for an Example Showing Evidence of Confounding but No Interaction (n = 400)

Model	Constant	SEX	AGE	SEX×AGE	Deviance	G
1	0.060	1.981			419.816	
2	−3.374	1.356	0.082		407.780	12.036
3	−4.216	4.239	0.103	−0.062	406.392	1.388

and l_1, $l_2 - l_1$. This difference is equal to the vertical distance between the two lines, which is the same for all ages.

Suppose instead that the logit for males is given by the line l_3. This line is steeper than the line l_1, for females, indicating that the relationship between age and CHD among males is different from that among females. When this occurs we say there is an interaction between age and sex. The estimate of the log-odds ratios for sex (males versus females) controlling for age is still given by the vertical distance between the lines, $l_3 - l_1$, but this difference now depends on the age at which the comparison is made. Thus, we cannot estimate the odds ratio for sex without first specifying the age at which the comparison is being made. In other words, age is an effect modifier.

Tables 3.12 and 3.13 present the results of fitting a series of logistic regression models to two different sets of hypothetical data. The variables in each of the data sets are the same: SEX, AGE, and the outcome variable CHD. In addition to the estimated coefficients, the deviance for each model is given. Recall that the change in the deviance may be used to test for the significance of coefficients for variables added to the model. An interaction is added to the model by creating a variable that is equal to the product of the value of the SEX and the value of AGE. Some programs have syntax that automatically creates interaction variables in a statistical model, while others require the user to create them through a data modification step.

Examining the results in Table 3.12 we see that the estimated coefficient for the variable SEX changed from 1.981 in model 1 to 1.356, a 46 percent decrease, when AGE was added in model 2. Hence, there is clear evidence of a confounding effect due to age. When the interaction term "SEX×AGE" is added in model 3 we see that the change in the deviance is only 1.388 which, when compared to the chi-square distribution with 1 degree of freedom, yields a p-value of 0.24, which is clearly not significant. Note that the coefficient for sex changed from

Table 3.13 Estimated Logistic Regression Coefficients, Deviance, and the Likelihood Ratio Test Statistic (*G*) for an Example Showing Evidence of Confounding and Interaction (n = 400)

Model	Constant	SEX	AGE	SEX×AGE	Deviance	*G*
1	0.201	2.386			376.712	
2	−6.672	1.274	0.166		338.688	38.024
3	−4.825	−7.838	0.121	0.205	330.654	8.034

1.356 to 4.239. This is not surprising since the inclusion of an interaction term, especially when it involves a continuous variable, usually produces fairly marked changes in the estimated coefficients of dichotomous variables involved in the interaction. Thus, when an interaction term is present in the model we cannot assess confounding via the change in a coefficient. For these data we would prefer to use model 2 that suggests age is a confounder but not an effect modifier.

The results in Table 3.13 show evidence of both confounding and interaction due to age. Comparing model 1 to model 2 we see that the coefficient for sex changes from 2.386 to 1.274, an 87 percent decrease. When the age by sex interaction is added to the model we see that the change in the deviance is 8.034 with a p-value of 0.005. Since the change in the deviance is significant, we prefer model 3 to model 2, and should regard age as both a confounder and an effect modifier. The net result is that any estimate of the odds ratio for sex should be made with reference to a specific age.

Hence, we see that determining whether a covariate, X, is an effect modifier and/or a confounder involves several issues. The plots of the logits shown in Figure 3.2 show us that determining effect modification status involves the parametric structure of the logit, while determination of confounder status involves two things. First the covariate must be associated with the outcome variable. This implies that the logit must have a nonzero slope in the covariate. Second the covariate must be associated with the risk factor. In our example this is characterized by having a difference in the mean age for males and females. However, the association may be more complex than a simple difference in means. The essence is that we have incomparability in our risk factor groups. This incomparability must be accounted for in the model if we are to obtain a correct, unconfounded, estimate of effect for the risk factor.

In practice, one method to check for the confounder status of a covariate is to compare the estimated coefficient for the risk factor variable from models containing and not containing the covariate. Any "clinically important" change in the estimated coefficient for the risk factor suggests that the covariate is a confounder and should be included in the model, regardless of the statistical significance of its estimated coefficient. As noted above, Bachand and Hosmer (1999) show that the change in coefficient method does not always provide evidence that a variable is a confounder and a more detailed evaluation may be required. We return to this point in Chapter 4.

On the other hand, we believe that a covariate is an effect modifier only when the interaction term added to the model is both clinically meaningful and statistically significant. When a covariate is an effect modifier, its status as a confounder is of secondary importance since the estimate of the effect of the risk factor depends on the specific value of the covariate.

The concepts of adjustment, confounding, interaction, and effect modification, may be extended to cover the situations involving any number of variables on any measurement scale(s). The dichotomous-continuous variables example illustrated in this section has the advantage that the results are easily shown graphically. This is not the case with more complicated models. The principles for identification and inclusion of confounder and interaction variables in the model are the same regardless of the number of variables and their measurement scales.

3.7 ESTIMATION OF ODDS RATIOS IN THE PRESENCE OF INTERACTION

In Section 3.6 we showed that when there was interaction between a risk factor and another variable, the estimate of the odds ratio for the risk factor depends on the value of the variable that is interacting with it. In this situation we may not be able to estimate the odds ratio by simply exponentiating an estimated coefficient. One approach that will always yield the correct model-based estimate is to (1) write down the expressions for the logit at the two levels of the risk factor being compared, (2) algebraically simplify the difference between the two logits and compute its value (3) exponentiate the value obtained in step 2.

As a first example, we develop the method for a model containing only two variables and their interaction. In this model, denote the risk factor as F, the covariate as X and their interaction as $F \times X$. The logit for this model evaluated at $F = f$ and $X = x$ is

$$g(f,x) = \beta_0 + \beta_1 f + \beta_2 x + \beta_3 f \times x \ . \tag{3.9}$$

Assume we want the odds ratio comparing two levels of F, $F = f_1$ versus and $F = f_0$, at $X = x$. Following the three step procedure first we evaluate the expressions for the two logits yielding

$$g(f_1,x) = \beta_0 + \beta_1 f_1 + \beta_2 x + \beta_3 f_1 \times x$$

and

$$g(f_0,x) = \beta_0 + \beta_1 f_0 + \beta_2 x + \beta_3 f_0 \times x \ .$$

Second we compute and simplify their difference to obtain the log-odds ratio yielding

$$\begin{aligned} \ln\left[\mathrm{OR}(F = f_1, F = f_0, X = x)\right] &= g(f_1,x) - g(f_0,x) \\ &= (\beta_0 + \beta_1 f_1 + \beta_2 x + \beta_3 f_1 \times x) \\ &\qquad - (\beta_0 + \beta_1 f_0 + \beta_2 x + \beta_3 f_0 \times x) \\ &= \beta_1(f_1 - f_0) + \beta_3 x(f_1 - f_0). \end{aligned} \tag{3.10}$$

Third we obtain the odds ratio by exponentiating the difference obtained at step 2 yielding

$$\mathrm{OR} = \exp\left[\beta_1(f_1 - f_0) + \beta_3 x(f_1 - f_0)\right] \ . \tag{3.11}$$

Note that the expression for the log-odds ratio in (3.10) does not simplify to a single coefficient. Instead, it involves two coefficients, the difference in the values of the risk factor and the interaction variable. The estimator of the log-odds ratio is obtained by replacing the parameters in (3.10) and (3.11) with their estimators.

We obtain the endpoints of the confidence interval estimator using the same approach used for models without interactions. We calculate the endpoints for the confidence interval for the log-odds ratio and then exponentiate the end points. The basic building block of the endpoints is the estimator of the variance of the estimator of the log-odds ratio in (3.10). Using methods for calculating the variance of a sum we obtain the following estimator,

$$\begin{aligned}\widehat{\text{Var}}\left\{\ln\left[\widehat{\text{OR}}(F=f_1,F=f_0,X=x)\right]\right\} &= (f_1-f_0)^2\times\widehat{\text{Var}}(\hat{\beta}_1)\\ &+\left[x(f_1-f_0)\right]^2\times\widehat{\text{Var}}(\hat{\beta}_3)+2x(f_1-f_0)^2\times\widehat{\text{Cov}}(\hat{\beta}_1,\hat{\beta}_3).\end{aligned} \tag{3.12}$$

Most logistic regression computer packages have the option to provide output showing estimates of the variances and covariances of the estimated parameters in the model. Substitution of these estimates into (3.12) obtains an estimate of the variance of the estimated log-odds ratio. The endpoints of a $100\times(1-\alpha)\%$ confidence interval estimator for the log-odds ratio are:

$$\begin{aligned}&\left[\hat{\beta}_1(f_1-f_0)+\hat{\beta}_3x(f_1-f_0)\right]\\ &\qquad\pm z_{1-\alpha/2}\widehat{\text{SE}}\left\{\ln\left[\widehat{\text{OR}}(F=f_1,F=f_0,X=x)\right]\right\},\end{aligned} \tag{3.13}$$

where the standard error in (3.13) is the positive square root of the variance estimator in (3.12). We obtain the endpoints of the confidence interval estimator for the odds ratio by exponentiating the endpoints in (3.13).

The estimators for the log-odds and its variance simplify in the case when F is a dichotomous risk factor. If we let $f_1=1$ and $f_0=0$ then the estimator of the log-odds ratio is

$$\ln\left[\widehat{\text{OR}}(F=1,F=0,X=x)\right]=\hat{\beta}_1+\hat{\beta}_3x, \tag{3.14}$$

the estimator of the variance is

$$\begin{aligned}&\widehat{\text{Var}}\left\{\ln\left[\widehat{\text{OR}}(F=1,F=0,X=x)\right]\right\}\\ &\qquad=\widehat{\text{Var}}(\hat{\beta}_1)+x^2\widehat{\text{Var}}(\hat{\beta}_3)+2x\widehat{\text{Cov}}(\hat{\beta}_1,\hat{\beta}_3)\end{aligned} \tag{3.15}$$

and the endpoints of the confidence interval are

$$\left(\hat{\beta}_1+\hat{\beta}_3x\right)\pm z_{1-\alpha/2}\widehat{\text{SE}}\left\{\ln\left[\widehat{\text{OR}}(F=1,F=0,X=x)\right]\right\}. \tag{3.16}$$

Table 3.14 Estimated Logistic Regression Coefficients, Deviance, the Likelihood Ratio Test Statistic (G), and the p-value for the Change for Models Containing LWD and AGE from the Low Birthweight Data (n = 189)

Model	Constant	LWD	AGE	LWD×AGE	$\ln[l(\boldsymbol{\beta})]$	G	p
0	−0.790				−117.34		
1	−1.054	1.054			−113.12	8.44	0.004
2	−0.027	1.010	−0.044		−112.14	1.96	0.160
3	0.774	−1.944	−0.080	0.132	−110.57	3.14	0.076

As an example, we consider a logistic regression model using the low birth weight data described in Section 1.6 containing the variables AGE and a dichotomous variable, LWD, based on the weight of the mother at the last menstrual period. This variable takes on the value 1 if LWT < 110 pounds, and is zero otherwise. The results of fitting a series of logistic regression models are given in Table 3.14.

Using the estimated coefficient for LWD in model 1 we estimate the odds ratio as $\exp(1.054) = 2.87$. The results shown in Table 3.14 indicate that AGE is not a strong confounder, $\Delta\hat{\beta}\% = 4.2$, but it does interact with LWD, $p = 0.076$. Thus, to assess the risk of low weight at the last menstrual period correctly we must include the interaction of this variable with the women's age because the odds ratio is not constant over age.

An effective way to see the presence of interaction is via a graph of the estimated logit under model 3 in Table 3.14. This is shown in Figure 3.3. The upper line in Figure 3.3 corresponds to the estimated logit for women with $LWD = 1$ and the lower line is for women with $LWD = 0$. Separate plotting symbols have been used for the two LWD groups. The estimated log-odds ratio for $LWD = 1$ versus $LWD = 0$ at $AGE = x$ from (3.14) is equal to the vertical distance between the two lines at $AGE = x$. We can see in Figure 3.3 that this distance is nearly zero at 15 years and progressively increases. Since the vertical distance is not constant we must choose a few specific ages for estimating the effect of low weight at the last menstrual period. We can see in Figure 3.3 that none of the women in the low weight group, $LWD = 1$, are older than about 33 years. Thus we should restrict our estimates of the effect of low weight to the range of 14 to 33 years. Based on these observations we estimate the effect of low weight at 15, 20, 25 and 30 years of age.

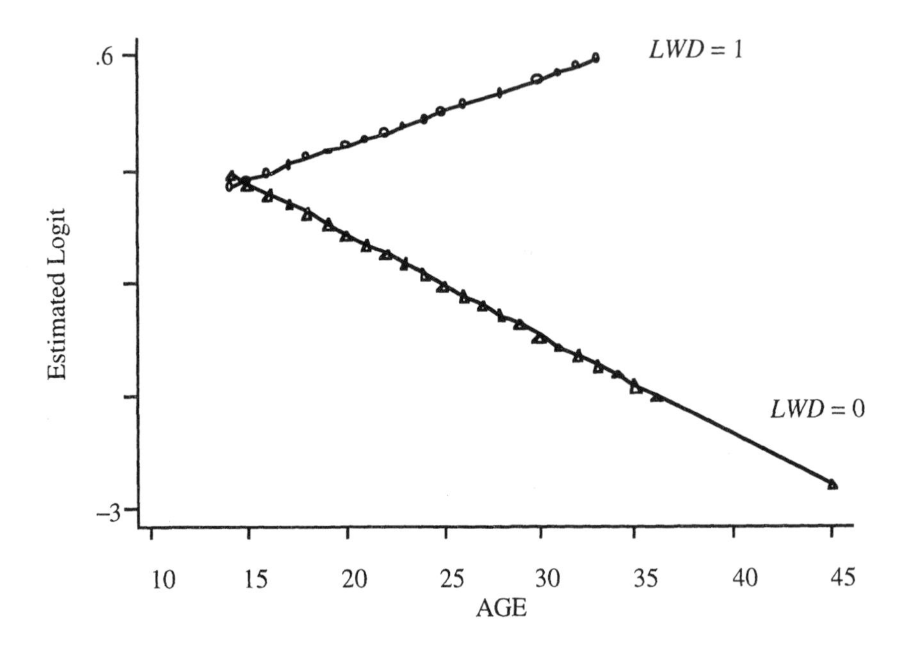

Figure 3.3 Plot of the estimated logit for women with $LWD = 1$ and for women with $LWD = 0$ from Model 3 in Table 3.17.

Using (3.14) and the results for model 3 the estimated log-odds ratio for low weight at the last menstrual period for a women of AGE = a is

$$\ln\left[\hat{\text{OR}}(\text{LWD}=1, \text{LWD}=0, \text{AGE}=a)\right] = -1.944 + 0.132a. \quad (3.17)$$

In order to obtain the estimated variance we must first obtain the estimated covariance matrix for the estimated parameters. Since this matrix is symmetric most logistic regression software packages print the

Table 3.15 Estimated Covariance Matrix for the Estimated Parameters in Model 3 of Table 3.14

Constant	0.828			
LWD	–0.828	2.975		
AGE	–0.353-02	–0.353-01	0.157-02	
LWD×AGE	–0.353-01	–0.128	–0.157-02	0.573-02
	Constant	LWD	AGE	LWD×AGE

Table 3.16 Estimated Odds Ratios and 95% Confidence Intervals for LWD, Controlling for AGE

Age	15	20	25	30
OR	1.04	2.01	3.90	7.55
95 % CIE	0.29, 3.79	0.91, 4.44	1.71, 8.88	1.95, 29.19

results in the form similar to that shown in Table 3.15.

The estimated variance of the log-odds ratio given (3.16) is obtained from (3.14) and is

$$\hat{\text{var}}\left\{\ln\left[\hat{\text{OR}}(\text{LWD}=1,\text{LWD}=0,\text{AGE}=a)\right]\right\}$$
$$=2.975+a^2\times 0.0057+2\times a\times(-0.128). \quad (3.19)$$

Values of the estimated odds ratio and 95% CI computed using (3.16) and (3.19) for several ages are given in Table 3.16. The results shown in Table 3.16 demonstrate that the effect of LWD on the odds of having a low birth weight baby increase exponentially with age. The results also show that the increase in risk is significant for low weight women 25 years and older. In particular low weight women of age 30 are estimated to have a risk that is about 7.5 times that of women of the same age who are not low weight. The increase in risk could be as little as two times or as much as 29 times with 95 percent confidence.

3.8 A COMPARISON OF LOGISTIC REGRESSION AND STRATIFIED ANALYSIS FOR 2 × 2 TABLES

Many users of logistic regression, especially those coming from a background in epidemiology, have performed stratified analyses of 2×2 tables to assess interaction and to control confounding. The essential objective of such analyses is to determine whether the odds ratios are constant, or homogeneous, over the strata. If the odds ratios are constant, then a stratified odds ratio estimator such as the Mantel-Haenszel estimator or the weighted logit-based estimator is computed. This same analysis may also be performed using the logistic regression modeling techniques discussed in Sections 3.6 and 3.7. In this section we compare these two approaches. An example from the low birth weight data illustrates the similarities and differences in the two approaches.

Table 3.17 Cross-Classification of Low Birth Weight by Smoking Status

		SMOKE		
		1	0	Total
LOW	1	30	29	59
	0	44	86	130
	Total	74	115	189

Consider an analysis of the risk factor smoking on low birth weight. The crude (or unadjusted) odds ratio computed from the 2×2 table shown in Table 3.17, cross-classifying the outcome variable LOW with SMOKE, is $\widehat{\text{OR}} = 2.02$.

Table 3.18 presents these data stratifying by the race of the mother. We can use these tables as the basis for computing either the Mantel-Haenszel estimate or the logit-based estimate of the odds ratio.

The Mantel-Haenszel estimator is a weighted average of the stratum specific odds ratios, $\widehat{\text{OR}}_i = (a_i \times d_i)/(b_i \times c_i)$, where a_i, b_i, c_i, and d_i are the observed cell frequencies in the 2×2 table for stratum i. For example, in stratum 1 $a_1 = 19$, $b_1 = 4$, $c_1 = 33$, and $d_1 = 40$ and the total number of subjects is $N_1 = 96$. The Mantel-Haenszel estimator of the odds ratio is defined in this case as follows:

$$\widehat{\text{OR}}_{\text{MH}} = \frac{\sum a_i \times d_i / N_i}{\sum b_i \times c_i / N_i}. \tag{3.20}$$

Evaluating (3.20) using the data in Table 3.18 yields the Mantel-Haenszel estimate

$$\widehat{\text{OR}}_{\text{MH}} = \frac{13.067}{4.234} = 3.09.$$

The logit-based summary estimator of the odds ratio is a weighted average of the stratum specific log-odds ratios where each weight is the inverse of the variance of the stratum specific log-odds ratio,

$$\hat{OR}_L = \exp\left[\sum w_i \ln\left(\hat{OR}_i\right) / \sum w_i\right]. \quad (3.21)$$

Table 3.19 presents the estimated odds ratio, log-odds ratio, estimate of the variance of the log-odds ratio and the weight, w.

The logit-based estimator based on the data in Table 3.18 is

$$\hat{OR}_L = \exp(7.109/6.582) = 2.95,$$

Table 3.18 Cross-Classification of Low Birth Weight by Smoking Status Stratified by RACE

White

		SMOKE 1	SMOKE 0	Total
LOW	1	19	4	23
	0	33	40	73
	Total	52	44	96

Black

		SMOKE 1	SMOKE 0	Total
LOW	1	6	5	11
	0	4	11	15
	Total	10	16	26

Other

		SMOKE 1	SMOKE 0	Total
LOW	1	5	20	25
	0	7	35	42
	Total	12	55	67

Table 3.19 Tabulation of the Estimated Odds Ratios, ln(Estimated Odds Ratios), Estimated Variance of the ln(Estimated Odds Ratios), and the Inverse of the Estimated Variance, *w*, for Smoking Status Within Each Stratum of RACE

	White	Black	Other
$\hat{OR}$	5.758	3.300	1.250
$\ln(\hat{OR})$	1.751	1.194	0.223
$\hat{var}[\ln(\hat{OR})]$	0.358	0.708	0.421
w	2.794	1.413	2.375

which is slightly smaller than the Mantel-Haenszel estimate. The high fluctuation in the odds ratio across the race strata suggests that there may be either confounding or effect modification due to RACE, or both. In general, the Mantel-Haenszel estimator and the logit based estimator are similar when the data are not too sparse within the strata. One considerable advantage of the Mantel-Haenszel estimator is that it may be computed when some of the cell entries are zero.

It is important to note that these estimators provide a correct estimate of the effect of the risk factor only when the odds ratio is constant across the strata. Thus, a crucial step in the stratified analysis is to assess the validity of this assumption. Statistical tests of this assumption are based on a comparison of the stratum specific estimates to an overall estimate computed under the assumption that the odds ratio is, in fact, constant. The simplest and most easily computed test of the homogeneity of the odds ratios across strata is based on a weighted sum of the squared deviations of the stratum specific log-odds ratios from their weighted mean. This test statistic, in terms of the current notation, is

$$X_H^2 = \sum \left\{ w_i \left[\ln\left(\hat{OR}_i\right) - \ln\left(\hat{OR}_L\right)\right]^2 \right\}. \tag{3.22}$$

Under the hypothesis that the odds ratios are constant, X_H^2 has a chi-square distribution with degrees-of-freedom equal to the number of strata minus 1. Thus, we would reject the homogeneity assumption when X_H^2 is large.

Using the data in Table 3.19 we have $X_H^2 = 3.017$ which, with 2 degrees-of-freedom, yields a p-value of 0.221. Thus, in spite of the apparent differences in the odds ratios seen in Table 3.19, the logit-based test of homogeneity indicates that they are within sampling variation of each other. It should be noted that the p-value calculated from the chi-square distribution is accurate only when the sample sizes are not too small within each stratum. This condition holds in this example.

Another test that also may be calculated by hand, but not as easily, is discussed in Breslow and Day (1980) and is corrected by Tarone (1985). This test compares the value of a_i to an estimated expected value, $\hat{e}_i$, if the odds ratio is constant. As noted by Breslow (1996) the correct formula for the test statistic is

$$X_{\text{BD}}^2 = \sum \frac{(a_i - \hat{e}_i)^2}{\hat{v}_i} - \frac{\left[\sum(a_i) - \sum(\hat{e}_i)\right]^2}{\sum(\hat{v}_i)}. \tag{3.23}$$

The quantity $\hat{e}_i$ is obtained as one of the solutions to a quadratic equation given by the following formula

$$\hat{e}_i = \frac{1}{2(\widehat{\text{OR}} - 1)} \left(\begin{array}{l} \widehat{\text{OR}}(n_{1i} + m_{1i}) + (n_{0i} - m_{1i}) \pm \\ \left\{ \left[\widehat{\text{OR}}(n_{1i} + m_{1i}) + (n_{0i} - m_{1i})\right]^2 - \left[4(\widehat{\text{OR}} - 1)\,\widehat{\text{OR}}\, n_{1i} m_{1i}\right] \right\}^{1/2} \end{array} \right), \tag{3.24}$$

where $n_{1i} = a_i + b_i$, $m_{1i} = a_i + c_i$ and $n_{0i} = c_i + d_i$. The quantity $\widehat{\text{OR}}$ in (3.24) is an estimate of the common odds ratio and either $\widehat{\text{OR}}_L$ or $\widehat{\text{OR}}_{\text{MH}}$ may be used. The quantity $\hat{v}_i$ is an estimate of the variance of a_i computed under the assumption of a common odds ratio and is

$$\hat{v}_i = \left(\frac{1}{\hat{e}_i} + \frac{1}{n_{1i} - \hat{e}_i} + \frac{1}{m_{1i} - \hat{e}_i} + \frac{1}{n_{0i} - m_{1i} + \hat{e}_i} \right)^{-1}. \tag{3.25}$$

If we use the value of the Mantel-Haenszel estimate, $\widehat{\text{OR}}_{\text{MH}} = 3.086$ in (3.23) then the resulting values of $\hat{e}$ and $\hat{v}$ are: $\hat{e}_1 = 17.01$, $\hat{v}_1 = 3.56$, $\hat{e}_2 = 5.91$, $\hat{v}_2 = 1.43$, $\hat{e}_3 = 7.16$, and $\hat{v}_3 = 2.33$. The value of the Breslow-Day statistic obtained is $X_{BD}^2 = 3.11 - 0.0081 = 3.10$, which is similar to

Table 3.20 Estimated Logistic Regression Coefficients for the Variable SMOKE, Log-Likelihood, the Likelihood Ratio Test Statistic (G), and Resulting p-Value for Estimation of the Stratified Odds Ratio and Assessment of Homogeneity of Odds Ratios Across Strata Defined by RACE

Model	SMOKE	Log-Likelihood	G	df	p
1	0.704	−114.90			
2	1.116	−109.99	9.83	2	0.007
3	1.751	−108.41	3.16	2	0.206

the value of the logit-based test. Some packages, for example SAS, report the value of the first term in (3.23) as the Breslow-Day test

The same analysis may be performed much more easily by fitting three logistic regression models. In model 1 we include only the variable SMOKE. We then add the two design variables for RACE to obtain model 2. For model 3 we add the two RACE×SMOKE interaction terms. The results of fitting these models are shown in Table 3.20. Since we are primarily interested in the estimates of the coefficient for SMOKE, the estimates of the coefficients for RACE and the RACE ×SMOKE interactions are not shown in Table 3.20.

Using the estimated coefficients in Table 3.20 we have the following estimated odds ratios. The crude odds ratio is $\widehat{\mathrm{OR}} = \exp(0.704) = 2.02$. Adjusting for RACE, the stratified estimate is $\widehat{\mathrm{OR}} = \exp(1.116) = 3.05$. This value is the maximum likelihood estimate of the estimated odds ratio, and it is similar in value to both the Mantel-Haenszel estimate, $\widehat{\mathrm{OR}}_{\mathrm{MH}} = 3.086$, and the logit-based estimate, $\widehat{\mathrm{OR}}_L = 2.95$. The change in the estimate of the odds ratio from the crude to the adjusted is 2.02 to 3.05, indicating considerable confounding due to RACE.

Assessment of the homogeneity of the odds ratios across the strata is based on the likelihood ratio test of model 2 versus model 3. The value of this statistic from Table 3.20 is $G = 3.156$. This statistic is compared to a chi-square distribution with 2 degrees-of-freedom, since two interaction terms were added to model 2 to obtain model 3. This test statistic is comparable to the ones from the logit-based test, X_H^2, and the Breslow-Day test, X_{BD}^2. If we had used the maximum likelihood estimate of the stratified odds ratio, exp(1.116), in computing the Breslow-Day test, then the resulting statistic would have been equal to the Pearson chi-square goodness-of-fit test of model 2, since model 3 is the saturated model.

The previously described analysis based on likelihood ratio tests may be used when the data have either been grouped into contingency tables in advance of the analysis, such as those shown in Table 3.17, or have remained in casewise form. When the data have been grouped it is possible to point out other similarities between classical analysis of stratified 2×2 tables and an analysis using logistic regression. Day and Byar (1979) have shown that the 1 degree of freedom Mantel-Haenszel test of the hypothesis that the stratum specific odds ratios are 1 is identical to the Score test for the exposure variable when added to a logistic regression model already containing the stratification variable. This test statistic may be easily obtained from a logistic regression package with the capability to perform Score tests such as the EGRET or SAS packages.

Thus, use of the logistic regression model provides a fast and effective way to obtain a stratified odds ratio estimator and to assess easily the assumption of homogeneity of odds ratios across strata.

3.9 INTERPRETATION OF THE FITTED VALUES

In previous sections in this chapter we discussed use of estimated coefficients to construct estimated odds in a number of settings typically encountered in practice. In our experience this accounts for the vast majority of the use of logistic regression modeling in applied settings. However there are situations where the fitted values from the model are equally, if not more, important. For example, Lemeshow, Teres, Klar, Avrunin, Gehlbach and Rapoport (1993) used logistic regression modeling methods to estimate a patient's probability of hospital mortality after admission to an intensive care unit.. We discussed in Section 1.4 and Section 2.5 the basic methods for computing the fitted values and confidence interval estimates. In this section, we expand on this work and include graphical presentation of fitted values and confidence bands. In addition we discuss prediction of the outcome for a subject not in the estimation sample.

As an example consider the model fit to the low birth weight data shown in Table 2.3. In Section 2.5 we illustrated the computations for a 150 pound white woman. A subject with these values was among the 189 subjects in the data set; thus estimates of the fitted value, logit and standard error of the logit are readily available from standard output.

Suppose instead that we wanted to present a graph illustrating the effect of weight of the mother at the last menstrual period on birth weight

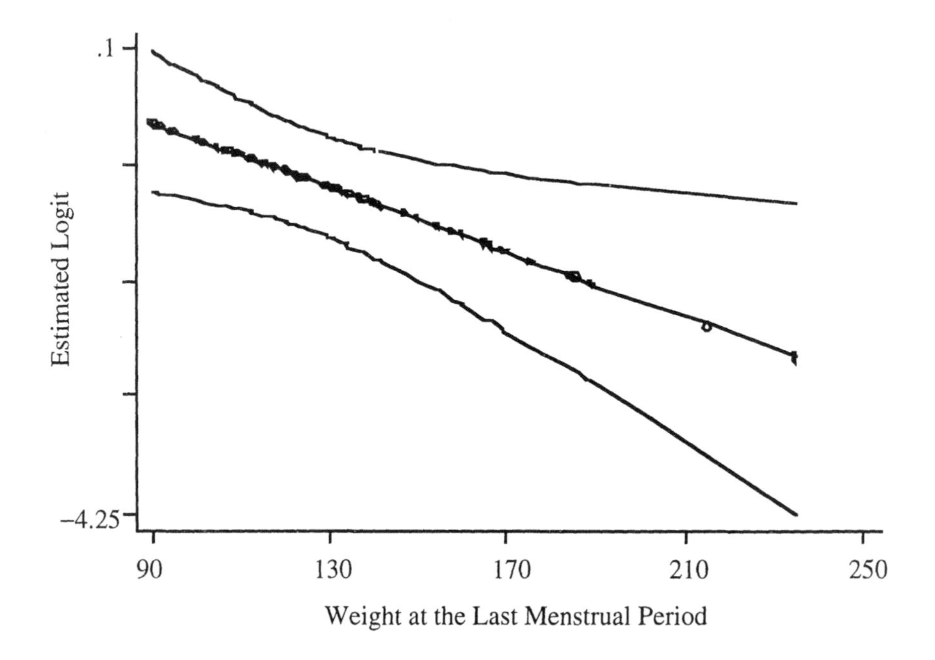

Figure 3.4 Graph of the estimated logit of low weight birth and 95 percent confidence intervals as a function of weight at the last menstrual period for white women.

holding race constant and equal to white. To accomplish this we take advantage of the fact that we can obtain the values of (2.6) and (2.7) for all subjects in the data set used to fit the model from standard logistic regression software. The graph for the estimated logit and its confidence bands is presented in Figure 3.4. The point and interval estimates for the logit are easily transformed to corresponding point and interval estimates for the logistic probability using the fundamental relationship between the two, see (1.19) and (1.21). These are presented in Figure 3.5. Note that we could have presented graphs for any of the three racial groups or for all three racial groups on the same graph. We arbitrarily chose the white mothers in order to keep the graph from getting unnecessarily complicated. The estimates in the figures are plotted at each observed value of LWT for the 96 white mothers. The estimated logit and probability decrease due to the fact that the estimated coefficient for LWT in Table 2.3 is negative. Note that the confidence bands in Figure 3.4 are narrowest near the mean value of LWT, approximately 130 pounds. The width increases in the same hyperbolic manner seen in similar plots from fitted linear regression models. The same pattern, transformed, can be seen in Figure 3.5.

Each point, and associated confidence interval, in Figure 3.5 is an estimate of the mean of the outcome, low birthweight, among white mothers of the specified value of LWT. Using the results in Section 2.5 at 150 pounds the point and interval estimates are 0.191 and (0.120, 0.289) respectively. The interpretation is that estimated proportion of low weight births among 150 pound white women is 0.191 and it could be as low as 0.12 or as high as 0.289 with 95 percent confidence. We would interpret estimates and confidence intervals at other values of LWT in a similar manner.

Suppose we wanted to use our fitted model to estimate the probability of low birthweight for a population of women not represented in the 189 in the estimation sample. As an example, suppose 150-pound black women. We obtain the value of the estimated logit from (2.6) using the estimated coefficients in Table 2.3 as follows

$$\begin{aligned}\hat{g}(LWT=150, RACE=Black) &= 0.806-0.015\times 150+1.081\times 1+0.481\times 0\\ &= -0.363\end{aligned}$$

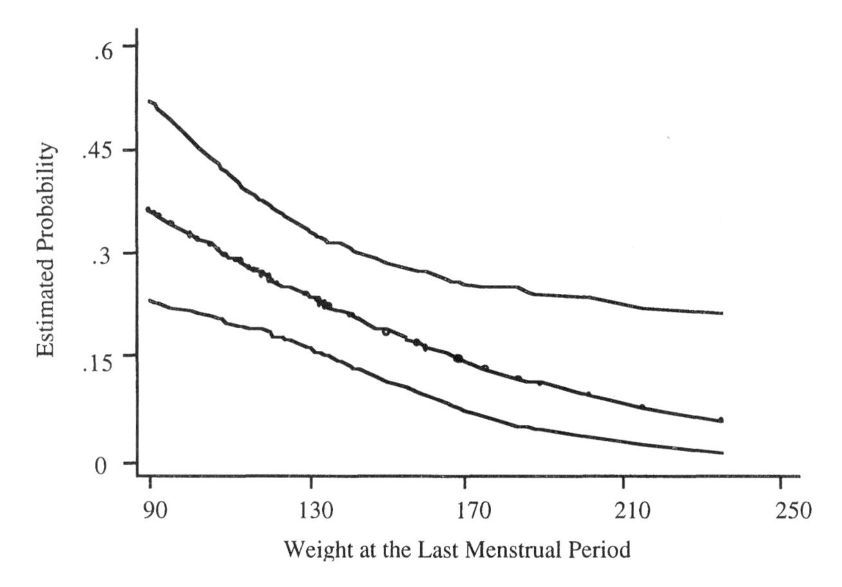

Figure 3.5 Graph of the estimated probability of low weight birth and 95 percent confidence intervals as a function of weight at the last menstrual period for white women.

and the estimated logistic probability is

$$\hat{\pi}(LWT=150, RACE=Black)=\frac{e^{-0.363}}{1+e^{-0.363}}=0.410\ .$$

The interpretation is the same as for patterns of data seen in the estimation sample. Namely, the model estimates that the 41 percent of 150 pound black women will have a low birthweight baby.

In order to obtain the confidence interval for this estimate we need to evaluate (2.7) or (2.9) using the covariance matrix in Table 2.4 with the data vector $\mathbf{x}'=(1,150,1,0)$. The resulting standard error from this computation is

$$\widehat{\text{SE}}\left[\hat{g}(LWT=150, RACE=Black)\right]=.1725,$$

yielding a 95 percent confidence interval for the probability of $(0.331,\ 0.494)$. The interpretation of this interval is that the proportion of 150 pound black women who give birth to a low weight baby could be as little as 0.331 or as high as 0.494 (with 95 percent confidence).

As is the case with any regression model we must take care not to extend model-based inferences out of the observed range of the data. The range of weight at the last menstrual period among the 26 black mothers is 98 to 241 pounds. We note that 150 pounds is well within this range. It is also important to keep in mind that any estimate is only as good as the model it is based on. In this section we did not attend to many of the important model building details that are discussed in Chapter 4. We have implicitly assumed that these steps have been performed.

EXERCISES

1. Consider the ICU data described in Section 1.6.1 and use as the outcome variable vital status (STA) and CPR prior to ICU admission (CPR) as a covariate.
   (a) Demonstrate that the value of the log-odds ratio obtained from the cross-classification of STA by CPR is identical to the estimated slope coefficient from the logistic regression of STA on CPR. Verify that the estimated standard error of the estimated slope coefficient for CPR obtained from the logistic regression package is identical to the square root of the sum of the inverse of the cell fre-

quencies from the cross-classification of STA by CPR. Use either set of computations to obtain 95% CI for the odds ratio. What aspect concerning the coding of the variable CPR makes the calculations for the two methods equivalent?

(b) For purposes of illustration, use a data transformation statement to recode, for this problem only, the variable CPR as follows: 4 = no and 2 = yes. Perform the logistic regression of STA on CPR (recoded). Demonstrate how the calculation of the logit difference of CPR = yes versus CPR = no is equivalent to the value of the log-odds ratio obtained in problem 1(a). Use the results from the logistic regression to obtain the 95% CI for the odds ratio and verify that they are the same limits as obtained in Exercise 1(a).

(c) Consider the ICU data and use as the outcome variable vital status (STA) and race (RACE) as a covariate. Prepare a table showing the coding of the two design variables for RACE using the value RACE = 1, white, as the reference group. Show that the estimated log-odds ratios obtained from the cross-classification of STA by RACE, using RACE = 1 as the reference group, are identical to estimated slope coefficients for the two design variables from the logistic regression of STA on RACE. Verify that the estimated standard errors of the estimated slope coefficients for the two design variables for RACE are identical to the square root of the sum of the inverse of the cell frequencies from the cross-classification of STA by RACE used to calculate the odds ratio. Use either set of computations to compute the 95% CI for the odds ratios.

(d) Create design variables for RACE using the method typically employed in ANOVA. Perform the logistic regression of STA on RACE. Show by calculation that the estimated logit differences of RACE = 2 versus RACE = 1 and RACE = 3 versus RACE = 1 are equivalent to the values of the log-odds ratio obtained in problem 1(c). Use the results of the logistic regression to obtain the 95% CI for the odds ratios and verify that they are the same limits as obtained in Exercise 1(c). Note that the estimated covariance matrix for the estimated coefficients is needed to obtain the estimated variances of the logit differences.

(e) Consider the variable AGE in the ICU data set. Prepare a table showing the coding of three design variables based on the empirical quartiles of AGE using the first quartile as the reference group. Fit the logistic regression of STA on AGE as recoded into these design variables and plot the three estimated slope coefficients versus the midpoint of the respective age quartile. Plot as a fourth point a value of zero at the midpoint of the first quartile of age. Does this plot suggest that the logit is linear in age?

(f) Consider the logistic regression of STA on CRN and AGE. Consider CRN to be the risk factor and show that AGE is a confounder of the association of CRN with STA. Addition of the interaction of AGE by CRN presents an interesting modeling dilemma. Examine the main effects only and interaction models graphically. Using the graphical results and any significance tests you feel are needed, select the best model (main effects or interaction) and justify your choice. Estimate relevant odds ratios. Repeat this analysis of confounding and interaction for a model that includes CPR as the risk factor and AGE as the potential confounding variable.

(g) Consider an analysis for confounding and interaction for the model with STA as the outcome, CAN as the risk factor, and TYP as the potential confounding variable. Perform this analysis using logistic regression modeling and Mantel-Haenszel analysis. Compare the results of the two approaches.

2. Use the data from the Prostatic Cancer Study described in Section 1.6.3 to answer the following questions:

(a) By fitting a series of logistic regression models show that RACE is not a confounder of the PSA CAPSULE odds ratio but is an effect modifier (at the 10 percent level).

(b) Graph the estimated logits from the interactions model versus PSA and interpret the two lines that appear on the graph. Use the graph to illustrate the log-odds of Black versus White for a subject with PSA = 7. Use the graph to illustrate the log-odds for a 5-unit increase in PSA for Whites and for Blacks.

(c) Estimate the point and 95 percent confidence interval estimates of the odds ratios corresponding to each of the log-odds illustrated in problem 2(b). Add the 95 percent confidence bands to the graph of the estimated logits from the interactions model in Exercise 2(b). Transform the lines and bands in this plot to obtain a plot of the estimated probability with its 95 percent confidence bands. Use the graph to estimate, point and interval, the probability of penetration for both a White and Black with PSA = 7. Interpret the two point and interval estimates.

CHAPTER 4

Model-Building Strategies and Methods for Logistic Regression

4.1 INTRODUCTION

In the previous chapters we focused on estimating, testing, and interpreting the coefficients in a logistic regression model. The examples discussed were characterized by having few independent variables, and there was perceived to be only one possible model. While there may be situations where this is the case, it is more typical that there are many independent variables that could potentially be included in the model. Hence, we need to develop a strategy and associated methods for handling these more complex situations.

The goal of any method is to select those variables that result in a "best" model within the scientific context of the problem. In order to achieve this goal we must have: (1) a basic plan for selecting the variables for the model and (2) a set of methods for assessing the adequacy of the model both in terms of its individual variables and its overall fit. We suggest a general strategy that considers both of these areas.

The methods to be discussed in this chapter are not to be used as a substitute for, but rather as an addition to, clear and careful thought. Successful modeling of a complex data set is part science, part statistical methods, and part experience and common sense. It is our goal to provide the reader with a paradigm that, when applied thoughtfully, yields the best possible model within the constraints of the available data.

4.2 VARIABLE SELECTION

The criteria for including a variable in a model may vary from one problem to the next and from one scientific discipline to another. The traditional approach to statistical model building involves seeking the most parsimonious model that still explains the data. The rationale for minimizing the number of variables in the model is that the resultant model is more likely to be numerically stable, and is more easily generalized. The more variables included in a model, the greater the estimated standard errors become, and the more dependent the model becomes on the observed data. Epidemiologic methodologists suggest including all clinically and intuitively relevant variables in the model, regardless of their "statistical significance." The rationale for this approach is to provide as complete control of confounding as possible within the given data set. This is based on the fact that it is possible for individual variables not to exhibit strong confounding, but when taken collectively, considerable confounding can be present in the data, see Rothman and Geenland (1998), Maldonado and Greenland (1993), Greenland (1989) and Miettinen (1976). The major problem with this approach is that the model may be "overfit," producing numerically unstable estimates. Overfitting is typically characterized by unrealistically large estimated coefficients and/or estimated standard errors. This may be especially troublesome in problems where the number of variables in the model is large relative to the number of subjects and/or when the overall proportion responding $(y=1)$ is close to either 0 or 1. In an excellent tutorial paper, Harrel, Lee and Mark (1996) discuss overfitting along with other model building issues.

There are several steps one can follow to aid in the selection of variables for a logistic regression model. The process of model building is quite similar to the one used in linear regression.

(1) The selection process should begin with a careful univariable analysis of each variable. For nominal, ordinal, and continuous variables with few integer values, we suggest this be done with a contingency table of outcome $(y=0,1)$ versus the k levels of the independent variable. The likelihood ratio chi-square test with $k-1$ degrees-of-freedom is exactly equal to the value of the likelihood ratio test for the significance of the coefficients for the $k-1$ design variables in a univariable logistic regression model that contains that single independent variable. Since the Pearson chi-square test is asymptotically equivalent to the likelihood ratio chi-square test, it may also be used. In addition to the

overall test, it is a good idea, for those variables exhibiting at least a moderate level of association, to estimate the individual odds ratios (along with confidence limits) using one of the levels as the reference group.

Particular attention should be paid to any contingency table with a zero cell. This yields a point estimate for one of the odds ratios of either zero or infinity. Including such a variable in any logistic regression program causes undesirable numerical outcomes to occur. These are addressed in the last section of this chapter. Strategies for handling the zero cell include: collapsing the categories of the independent variable in some sensible fashion to eliminate the zero cell; eliminating the category completely; or, if the variable is ordinal scaled, modeling the variable as if it were continuous.

For continuous variables, the most desirable univariable analysis involves fitting a univariable logistic regression model to obtain the estimated coefficient, the estimated standard error, the likelihood ratio test for the significance of the coefficient, and the univariable Wald statistic. An alternative analysis, which is equivalent at the univariable level, may be based on the two-sample t-test. Descriptive statistics available from a two-sample t-test analysis generally include group means, standard deviations, the t statistic, and its p-value. The similarity of this approach to the logistic regression analysis follows from the fact that the univariable linear discriminant function estimate of the logistic regression coefficient is

$$\frac{(\bar{x}_1 - \bar{x}_0)}{s_p^2} = \frac{t}{s_p}\sqrt{\frac{1}{n_1} + \frac{1}{n_0}}$$

and that the linear discriminant function and the maximum likelihood estimate of the logistic regression coefficient are usually quite close when the independent variable is approximately normally distributed within each of the outcome groups, $y = 0,1$, [see Halpern, Blackwelder, and Verter (1971)]. Thus, univariable analysis based on the t-test should be useful in determining whether the variable should be included in the model, since the p-value should be of the same order of magnitude as that of the Wald statistic, Score test, or likelihood ratio test from logistic regression.

For continuous covariates, we may wish to supplement the evaluation of the univariable logistic fit with some sort of smoothed scatterplot. This plot is helpful, not only in ascertaining the potential importance of the variable and possible presence and effect of extreme (large or small) observations, but also its appropriate scale. One simple and easily computed form of a smoothed scatterplot was illustrated in Figure 1.2 using the data in Table 1.2. Other more complicated methods that have greater precision are available.

Kay and Little (1987) illustrate the use of a method proposed by Copas (1983). This method requires computing a smoothed value for the response variable for each subject that is a weighted average of the values of the outcome variable over all subjects. The weight for each subject is a continuous decreasing function of the distance of the value of the covariate for the subject under consideration from the value of the covariate for all other cases. For example, for covariate x for the ith subject we compute

$$\bar{y}_i = \frac{\sum_{j=i_l}^{i_u} w(x_i, x_j) y_j}{\sum_{j=i_l}^{i_u} w(x_i, x_j)},$$

where $w(x_i, x_j)$ represents a particular weight function. For example if we use STATA's scatterplot smooth command, ksm, with the weight option and band width k, then

$$w(x_i, x_j) = \left[1 - \left(\frac{|x_i - x_j|^3}{\Delta}\right)\right]^3,$$

where Δ is defined so that the maximum value for the weight is ≤ 1 and the two indices defining the summation, i_l and i_u, include the k percent of the n subjects with x values closest to x_i. Other weight functions are possible as well as additional smoothing using locally weighted least squares regression, called *lowess* in some packages. See Cleveland (1993) for a more complete discussion of scatterplot smoothing methods. In general, when using STATA, we prefer to use the lowess option with a band width of $k = 80$. We plot the triplet $(x_i, y_i, \bar{y}_i)$, i.e., observed and smoothed values of y on the same set of axes. The shape of the smoothed plot should provide some idea about the parametric relationship between the outcome and the covariate. Some packages, including

STATA, provide the option for plotting the smoothed values on the logit scale, thus making it a little easier to make decisions about the possible scale of the covariate.

We discuss and illustrate methods for identification of the scale of continuous covariates in the logit later in this section.

(2) Upon completion of the univariable analyses, we select variables for the multivariable analysis. Any variable whose univariable test has a p-value < 0.25 is a candidate for the multivariable model along with all variables of known clinical importance. Once the variables have been identified, we begin with a model containing all of the selected variables.

Our recommendation that 0.25 level be used as a screening criterion for variable selection is based on the work by Bendel and Afifi (1977) on linear regression and on the work by Mickey and Greenland (1989) on logistic regression. These authors show that use of a more traditional level (such as 0.05) often fails to identify variables known to be important. Use of the higher level has the disadvantage of including variables that are of questionable importance at the model building stage. For this reason, it is important to review all variables added to a model critically before a decision is reached regarding the final model.

One problem with any univariable approach is that it ignores the possibility that a collection of variables, each of which is weakly associated with the outcome, can become an important predictor of outcome when taken together. If this is thought to be a possibility, then we should choose a significance level large enough to allow the suspected variables to become candidates for inclusion in the multivariable model. The best subsets selection technique, discussed briefly below and in greater detail in Section 4.4, is an effective model-building strategy for identifying collections of variables having this type of association with the outcome variable.

As noted above, the issue of variable selection is made more complicated by different analytic philosophies as well as by different statistical methods. One school of thought argues for the inclusion of all scientifically relevant variables into the multivariable model regardless of the results of univariable analyses. In general, the appropriateness of the decision to begin the multivariable model with all possible variables depends on the overall sample size and the number in each outcome group relative to the total number of candidate variables. When the data are adequate to support such an analysis it may be useful to begin the multivariable modeling from this point. However, when the data are inadequate, this approach can produce a numerically unstable multivariable model, discussed in greater detail in Section 4.5. In this case the Wald statistics should not be used to select variables because of the unstable nature of the results. Instead, we should select a subset of variables based on results of the univariable analyses and refine the definition of "scientifically relevant."

Another approach to variable selection is to use a stepwise method in which variables are selected either for inclusion or exclusion from the model in a sequential fashion based solely on statistical criteria. There are two main versions of the stepwise procedure: (a) forward selection with a test for backward elimination and (b) backward elimination followed by a test for forward selection. The algorithms used to define these procedures in logistic regression are discussed in Section 4.3. The stepwise approach is useful and intuitively appealing in that it builds models in a sequential fashion and it allows for the examination of a collection of models which might not otherwise have been examined.

"Best subsets selection" is a selection method that has not been used extensively in logistic regression. With this procedure a number of models containing one, two, three variables, and so on, are examined to determine which are considered the "best" according to some specified criteria. Best subsets linear regression software has been available for a number of years. A parallel theory has been worked out for nonnormal errors models [Lawless and Singhal (1978, 1987a, 1987b)]. We show in Section 4.4 how logistic regression may be performed using any best subsets linear regression program.

Stepwise, best subsets, and other mechanical selection procedures have been criticized because they can yield a biologically implausible model [Greenland (1989)] and can select irrelevant, or noise, variables [Flack and Chang (1987), Griffiths and Pope (1987)]. The problem is not the fact that the computer can select such models, but rather that the analyst fails to scrutinize the resulting model carefully, and reports such results as the final, best model. The wide availability and ease with which stepwise methods can be used has undoubtedly reduced some analysts to the role of assisting the computer in model selection rather than the more appropriate alternative. It is only when the analyst understands the strengths, and especially the limitations, of the methods that these methods can serve as useful tools in the model-building process. The analyst, not the computer, is ultimately responsible for the review and evaluation of the model.

(3) Following the fit of the multivariable model, the importance of each variable included in the model should be verified. This should

include (a) an examination of the Wald statistic for each variable and (b) a comparison of each estimated coefficient with the coefficient from the model containing only that variable. Variables that do not contribute to the model based on these criteria should be eliminated and a new model should be fit. The new model should be compared to the old, larger, model using the likelihood ratio test. Also, the estimated coefficients for the remaining variables should be compared to those from the full model. In particular, we should be concerned about variables whose coefficients have changed markedly in magnitude. This indicates that one or more of the excluded variables was important in the sense of providing a needed adjustment of the effect of the variable that remained in the model. This process of deleting, refitting, and verifying continues until it appears that all of the important variables are included in the model and those excluded are clinically and/or statistically unimportant.

At this point, we suggest that any variable not selected for the original multivariable model be added back into the model. This step can be helpful in identifying variables that, by themselves, are not significantly related to the outcome but make an important contribution in the presence of other variables.

We refer to the model at the end of step (3) as the *preliminary main effects model.*

(4) Once we have obtained a model that we feel contains the essential variables, we should look more closely at the variables in the model. The question of the appropriate categories for discrete variables should have been addressed at the univariable stage. For continuous variables we should check the assumption of linearity in the logit.

Assuming linearity in the logit at the variable selection stage is a common practice and is consistent with the goal of determining whether a particular variable should be in the model. The graphs for several different relationships between the logit and a continuous independent variable are shown in Figure 4.1. The figure illustrates the case when the logit is (a) linear, (b) quadratic, (c) some other nonlinear continuous relationship, and (d) binary where there is a cutpoint above and below which the logit is constant. In each of the situations described in Figure 4.1 fitting a linear model would yield a significant slope. Once the variable is identified as important, we can obtain the correct parametric relationship or scale in the model refinement stage. The exception to this would be the rare instance where the function is U-shaped. Specific methods to assess scale of continuous variables are discussed in detail

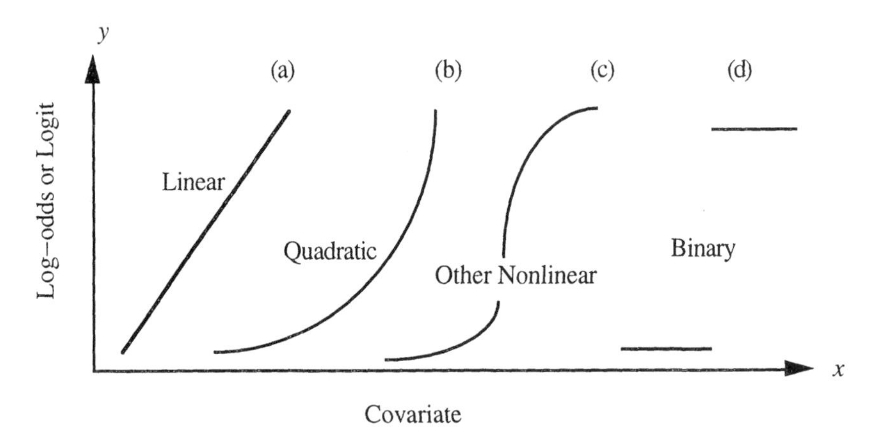

Figure 4.1 Different types of models for the relationship between the logit and a continuous variable.

later in this section. We refer to the model at the end of step (4) as the *main effects model.*

(5) Once we have refined the main effects model and ascertained that each of the continuous variables is scaled correctly, we check for interactions among the variables in the model. In any model an interaction between two variables implies that the effect of one of the variables is not constant over levels of the other. For example, an interaction between sex and age implies that the slope coefficient for age is different for males and females. The final decision as to whether an interaction term should be included in a model should be based on statistical as well as practical considerations. Any interaction term in the model must make sense from a clinical perspective.

We address the clinical plausibility issue by creating a list of possible pairs of variables in the model that have some scientific basis to interact with each other. The interaction variables are created as the arithmetic product of the pairs of main effect variables. We add the interaction variables, one at a time, to the model containing all the main effects and assess their significance using a likelihood ratio test. We feel that interactions must contribute to the model at traditional levels of statistical significance. Inclusion of an interaction term in the model that is not significant typically increases the estimated standard errors without changing the point estimates. In general, for an interaction term to alter both point and interval estimates, the estimated coefficient for the interaction term must be statistically significant.

We refer to the model at the conclusion of step (5) as the *preliminary final model.* Before we use any model for inferences we must assess its adequacy and check its fit. We discuss these methods in Chapter 5.

As noted in step (4), an important step in refining the main effects model is to determine whether the model is linear in the logit for continuous variables. We discuss two methods to address this problem: (1) design variables and (2) fractional polynomials.

In step (1) we mentioned that one way to examine the scale of the covariate is to use a scatterplot smooth, plotting the results on the logit scale. Unfortunately scatterplot smoothing methods are not easily extended to multivariable models and thus have limited applicability in the model refinement step. However, it is possible to extend the grouping type smooth shown in Figure 1.2 to multivariable models.

The procedure is easily implemented within any logistic regression package and is based on the following observation. The difference, adjusted for other model covariates, between the logits for two different groups is equal to the value of an estimated coefficient from a fitted logistic regression model that treats the grouping variable as categorical. We have found that the following implementation of the grouped smooth is usually adequate for purposes of visually checking the scale of a continuous covariate.

First, using the descriptive statistics capabilities of most any statistical package, obtain the quartiles of the distribution of the variable. Next create a categorical variable with 4 levels using three cutpoints based on the quartiles. Other grouping strategies can be used but one based on quartiles seems to work well in practice. Fit the multivariable model replacing the continuous variable with the 4-level categorical variable. To do this, 3 design variables must be used with the lowest quartile serving as the reference group. Following the fit of the model, plot the estimated coefficients versus the midpoints of the groups. In addition, plot a coefficient equal to zero at the midpoint of the first quartile. To aid in the interpretation we connect the four plotted points. Visually inspect the plot and choose the most logical parametric shape(s) for the scale of the variable.

The next step is to refit the model using the possible parametric forms suggested by the plot and choose one that is significantly different from the linear model and makes clinical sense. It is possible that two or more different parametrizations of the covariate will yield similar models in the sense that they are significantly different from the linear model. However, it is our experience that one of the possible models will be more appealing clinically, thus yielding more easily interpreted parameter estimates.

Another more analytic approach is to use the method of fractional polynomials, developed by Royston and Altman (1994), to suggest transformations. We wish to determine what value of x^p yields the best model for the covariate. In theory, we could incorporate the power, p, as an additional parameter in the estimation procedure. However, this greatly increases the complexity of the estimation problem. Royston and Altman propose replacing full maximum likelihood estimation of the power by a search through a small but reasonable set of possible values. Hosmer and Lemeshow (1999) provide a brief introduction to the use of fractional polynomials when fitting a proportional hazards regression model. This material provides the basis for our discussion of its application to logistic regression.

The method of fractional polynomials may be used with a multivariable logistic regression model, but, for sake of simplicity, we describe the procedure using a model with a single continuous covariate. The logit, that is linear in the covariate, is

$$g(x,\boldsymbol{\beta}) = \beta_0 + x\beta_1,$$

where $\boldsymbol{\beta}$ denotes the vector of model coefficients. One way to generalize this function is to specify it as

$$g(x,\boldsymbol{\beta}) = \beta_0 + \sum_{j=1}^{J} F_j(x)\beta_j \ .$$

The functions $F_j(x)$ are a particular type of power function. The value of the first function is $F_1(x) = x^{p_1}$. In theory, the power, p_1, could be any number, but in most applied settings it makes sense to try to use something simple. Royston and Altman (1994) propose restricting the power to be among those in the set $\wp = \{-2,-1,-0.5,0,0.5,1,2,3\}$, where $p_1 = 0$ denotes the log of the variable. The remaining functions are defined as

$$F_j(x)=\begin{cases} x^{p_j}, p_j \neq p_{j-1} \\ F_{j-1}(x)\ln(x), p_j = p_{j-1} \end{cases}$$

for $j=2,\ldots,J$ and restricting powers to those in $\wp$. For example, if we chose J = 2 with $p_1=0$ and $p_2=-0.5$, then the logit is

$$g(x,\boldsymbol{\beta})=\beta_0+\beta_1\ln(x)+\beta_2\frac{1}{\sqrt{x}}.$$

As another example, if we chose J = 2 with $p_1=2$ and $p_2=2$, then the logit is

$$g(x,\boldsymbol{\beta})=\beta_0+\beta_1x^2+\beta_2x^2\ln(x).$$

The model is quadratic in x when $J=2$ with $p_1=1$ and $p_2=2$. Again, we could allow the covariate to enter the model with any number of functions, J; but in most applied settings an adequate transformation may be found if we use $J=1$ or 2.

Implementation of the method requires, for $J=1$, fitting 8 models, that is, $p_1 \in \wp$. The best model is the one with the largest log likelihood. The process is repeated with $J=2$ by fitting the 36 models obtained from the distinct pairs of powers, that is, $(p_1,p_2)\in\wp\times\wp$, and the best model is again the one with the largest log likelihood.

The relevant question is whether either of the two best models is significantly better than the linear model. Let $L(1)$ denote the log likelihood for the linear model, that is, $J=1$ and $p_1=1$, and $L(p_1)$ denote the log likelihood for the best $J=1$ model and $L(p_1,p_2)$ denote the log likelihood for the best $J=2$ model. Royston and Altman (1994) suggest, and verify with simulations, that each term in the fractional polynomial model contributes approximately 2 degrees of freedom to the model, effectively one for the power and one for the coefficient. Thus, the partial likelihood ratio test comparing the linear model to the best $J=1$ model,

$$G(1,p_1)=-2\{L(1)-L(p_1)\},$$

is approximately distributed as chi-square with 1 degree of freedom under the null hypothesis of linearity in x. The partial likelihood ratio test comparing the best $J=1$ model to the best $J=2$ model,

$$G[p_1,(p_1,p_2)]=-2\{L(p_1)-L(p_1,p_2)\},$$

is approximately distributed as chi-square with 2 degrees of freedom under the null hypothesis that the second function is equal to zero. Similarly, the partial likelihood ratio test comparing the linear model to the best $J=2$ model is distributed approximately as chi-square with 3 degrees of freedom. Note that to keep the notation simple, we use p_1 to denote the best power both when $J=1$ and as the first of the two powers for $J=2$. These are not likely to be the same numeric value in practice.

The only software package that has fully implemented the method of fractional polynomials is STATA. In addition to the method described above, STATA's fractional polynomial routine offers the user considerable flexibility in expanding the set of powers searched; however, in most settings the default set of values should be adequate.

The previous discussion introduced the basic approach to the use of fractional polynomials in the setting of a simple univariable logistic regression model. In practice most models are multivariable and can contain numerous continuous covariates, each of which must be checked for linearity. Recently Royston and Ambler (1998, 1999) extended the original fractional polynomial software in STATA to incorporate an iterative examination for scale with multivariable models. The default method incorporates recommendations discussed in detail in Sauerbrei and Royston (1999). In our discussion of the method we assume that the fractional polynomial analysis will not incorporate any variable selection. That is, all the covariates remain in the model. The variables are ordered by the Wald statistics with the most significant (i.e., smallest p-value) first. The following two step procedure is performed on each covariate that is being checked for the scale in the logit. The first step is the 3 degree-of-freedom test of the best $J=2$ versus the linear model, $G[1,(p_1,p_2)]$. If this test is not significant at a user-specified level of significance, α_{fp}, then the covariate is modeled as linear in the logit. If the test is significant, the second step is the 2 degree-of-freedom test of the best $J=2$ versus the best $J=1$ fractional polynomial models, $G[p_1,(p_1,p_2)]$. If this test is significant at the α_{fp} level then the best $J=2$ model is chosen, otherwise the best $J=1$ model is chosen. After checking each covariate, the process recycles through each covariate using the results of the first cycle in the sense that covariates not being checked are included in the model using the results from the first cycle. The purpose of recycling is to ascertain whether a transformation of one covariate changes the transformation of one or more of the other covariates. The process keeps cycling until no further transformations are indicated. It is rare for the method to require more than two cycles and it usually stops after one cycle.

In an applied setting, we recommend that if a more complicated model is selected for use then it should provide a statistically significant improvement over the linear model, and it is vital that the transformation make clinical sense.

Kay and Little (1987) suggest another method for examining the scale of continuous covariates. They illustrate how examination of the marginal distribution of the continuous covariates within outcome groups may help suggest the appropriate scale. For example, they show that if the distribution of a particular covariate, X, is normal within each outcome group but with different means and variances, then a linear term, X, and a quadratic term, X^2, are needed in the logit. This parametrization corresponds to a fractional polynomial with $J = 2$ and $p_1 = 1$ and $p_2 = 2$. If X follows a gamma distribution, skewed right, then we need to include X and $\ln(X)$ in the model, equivalently a fractional polynomial with $J = 2$ and $p_1 = 1$ and $p_2 = 0$. If X follows a beta distribution with different parameters within outcome groups, then inclusion of $\ln(X)$ and $\ln(1 - X)$ is necessary to correctly model the covariate. Due to the need for $\ln(1 - X)$ this parametrization cannot be expressed in terms of a fractional polynomial. This should be kept in mind as occasionally one encounters a covariate with a fixed range. For example we may use the logistic probability from one model as a covariate in a second model. The approach of Kay and Little may be most useful for continuous variables when there are enough observations within each outcome group to obtain an accurate approximation to the distribution of the covariate.

Another approach to scale selection that can be thought of as a non-parametric generalization of fractional polynomials is to fit a generalized additive model. Hastie and Tibshirani (1986, 1987, and 1990) discuss the use of a generalized additive model for the analysis of binary data. The results of fitting such a model may be used to plot an adjusted non-parametrically smoothed estimate of the effect of a covariate. The plot can then be visually checked for nonlinearity and a possible parametric transformation. The models are quite sophisticated and require considerable experience to be used effectively. For these reasons we do not consider them in any detail. In addition, their use requires special software not typically available in most packages. Royston and Ambler (1998) have written a module that may be used in conjunction with STATA to fit additive models.

Before proceeding to an example illustrating the proposed method for building a logistic regression model we need to discuss one special type of variable that does occur reasonably often in practice. Consider a study in which subjects are asked to report their lifetime use of cigarettes. All the non-smokers report a value of zero. A one-half pack-a-day smoker for 20 years has a value of approximately 73,000 cigarettes. What makes this covariate unusual is the fact that the zero value occurs with a frequency much greater than expected for a fully continuous distribution. In addition, the non-zero values typically exhibit right skewness. Robertson, Boyle, Hsieh, Macfarlane, and Maisonneuve (1994) show that the correct way to model such a covariate is to include two terms, one that is dichotomous recording zero versus non-zero and one for the actual recorded value. Thus, the logit for a univariable model is

$$g(x,\boldsymbol{\beta}) = \beta_0 + \beta_1 d + \beta_2 x,$$

where $d = 0$ if $x = 0$ and $d = 1$ if $x > 0$. The advantage of this parameterization is that it allows us to model two different odds ratios,

$$\text{OR}\left(x = x^*, x = 0\right) = e^{\beta_1 + \beta_2 x^*}$$

and

$$\text{OR}(x = x + c, x = x) = e^{\beta_2 c}.$$

Note that during the modeling process we still need to check the scale in the logit for the positive values of the covariate.

Example

As an example of the model-building process, consider the analysis of the UMARU IMPACT study (UIS). The study is described in Section 1.6.4 and a code sheet for the data is shown in Table 1.8. Briefly, the goal of the analysis is to determine whether there is a difference between the two treatment programs after adjusting for potential con-

Table 4.1 Univariable Logistic Regression Models for the UIS (n = 575)

Variable	Coeff.	Std. Err.	$\widehat{OR}$	95 % CI	G	p
AGE	0.018	0.0153	1.20*	(0.89, 1.62)	1.40	0.237
BECK	−0.008	0.0103	0.96+	(0.87, 1.06)	0.64	0.425
NDRGTX	−0.075	0.0247	0.93	(0.88, 0.97)	11.84	<0.001
IVHX_2	−0.481	0.2657	0.62	(0.37, 1.04)		
IVHX_3	−0.775	0.2166	0.46	(0.30, 0.70)	13.35	0.001
RACE	0.459	0.2110	1.58	(1.04, 2.39)	4.62	0.032
TREAT	0.437	0.1931	1.55	(1.06, 2.26)	5.18	0.023
SITE	0.264	0.2034	1.30	(0.87, 1.94)	1.67	0.197

*: Odds Ratio for a 10 year increase in AGE

+: Odds Ratio for a 5 point increase in BECK

founding and interaction variables. One outcome of considerable public health interest is whether or not a subject remained drug free for at least one year from randomization to treatment (DFREE in Table 1.8). A total of 147 of the 575 subjects (25.57%), considered in the analyses presented in this text, remained drug free for at least one year. The analyses in this chapter are primarily designed to demonstrate specific aspects of logistic model building. Hosmer and Lemeshow (Chapter 5, 1999) present an analysis based on the actual length of time to return to drug use. The analyses in this chapter and in Hosmer and Lemeshow (1999) should not be considered definitive. One should see the papers written by our colleagues cited in Section 1.6.4 for more detailed analyses of the UIS data and a discussion of study results.

The results of fitting the univariable logistic regression models to these data are given in Table 4.1. In this table we present, for each variable listed in the first column, the following information. (1) The estimated slope coefficient(s) for the univariable logistic regression model containing only this variable, (2) The estimated standard error of the estimated slope coefficient, (3) The estimated odds ratio, which is obtained by exponentiating the estimated coefficient. For the variable AGE the odds ratio is for a 10-year increase and for Beck depression score (BECK) the odds ratio is for a 5-point increase. This was done since a change of 1 year or 1 point would not be clinically meaningful. (4) The 95% CI for the odds ratio. (5) The likelihood ratio test statistic, G, for the hypothesis that the slope coefficient is zero. Under the null

Table 4.2 Results of Fitting a Multivariable Model Containing the Covariates Significant at the 0.25 Level in Table 4.1

Variable	Coeff.	Std. Err.	z	P>|z|
AGE	0.050	0.0173	2.91	0.004
NDRGTX	−0.062	0.0256	−2.40	0.016
IVHX_2	−0.603	0.2873	−2.10	0.036
IVHX_3	−0.733	0.2523	−2.90	0.004
RACE	0.226	0.2233	1.01	0.311
TREAT	0.443	0.1993	2.22	0.026
SITE	0.149	0.2172	0.68	0.494
Constant	−2.405	0.5548	−4.34	<0.001

Log likelihood = −309.6241

hypothesis, this quantity follows the chi-square distribution with 1 degree of freedom, except for the variable IVHX, where it has 2 degrees of freedom. (6) The significance level for the likelihood ratio test.

With the exception of Beck score there is evidence that each of the variables has some association ($p<0.25$) with the outcome, remaining drug free for at least one year (DFREE). The covariate recording history of intravenous drug use (IVHX) is modeled via two design variables using "1 = Never" as the reference code. Thus its likelihood ratio test has two degrees-of-freedom. We begin the multivariable model with all but BECK. The results of fitting the multivariable model are given in Table 4.2.

The results in Table 4.2, when compared to Table 4.1, indicate weaker associations for some covariates when controlling for other variables. In particular, the significance level for the Wald test for the coefficient for SITE is $p = 0.494$ and for RACE is $p = 0.311$. Strict adherence to conventional levels of statistical significance would dictate that we consider a smaller model deleting these two covariates. However, due to the fact that subjects were randomized to treatment within site we keep SITE in the model. On consultation with our colleagues we were advised that race is an important control variable. Thus on the basis of subject matter considerations we keep RACE in the model.

The next step in the modeling process is to check the scale of the continuous covariates in the model, AGE and NDRGTX in this case. One approach to deciding the order in which to check for scale is to

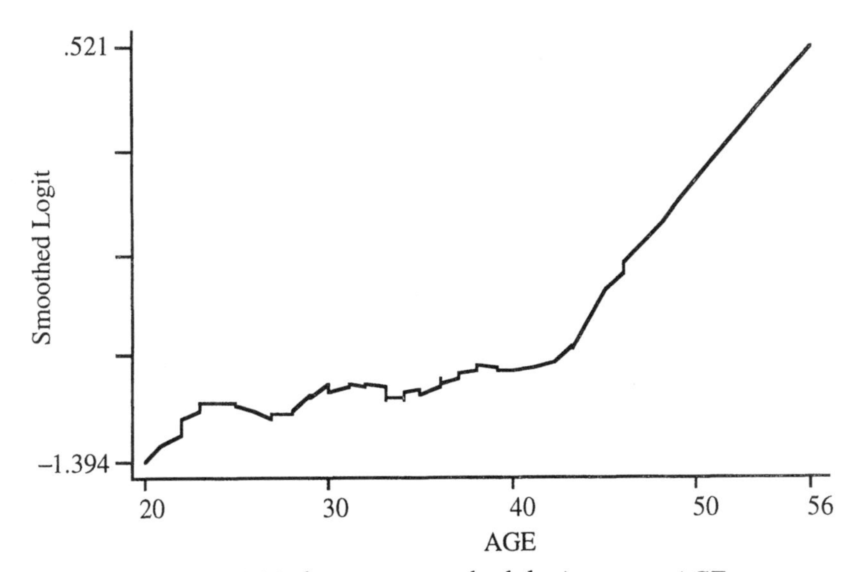

Figure 4.2 Univariable lowess smoothed logit versus AGE.

rank the continuous variables by their respective significance levels. Results in Table 4.2 suggest that we consider AGE and then NDRGTX.

To explore the scale of AGE we use three different methods: (1) a univariable smoothed scatterplot on the logit scale, (2) design variables based on the quartiles of the distribution and (3) the method of fractional polynomials. We begin with the lowess smoothed univariable logit shown in Figure 4.2. This plot shows a linear increase from age 20 to about age 40 and then a steeper linear increase to age 56. Overall, the plot supports treating AGE as linear in the logit.

Results from an analysis using design variables with the first quar-

Table 4.3 Results of the Quartile Analyses of AGE from the Multivariable Model Containing the Variables Shown in the Model in Table 4.2

Quartile	1	2	3	4
Midpoint	24	30.5	35.5	47.5
Number	148	144	166	117
Coeff.	0.0	−0.166	0.469	0.596
95 % CI		(−0.74, 0.40)	(−0.06, 1.00)	(−0.02, 1.21)

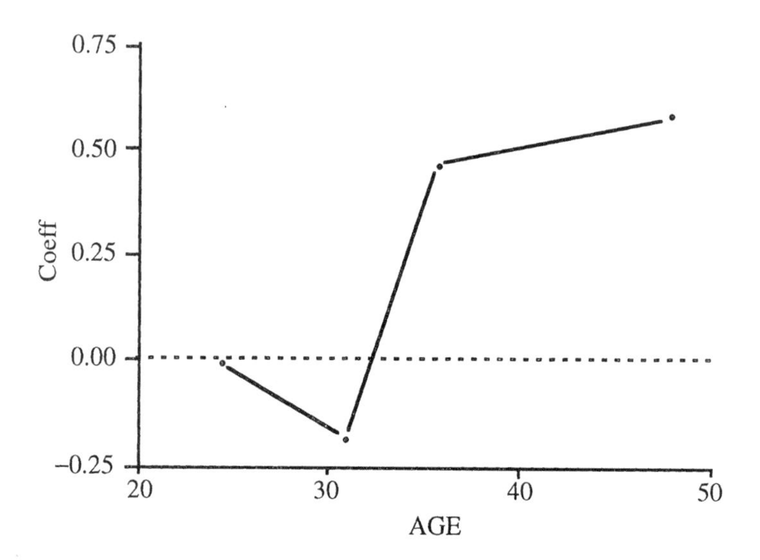

Figure 4.3 Plot of estimated logistic regression coefficients versus approximate quartile midpoints of AGE.

tile as the reference group are shown in Table 4.3 and Figure 4.3. The coefficients in Table 4.3 show an initial decrease followed by an increase in the log odds. The results do not conclusively support linearity in the logit for AGE nor do they rule it out. One possible scaling suggested by these results is to create a dichotomous covariate at the median. This suggestion is based on the observation that zero is well within the confidence interval for the coefficient for the second quartile and zero is barely contained in the confidence intervals for the other two coefficients. In addition, there is considerable overlap in the confidence intervals for the coefficients for the third and fourth quartiles. Wald tests, left as an exercise, support these observations. However, the quartile analysis is not an especially powerful diagnostic tool. Its strength is that it is easily explained to and understood by subject matter scientists.

The results of the fractional polynomial analysis are presented in Table 4.4. A detailed description of the results is as follows:

(1) The significance level in the second line of Table 4.4, $p = 0.003$ is for the single degree-of-freedom likelihood ratio test of a model not containing AGE versus the model containing AGE as a single linear term, e.g. the model in Table 4.2.

(2) The significance level in the third line of Table 4.4, $p = 0.545$, is for the single degree-of-freedom likelihood ratio test of a

model containing AGE as a single linear term versus the model containing AGE^3, i.e., $G = 0.366$ and $\Pr[\chi^2(1) \geq 0.366] = 0.545$.

(3) The significance level in the fourth line of Table 4.4, p = 0.945 is for the two degree-of-freedom likelihood ratio test of a model containing AGE^3 versus the model containing AGE^{-2} and AGE^3, i.e., $G = 618.882 - 618.769 = 0.113$ and $\Pr[\chi^2(2) \geq 0.113] = 0.945$.

(4) The likelihood ratio test statistic for the best $J = 2$ model versus the linear model is $G = 619.248 - 618.769$ = 0.479 and its p-value (not shown in Table 4.4) is $\Pr[\chi^2(3) \geq 0.479] = 0.923$.

The best non-linear transformations are not significantly different from the linear model and thus the fractional polynomial analysis supports treating age as linear in the logit.

In summary, the smoothed logit and fractional polynomial analysis support treating age as continuous and linear in the logit. The quartile based design variable analysis, while not conclusive, suggests using a dichotomous variable with the median age as the cutpoint. Results not shown, left as an exercise, indicate that modeling age as a dichotomous variable provides a model that is not better than one treating age as continuous and linear in the logit. Hence we choose to treat age as continuous and linear in the logit.

We use the same three methods to assess the scale of NDRGTX and begin with the univariable smoothed logit in Figure 4.4. The plot shows an initial increase in the logit at 1 and 2 previous treatments. This is followed by a nearly linear decrease in the range of 3 to about 15 previous treatments. The logit appears to have no consistent trend in the

Table 4.4 Summary of the Use of the Method of Fractional Polynomials for AGE

	df	Deviance	G for Model vs. Linear	Approx. p-Value	Powers
Not in model	0	627.801			
Linear	1	619.248	0.000	0.003*	1
$J = 1$	2	618.882	0.366	0.545+	3
$J = 2$	4	618.769	0.479	0.945#	−2, 3

* Compares linear model to model without AGE
+ Compares the $J = 1$ model to the linear model
# Compares the $J = 2$ model to the $J = 1$ model

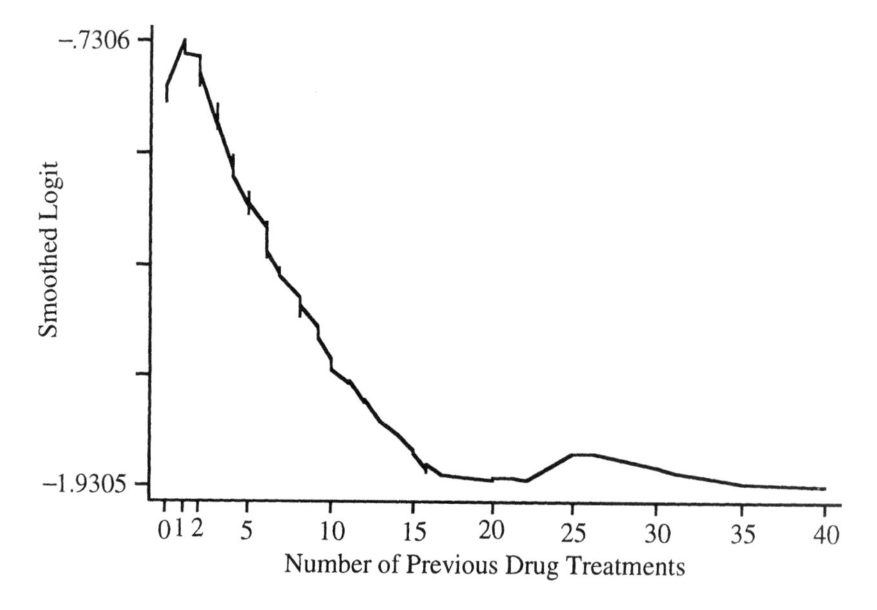

Figure 4.4 Univariable lowess smoothed logit versus number of previous drug treatments (NDRGTX).

range from 15 to 40 previous treatments. It is of particular interest to determine whether the initial increase in the logit is significant or simply a numerical artifact of the smoothing. We explore this possibility using a design variable analysis. We choose cutpoints using Figure 4.4 as a guide. The results of the design variable analysis are shown in Table 4.5 and Figure 4.5.

The results in Table 4.5 and Figure 4.5 agree with the pattern seen in Figure 4.4 of an increase followed by a progressive decrease in the logit. Since zero is contained in each of the confidence intervals none

Table 4.5 Results of the Design Variable Analysis of Number of Previous Drug Treatments (NDRGTX) from the Multivariable Model Containing the Variables Shown in the Model in Table 4.2

Group	1	2	3	4
Interval	0	1-2	3-15	16-40
Number	79	173	294	29
Coeff.	0.0	0.406	−0.154	−0.585
95 % CI		(−0.20, 1.01)	(−0.76, 0.46)	(−1.80, 0.63)

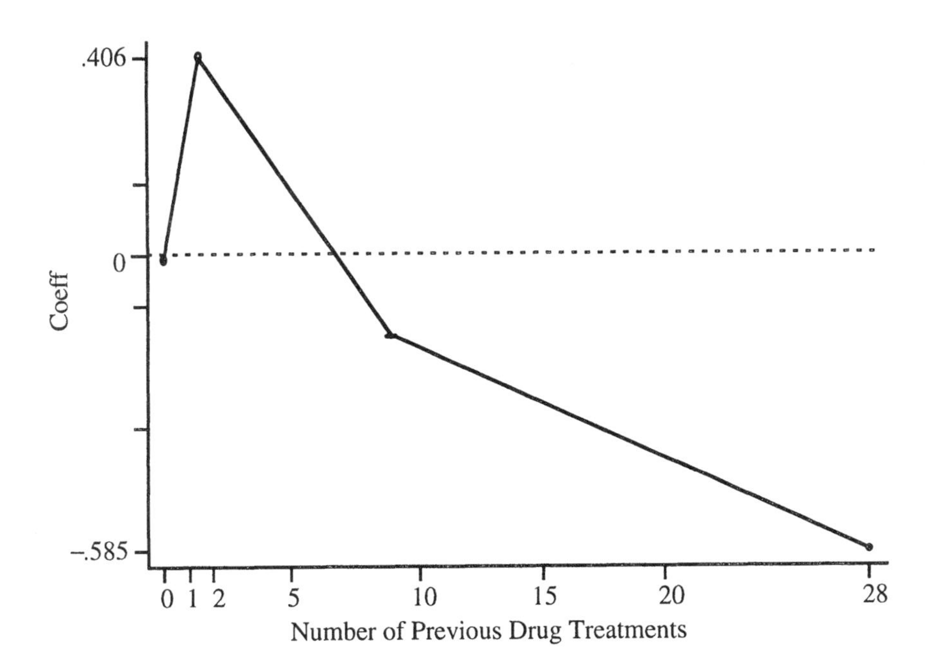

Figure 4.5 Plot of estimated logistic regression coefficients from Table 4.4 versus the midpoints of number of previous drug treatment groups.

of the coefficients is significantly different from zero. This result does not agree with the fact that NDRGTX is significant in Table 4.2.

We next use the method of fractional polynomials. The results of this analysis, presented in Table 4.6, (see Addendum, page 352) suggest that we consider the $J = 2$ model as a possible non-linear transformation of NDRGTX. This model is significantly different from the best $J = 1$ model, at the 10 percent level. The significance level of the $J = 2$ model versus the linear model is $\Pr[\chi^2(3) \geq 5.797] = 0.122$, which indicates that the model offers a small improvement over the linear model. Recall that the goals are to see whether a fractional polynomial transformation of NDRGTX is able to provide a parametric model similar in shape to the smoothed logit in Figure 4.4, and to determine whether such a model is significantly better than the linear model from both a statistical and a clinical perspective.

A graph of the univariable smoothed logit and the two term fractional polynomial model, adjusted for other model covariates, is shown in Figure 4.6. We explain how the graph was obtained shortly. The

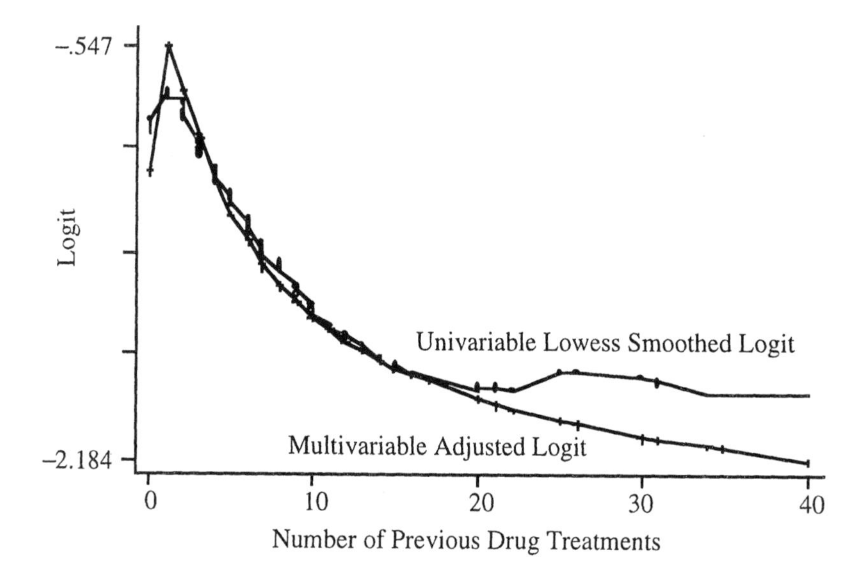

Figure 4.6 Plot of the univariable lowess smoothed logit (o) and the multivariable adjusted logit (+) from the $J = 2$ fractional polynomial model versus number of previous drug treatments (NDRGTX).

fractional polynomial model has the desired shape, a sharp rise followed by a gradual fall and there is good agreement between the two sets of plotted points. At this point, we consulted our colleagues to see if the shape of the two term fractional polynomial model made any clinical sense. They advised us that, in fact, the basic shape makes sense from a clinical point of view. Namely, subjects with no previous drug treatments tend to be less likely to return to drug use than subjects with one or two previous treatments. The rate of return to drug use tends to decrease thereafter for larger numbers of treatments. They also found the fact that this pattern could be described by a parametric function quite an interesting result. Hence we proceed with the two term fractional polynomial model. The results of fitting this model are shown in Table 4.7.

The two fractional polynomial transformations are

$$NDRGFP1 = \left[\frac{(NDRGTX + 1)}{10}\right]^{-1}$$

and

Table 4.7 Results of Fitting the Multivariable Model with the Two Term Fractional Polynomial Transformation of NDRGTX

Variable	Coeff.	Std. Err.	z	P>|z|
AGE	0.054	0.0175	3.11	0.002
NDRGFP1	0.981	0.2888	3.40	0.001
NDRGFP2	0.361	0.1099	3.29	0.001
IVHX_2	−0.609	0.2911	−2.09	0.036
IVHX_3	−0.724	0.2556	−2.83	0.005
RACE	0.248	0.2242	1.11	0.269
TREAT	0.422	0.2004	2.11	0.035
SITE	0.173	0.2210	0.78	0.433
Constant	−2.928	0.5867	−4.99	<0.001

Log likelihood = −306.7256

$$NDRGFP2 = NDRGFP1 \times \ln\left[\frac{(NDRGTX+1)}{10}\right].$$

It should be noted that there are several alternative ways of incorporating fractional polynomial transformed variables. STATA uses a scaled and centered version. Alternatively, one could use an uncentered transformation or a transformation that is neither centered nor scaled. In this example, we chose to use the scaled but uncentered transformation shown above. It should be noted that, in the equations above, 1 is added to NDRGTX to avoid problems with $NDRGTX = 0$ and to control the range of the transformed covariate. We note that the Wald statistics for the coefficients of NDRGFP1 and NDRGFP2 are statistically significant.

We obtained the plot presented in Figure 4.6 using the following procedure. First we requested that STATA create and save a new variable containing the values of the smoothed logit plotted in Figure 4.4. Suppose we named this variable LGTSM. Next, we used the coefficients in Table 4.7 to compute an estimated logit as a function of NDRGFP1 and NDRGFP2 as follows:

$$LGTFP = -4.314 + 0.981 \times NDRGFP1 + 0.361 \times NDRGFP2.$$

Before plotting, we added the difference between the mean of LGTSM and LGTFP to the values of LGTFP so that the two plotted variables would have the same mean, namely,

$$LGTFP^* = LGTFP + \left(\overline{LGTSM} - \overline{LGTFP}\right).$$

We plotted LGTSM and LGTFP* versus the values of NDRGTX to obtain Figure 4.6.

The two curves in Figure 4.6 are quite similar after adjusting for the difference in the means. This may not always be the case. The lowess scatterplot smooth does not take into account all the covariates in the model whereas the plot of the fractional polynomial does. In addition, we note that we get -4.314 from $\hat{\beta}_0 + \sum_{i=1}^{k} \hat{\beta}_i \bar{x}_i$, where the summation is over all but the fracploy covariate. We could have computed *LGTFP* using a different set of typical values for the other covariates (e.g., the median age and some choice of zero and one for each of the dichotomous covariates). The net effect would have been to change the value of the constant from −4.314 to some other value. This would have produced a different mean value for *LGTFP* and, therefore, a different value for $\overline{LGTSM} - \overline{LGTFP}$; however, once done, the plot would be identical to that seen in Figure 4.6.

The current model building example demonstrates that the family of fractional polynomial transformations provides the ability to model a surprisingly complex function parametrically. This method is an extremely powerful analysis tool, the results of which must be examined carefully using statistical and clinical criteria. It is absolutely vital that any fractional polynomial transformation makes clinical sense.

Before moving on to assess interactions we check to make sure that any main effects not included in the model still are neither significant nor important confounding variables. In this case the only variable not included in the model is BECK. It was neither significant (p = 0.932) nor an important confounder when added to the main effects model in Table 4.7. We used the iterative multivariable fractional polynomial procedure as a final check on the scale of AGE and NDRGTX. This analysis yielded the same results: no need to transform AGE and use the (−1, −1) transformation for NDRGTX.

The model in Table 4.7 is our main effects model. The next step in the model building process is to assess the need to include interactions. At this point we recommend that a list be prepared of the clinically plausible interactions that can be formed from the main effects in the model. This list may or may not consist of all possible interactions. Our main effects model contains six covariates, hence there are 15 possible pairwise interactions. We think that each one offers the possibility

Table 4.9 Preliminary Final Model Containing Significant Main Effects and Interactions

Variable	Coeff.	Std. Err.	z	P>|z|
AGE	0.117	0.0289	4.04	<0.001
NDRGFP1	1.669	0.4072	4.10	<0.001
NDRGFP2	0.434	0.1169	3.71	<0.001
IVHX_2	−0.635	0.2987	−2.13	0.034
IVHX_3	−0.705	0.2616	−2.70	0.007
RACE	0.684	0.2641	2.59	0.010
TREAT	0.435	0.2038	2.14	0.033
SITE	0.516	0.2549	2.03	0.043
AGE×NDRGFP1	−0.015	0.0060	−2.53	0.011
RACE×SITE	−1.429	0.5298	−2.70	0.007
Constant	−6.844	1.2193	−5.61	<0.001

Log likelihood = −298.9814

of a clinically plausible modification of the covariate effects. The results of adding each of the 15 interactions one at a time to the main effects model in Table 4.7 are presented in Table 4.8 (see Addendum, page 352).

The results in Table 4.8 show that only the AGE×NDRGTX, AGE×TREAT and RACE×SITE interactions are significant at the 10 percent level. Next we add these three interactions (i.e., four terms) to the main effects model. The fit of this model (not presented) yields Wald statistics for the two coefficients for the AGE×NDRGTX interaction that are not significant yet the two degree of freedom likelihood ratio test for exclusion of the two interaction terms is highly significant, $p = 0.026$. We attribute this conflict in results to high correlation between NDRGFP1 and NDRGFP2. To explore this further we fit a model (not shown) containing three interaction terms: AGE×NDRGFP1, AGE×TREAT and RACE×SITE. In this model the Wald statistic for the AGE×TREAT interaction term is not significant, $p = 0.113$, and the likelihood ratio test for its exclusion has $p = 0.111$. Thus we feel that this interaction term should not be included in the model. We present in Table 4.9 the results of fitting the model containing the main effects in Table 4.7 and the two significant interaction terms.

We refer to the model in Table 4.9 as the preliminary final model as we have not checked its goodness of fit or other diagnostic statistics of model adequacy. These methods are presented in detail in Chapter 5. Following the assessment of fit we discuss, in Section 5.5, the interpretation and presentation of the results from a fitted model using our final model from the UIS as an example.

4.3 STEPWISE LOGISTIC REGRESSION

Stepwise selection of variables is widely used in linear regression. All the major software packages have either a separate program or an option to perform this type of analysis. Currently, most, if not all, major software packages offer an option for stepwise logistic regression. At one time, stepwise regression was an extremely popular method for model building. In recent years there has been a shift away from deterministic methods for model building to purposeful selection of variables. However, we feel that the procedure provides a useful and effective data analysis tool. In particular, there are times when the outcome being studied is relatively new and the important covariates may not be known and associations with the outcome not well understood. In these instances most studies collect many possible covariates and screen them for significant associations. Employing a stepwise selection procedure can provide a fast and effective means to screen a large number of variables, and to fit a number of logistic regression equations simultaneously.

Any stepwise procedure for selection or deletion of variables from a model is based on a statistical algorithm that checks for the "importance" of variables, and either includes or excludes them on the basis of a fixed decision rule. The "importance" of a variable is defined in terms of a measure of the statistical significance of the coefficient for the variable. The statistic used depends on the assumptions of the model. In stepwise linear regression an F-test is used since the errors are assumed to be normally distributed. In logistic regression the errors are assumed to follow a binomial distribution, and significance is assessed via the likelihood ratio chi-square test. Thus, at any step in the procedure the most important variable, in statistical terms, is the one that produces the greatest change in the log-likelihood relative to a model not containing the variable (i.e., the one that would result in the largest likelihood ratio statistic, G).

We discussed in Chapter 3 that a polychotomous variable with k levels is appropriately modeled through its $k-1$ design variables. Since the magnitude of G depends on its degrees-of-freedom, any procedure based on the likelihood ratio test statistic, G, must account for possible differences in degrees-of-freedom between variables. This is done by assessing significance through the p-value for G.

We describe and illustrate the algorithm for forward selection followed by backward elimination in stepwise logistic regression. Any variants of this algorithm are simple modifications of this procedure. The method is described by considering the statistical computations that the computer must perform at each step of the procedure.

Step (0): Suppose we have available a total of p possible independent variables, all of which are judged to be of plausible "clinical" importance in studying the outcome variable. Step (0) begins with a fit of the "intercept only model" and an evaluation of its log-likelihood, L_0. This is followed by fitting each of the p possible univariable logistic regression models and comparing their respective log-likelihoods. Let the value of the log-likelihood for the model containing variable x_j at step zero be denoted by $L_j^{(0)}$. The subscript j refers to the variable that has been added to the model, and the superscript (0) refers to the step. This notation is used throughout the discussion of stepwise logistic regression to keep track of both step number and variables in the model.

Let the value of the likelihood ratio test for the model containing x_j versus the intercept only model be denoted by $G_j^{(0)} = -2\left(L_0 - L_j^{(0)}\right)$, and its p-value be denoted by $p_j^{(0)}$. Hence, this p-value is determined by the tail probability $\Pr\left[\chi^2(\nu) > G_j^{(0)}\right] = p_j^{(0)}$, where $\nu = 1$ if x_j is continuous and $\nu = k-1$ if x_j is polychotomous with k categories.

The most important variable is the one with the smallest p-value. If we denote this variable by x_{e_1}, then $p_{e_1}^{(0)} = \min\left(p_j^{(0)}\right)$, where "min" stands for selecting the minimum of the quantities enclosed in the brackets. The subscript "e_1" is used to denote that the variable is a candidate for entry at step 1. For example, if variable x_2 had the smallest p-value, then $p_2^{(0)} = \min\left(p_j^{(0)}\right)$, and $e_1 = 2$. Just because x_{e_1} is the most important variable, there is no guarantee that it is "statistically significant." For example, if $p_{e_1}^{(0)} = 0.83$, we would probably conclude that there is little point in continuing this analysis because the "most important" variable is not related to the outcome. On the other hand, if $p_{e_1}^{(0)} = 0.003$, we would like to look at the logistic regression containing this variable and then see if there are other variables that are important given that x_{e_1} is in the model.

A crucial aspect of using stepwise logistic regression is the choice of an "alpha" level to judge the importance of variables. Let p_{E} denote our choice where the "E" stands for entry. The choice for p_{E} determines how many variables eventually are included in the model. Bendel and Afifi (1977) have studied the choice of p_{E} for stepwise linear regression, and Costanza and Afifi (1979) have studied the choice for stepwise discriminant analysis. More recently Lee and Koval (1997) examined the issue of significance level for forward stepwise logistic regression. The results of this research have shown that the choice of $p_{\text{E}} = 0.05$ is too stringent, often excluding important variables from the model. Choosing a value for p_{E} in the range from 0.15 to 0.20 is highly recommended.

Sometimes the goal of the analysis may be broader, and models containing more variables are sought to provide a more complete picture of possible models. In these cases, use of $p_{\text{E}} = 0.25$ or even larger might be a reasonable choice. Whatever the choice for p_{E}, a variable is judged important enough to include in the model if the p-value for G is less than p_{E}. Thus, the program proceeds to step (1) if $p_{e_1}^{(0)} < p_{\text{E}}$; otherwise, it stops.

Step (1): Step (1) commences with a fit of the logistic regression model containing x_{e_1}. Let $L_{e_1}^{(1)}$ denote the log-likelihood of this model. To determine whether any of the remaining $p-1$ variables are important once the variable x_{e_1} is in the model, we fit the $p-1$ logistic regression models containing x_{e_1} and x_j, $j = 1,2,3,\ldots,p$ and $j \neq e_1$. For the model containing x_{e_1} and x_j let the log-likelihood be denoted by $L_{e_1 j}^{(1)}$, and let the likelihood ratio chi-square statistic of this model versus the model containing only x_{e_1} be denoted by $G_j^{(1)} = -2\left(L_{e_1}^{(1)} - L_{e_1 j}^{(1)}\right)$. The p-value for this statistic is denoted by $p_j^{(1)}$. Let the variable with the smallest p-value at step (1) be x_{e_2} where $p_{e_2}^{(1)} = \min\left(p_j^{(1)}\right)$. If this value is less than p_{E} then we proceed to step (2); otherwise we stop.

Step (2): Step (2) begins with a fit of the model containing both x_{e_1} and x_{e_2}. It is possible that once x_{e_2} has been added to the model, x_{e_1} is no longer important. Thus, step (2) includes a check for backward elimination. In general this is accomplished by fitting models that delete one of the variables added in the previous steps and assessing the continued importance of the variable removed. At step (2) let $L_{-e_j}^{(2)}$ denote the log-likelihood of the model with x_{e_j} removed. In similar fashion let the likelihood ratio test of this model versus the full model at step (2) be $G_{-e_j}^{(2)} = -2\left(L_{-e_j}^{(2)} - L_{e_1e_2}^{(2)}\right)$ and $p_{-e_j}^{(2)}$ be its p-value.

To ascertain whether a variable should be deleted from the model the program selects that variable which, when removed, yields the maximum p-value. Denoting this variable as x_{r_2}, then $p_{r_2}^{(2)} = \max\left(p_{-e_j}^{(2)}, p_{-e_2}^{(2)}\right)$. To decide whether x_{r_2} should be removed, the program compares $p_{r_2}^{(2)}$ to a second pre-chosen "alpha" level, p_{R}, which indicates some minimal level of continued contribution to the model where "R" stands for remove. Whatever value we choose for p_{R}, it must exceed the value of p_{E} to guard against the possibility of having the program enter and remove the same variable at successive steps.

If we do not wish to exclude many variables once they have entered then we might use $p_{\text{R}} = 0.9$. A more stringent value would be used if a continued "significant" contribution were required. For example, if we used $p_{\text{E}} = 0.15$, then we might choose $p_{\text{R}} = 0.20$. If the maximum p-value to remove, $p_{r_2}^{(2)}$, exceeds p_{R} then x_{r_2} is removed from the model. If $p_{r_2}^{(2)}$ is less than p_{R} then x_{r_2} remains in the model. In either case the program proceeds to the variable selection phase.

At the forward selection phase each of the $p-2$ logistic regression models are fit containing x_{e_1}, x_{e_2} and x_j for $j = 1,2,3,...p,\ j \neq e_1, e_2$. The program evaluates the log-likelihood for each model, computes the likelihood ratio test versus the model containing only x_{e_1} and x_{e_2} and determines the corresponding p-value. Let x_{e_3} denote the variable with the minimum p-value, that is, $p_{e_3}^{(2)} = \min\left(p_j^{(2)}\right)$. If this p-value is smaller than p_{E}, $p_{e_3}^{(2)} < p_{\text{E}}$, then the program proceeds to step (3); otherwise, it stops.

Step (3): The procedure for step (3) is identical to that of step (2). The program fits the model including the variable selected during the previous step, performs a check for backward elimination followed by forward selection. The process continues in this manner until the last step, step (S).

Step (S): This step occurs when: (1) all p variables have entered the model or (2) all variables in the model have p-values to remove that are less than p_{R}, and the variables not included in the model have p-values to enter that exceed p_{E}. The model at this step contains those variables that are important relative to the criteria of p_{E} and p_{R}. These may or may not be the variables reported in a final model. For instance, if the chosen values of p_E and p_R correspond to our belief for statistical significance, then the model at step S may well contain the significant variables. However, if we have used values for p_E and p_R which are less stringent, then we should select the variables for a final model from a table that summarizes the results of the stepwise procedure.

There are two methods that may be used to select variables from a summary table; these are comparable to methods commonly used in stepwise linear regression. The first method is based on the p-value for entry at each step, while the second is based on a likelihood ratio test of the model at the current step versus the model at the last step.

Let "q" denote an arbitrary step in the procedure. In the first method we compare $p_{e_q}^{(q-1)}$ to a pre-chosen significance level such as $\alpha = 0.15$. If the value $p_{e_q}^{(q-1)}$ is less than α, then we move to step q. We stop at the step when $p_{e_q}^{(q-1)}$ exceeds α. We consider the model at the previous step for further analysis. In this method the criterion for entry is based on a test of the significance of the coefficient for x_{e_q} conditional on $x_{e_1}, x_{e_2}, \ldots, x_{e_{q-1}}$ being in the model. The degrees-of-freedom for the test are 1 or $k-1$ depending on whether x_{e_q} is continuous or polychotomous with k categories.

In the second method, we compare the model at the current step, step q, not to the model at the previous step, step $q-1$ but to the model at the last step, step (S). We evaluate the p-value for the likelihood ratio test of these two models and proceed in this fashion until this p-value exceeds α. This tests that the coefficients for the variables added to the model from step q to step (S) are all equal to zero. At any given step it has more degrees-of-freedom than the test employed in the first method. For this reason the second method may possibly select a larger number of variables than the first method.

It is well known that the p-values calculated in stepwise selection procedures are not p-values in the traditional hypothesis testing context. Instead, they should be thought of as indicators of relative importance among variables. We recommend that one err in the direction of selecting a relatively rich model following stepwise selection. The variables so identified should then be subjected to the more intensive analysis described in the previous section.

A common modification of the stepwise selection procedure just described is to begin with a model at step zero which contains known important covariates. Selection is then performed from among other variables. One instance when this approach may be useful is to select interactions from among those possible from a main effects model.

One disadvantage of the stepwise selection procedures just described is that the maximum likelihood estimates for the coefficients of all variables not in the model must be calculated at each step. For large data files with large numbers of variables this can be quite time consuming. An alternative to a full maximum likelihood analysis that is available in some packages, for example SAS, selects new variables based on the Score tests for the variables not included in the model. Another less time consuming method available in some packages, for example STATA, is based on a multivariable Wald test first suggested by Peduzzi, Hardy, and Holford (1980). To date there has been no work published which has compared these different selection methods although it does seem likely that an important variable would be identified, regardless of method used. For comparison purposes we present, in the example, results based on using full maximum likelihood, the Score test and the Wald test.

Freedman (1983) urges caution when considering a model with many variables, noting that significant linear regressions may be obtained from "noise" variables, completely unrelated to the outcome variable. Flack and Chang (1987) have shown similar results regarding the frequency of selection of "noise" variables. Thus, a thorough analysis that examines statistical and clinical significance is especially important following any stepwise method.

As an example, we apply the stepwise variable selection procedure to the UIS data analyzed using purposeful selection earlier in this chapter. The reader is reminded that this procedure should be viewed as a first step in the model building process – basic variable selection. Subsequent steps such as determination of scale as described in Section 4.2 would follow. The results of this process are presented in Table 4.10 (see Addendum, page 353) in terms of the p-values to enter and remove calculated at each step. These p-values are those of the relevant likelihood ratio test described previously. The order of the variables given column-wise in the table is the order in which they were selected. In each column the values below the horizontal line are p_E values and values above the horizontal lines are p_R values. The program was run using $p_E = 0.15$ and $p_R = 0.20$. This particular analysis was performed using program BMDPLR. One reason we used BMDPLR is it is one of the few stepwise logistic regression programs that correctly considers the design variables formed from a polychotomous variable together for entry or removal from a model.

Step (0): At step (0) the program selects as a candidate for entry at step (1) the variable with the smallest p-value in the first column of Table 4.10. This is the variable NDRGTX with a p-value of 0.0006. Since this p-value is less than 0.15, the program proceeds to step (1).

Step (1): At step (1) a model is fit containing NDRGTX. The program does not remove the variable just entered since $p_R > p_E$ and the p-value to remove at step (1) is equal to the p-value to enter at step (0). This is true for the variable entered at any step - not just the first step. The variable with the smallest p-value to enter at step (1) is TREAT with a value of 0.0249, which is less than 0.15 so the program moves to step (2).

Step (2): The p-values to remove appear first in each row. The largest p-value to remove is indicated with an "*". The model con taining both NDRGTX and TREAT is fit at step (2). The largest p-value to remove is 0.0249, which does not exceed 0.20, thus the program moves to the variable selection phase. The smallest p-value to enter among the remaining variables not in the model is for the variable IVHX and is 0.0332. This value is less than 0.15 so the program proceeds to step (3).

Step (3) At step (3) the largest p-value to remove is 0.0332, which does not exceed 0.20, thus the program moves to the variable selection phase. The smallest p-value to enter among the remaining variables not in the model is for the variable AGE and is 0.0021. This value is less than 0.15 so the program proceeds to step (4).

Step (4): At step (4) the program finds that the maximum p-value to remove is 0.0224 for TREAT. This value is less than 0.20, so TREAT is not removed from the model. In the selection phase the program finds that the minimum p-value for entry is 0.350 for the variable RACE. Since this value exceeds 0.15, no further variables may be entered into the model, and the program stops.

Since the program was run with $p_E = 0.15$, a value we believe selects variables with significant coefficients, it is not strictly necessary to go to the summary table to select the variables to be used in a final model. However, we illustrate the calculations for the two methods of variable selection from the summary table in Table 4.11.

For method 1, we compare the p-value for entry at each step to our chosen level of significance. For purposes of illustration only we use the value of 0.03, even though we noted earlier in this section that it is too stringent for actual practice. The information for method 1 is in the second panel of Table 4.11.

Table 4.11 Log-Likelihood for the Model at Each Step and Likelihood Ratio Test Statistics (G), Degrees-of-freedom (df), and p-Values for Two Methods of Selecting Variables for a Final Model from a Summary Table

	Variable		Method 1			Method 2		
Step	Entered	Log Like.	G	df	p	G	df	p
0		−326.864				33.14	5	<0.001
1	NDRGTX	−320.945	11.84	1	<0.001	21.30	4	<0.001
2	TREAT	−318.430	5.03	1	0.025	16.27	3	<0.001
3	IVHX	−315.025	6.81	2	0.033	9.46	1	0.002
4	AGE	−310.293	9.46	1	0.002			

The value of the likelihood ratio test for the model at step (0) compared to that containing NDRGTX at step (1) is

$$G = -2[-326.864 - (-320.945)] = 11.84.$$

The p-value for G is <0.001, which is less than 0.03, so we conclude that the coefficient for NDRGTX is significant and move to step (2). The p-value for the variable, TREAT, entered at step (2) is 0.025. This is the p-value for the likelihood ratio test of the significance of the coefficient for TREAT, given that NDRGTX is in the model. The value of the test statistic is

$$G = -2[-320.945 - (-318.430)] = 5.03.$$

Since the p-value for G is less than 0.03 we move to step (3). At step (3) we find that the value of the likelihood ratio test of the model at step (3) versus that at step (2) is

$$G = -2[-318.430 - (-315.025)] = 6.81,$$

resulting in a p-value of 0.033. This value is greater than 0.03 so we conclude that TREAT does not provide a significant addition to the variables already selected at step (2). Hence, the final model would be the one with all variables entered through step (2) even though the variable entered at step (4), AGE, has a p-value of less than 0.03.

The information for method 2 is in the last panel of Table 4.11. In the second method the model at each step is compared to the model at the last step via a likelihood ratio test. This is a test of the joint significance of variables added at subsequent steps. We again proceed until the p-value for the test exceeds the chosen significance level. For purposes of illustration only we use 0.03 again. The value of G at step (0) is

$$G = -2[-326.864 - (-310.293)] = 33.14$$

with a p-value of <0.001 based on 5 degrees-of-freedom. Since this p-value is less than 0.03 we proceed to step (1). At step (1) the test of this model versus that at the last step is

$$G = -2[-320.945 - (-310.293)] = 21.304$$

with a p-value of <0.001 based on 4 degrees-of-freedom. Since the p-value is less than 0.03 we proceed to step (2).

At step (2) the test of this model versus that at the last step is

$$G = -2[-318.430 - (-310.293)] = 16.27$$

with a p-value of <0.001 based on 3 degrees-of-freedom. Since the p-value is less than 0.03 we proceed to step (3).

At step (3) the test of this model versus that at the last step is

$$G = -2[-315.025 - (-310.293)] = 9.46$$

with a p-value of 0.002 based on 1 degree of freedom. This value is less than 0.03, so we use the model at step (4).

In this example methods 1 and 2 have identified different sets of variables. Each method provides a test of a different hypothesis at each step. The number of parameters being tested in method 2 is, except for the last step, larger than that for method 1. Thus, method 2 may select, as it does in this example, more variables than method 1. In cases where this occurs, one should carefully examine the additional variables and include them if they seem clinically relevant. In this case we would undoubtedly opt for the richer model selected by method 2. Again we wish to emphasize that had we used the recommended level of significance of 0.15 both methods suggest the same model.

At the conclusion of the stepwise selection process we have only identified a collection of variables which seem to be statistically important. Thus, any known clinically important variables, for example RACE, or variables that must be controlled for due to the design of the study, such as SITE, should be added before proceeding with the steps necessary to obtain the final main effects model. As noted earlier, this should include determining the appropriate scale of continuous covariates.

Once the scale of the continuous covariates has been examined, and corrected if necessary, we may consider applying stepwise selection to identify interactions. The candidate interaction terms are those that seem clinically reasonable given the main effect variables in the model. We begin at step (0) with the main effects model and sequentially select from among the possible interactions. We use the summary table to select the significant interactions using either method 1 or method 2. Consequently the final model contains previously identified main effects and significant interaction terms.

As we noted, some packages use various combinations of the computationally more efficient Score and Wald tests. For example, SPSS uses the score test for selection and the likelihood ratio test for removal of covariates. SAS also uses the score test to select covariates but uses individual Wald statistics to check for removal of covariates. STATA has the option to use the Wald test for both entry and removal of covariates. The results shown in Table 4.12 are from SPSS. At each step, the p-values from the score test to enter are below the horizontal line. The p-values for the likelihood ratio test to remove are above the horizontal line. Since we use these p-values in exactly the same manner as discussed in detail when selection for entry and removal is based on the likelihood ratio test we do not repeat it. Instead we focus on comparing similarities and differences between the results in Table 4.10 and Table 4.12.

We note that the order of variable entry into the models in Tables 4.10 and 4.12 is different. SPSS (Table 4.12) selects IVHX first, followed by AGE, NDRGTX and TREAT. BMDP (Table 4.10) selected NDRGTX first, followed by TREAT, IVHX and AGE. After four steps, the variables selected by both methods were the same and no other variables were significant at the 0.15 level and so were unable to enter the model at step 5. Thus the composition of the final model selected is the same using the two approaches. The p-values for entry for the likelihood ratio test in Table 4.10 are quite similar to those for the Score test in Table 4.12. This is in agreement with the discussion of these two tests

Table 4.12 Results of Applying Stepwise Variable Selection Using the Score Test to Select and Maximum Likelihood Test to Remove Covariates at Each Step to the UIS Data. Results Are Presented at Each Step in Terms of the p-Values to Enter (Below the Horizontal Line), and the p-Value to Remove (Above the Horizontal Line) in Each Column. The Asterisk Denotes the Maximum p-Value to Remove at Each Step

Variable/Step	0	1	2	3	4
IVHX+	0.0012	0.0012	0.0001	0.0024	0.0027
AGE	0.2357	0.0068	0.0070*	0.0029	0.0021
NDRGTX	0.0018	0.0264	0.0127	0.0069*	0.0062
TREAT	0.0231	0.0329	0.0253	0.0226	0.0224*
RACE	0.0288	0.1462	0.2330	0.2725	0.3470
SITE	0.1933	0.4549	0.4762	0.6231	0.5676
BECK	0.4262	0.7738	0.9630	0.9739	0.9948

+ IVHX is entered with 2 degrees-of-freedom corresponding to its 2 design variables.

in Chapters 1 and 2. Use of SAS and STATA yields results similar to those shown in Table 4.12 (results are not presented).

Given a choice, we prefer to use the likelihood ratio test for both entry and removal as research has shown it has the best statistical properties. However the results in Table 4.10 and Table 4.12 are typical of our experience in using stepwise methods. Namely, with different tests there may be some swapping in the order selected but the total set is usually the same.

The variables identified using stepwise selection are the same as those identified earlier by purposeful selection. Therefore, the work necessary to check the scale of the continuous covariates, NDRGTX and AGE, is not repeated and we begin stepwise selection of the interactions listed in Table 4.8.

In order to simplify the presentation somewhat we use the fact that all sets of interactions involving the two fractional polynomial transformations of NDRGTX are quite highly correlated. Thus, as was the case in purposeful selection, we only consider interactions with NDRGFP1. We note that interactions involving history of IV drug use, IVHX, are computed using the design variables, IVHX_2 and IVHX_3.

The same software may be used for stepwise selection of interactions as was used for the selection of main effects. The difference is that all main effect variables are forced into the model at Step (0) and selection is restricted to interactions. In total there are 15 possible interac-

Table 4.13 Results of Applying Stepwise Variable Selection to Interactions from the Main Effects Model from the UIS Data, Using the Maximum Likelihood Method Presented at Each Step in Terms of the p-Values to Enter (Below the Horizontal Line), and the p-Values to Remove (Above the Horizontal Line) in Each Column. The Asterisk Denotes the Maximum p-Value to Remove at Each Step

Variable/Step	0	1	2	3
RACExSITE	0.0035	0.0035	0.0053	0.0064
AGExNDRGFP1	0.0055	0.0084	0.0084*	0.0080
AGExTREAT	0.0693	0.1174	0.1110	0.1110*
AGExIVHX	0.6910	0.6075	0.2341	0.2577
NDRGFP1xTREAT	0.0900	0.1816	0.1376	0.2837
RACExTREAT	0.3315	0.3449	0.2885	0.2871
IVHXxRACE	0.9798	0.5902	0.4139	0.3844
IVHXxSITE	0.6475	0.4097	0.4487	0.4610
NDRGFP1xIVHX	0.3291	0.3726	0.6063	0.5682
AGExRACE	0.6568	0.5441	0.4489	0.5734
IVHXxTREAT	0.9798	0.9223	0.9779	0.6271
TREATxSITE	0.8542	0.9591	0.8706	0.7962
NDRGFP1xSITE	0.9587	0.6876	0.7664	0.8260
AGExSITE	0.2062	0.2908	0.6643	0.8583
NDRGFP1xRACE	0.3679	0.8855	0.9521	0.8815

tions listed in Table 4.13 where they are inverse rank ordered by the p-values at the last step.

The results in Table 4.13 indicate that only 3 interactions entered the model using the 15 percent level of significance and none were removed at the 20 percent level of significance. These are the same 3 interactions identified previously via purposeful selection. At this point the analysis proceeds in exactly the same manner as discussed in detail earlier in this chapter. In that analysis we concluded that only the RACE by SITE and AGE by NDRGFP1 interactions should be in the model. Hence the preliminary final model obtained via stepwise selection turns out to be the same as the one obtained via purposeful selection. This may not always be the case. In our experience models obtained by these two approaches rarely differ by more than a couple of variables. In a situation where different approaches yield different models we recommend proceeding with a combined larger model via purposeful selection using both confounding and statistical significance as criteria for model simplification.

We note that the same three interactions were identified as significant when both the Score and Wald test options were used. Thus we do not present this output or discuss the results.

In conclusion, we emphasize that stepwise selection identifies variables as candidates for a model solely on statistical grounds. Thus, following stepwise selection of main effects all variables should be carefully scrutinized for clinical plausibility. In general, interactions must attain statistical significance to alter the point and interval estimates from a main effects model. Thus, stepwise selection of interactions using statistical significance can provide a valuable contribution to model identification, especially when there are large numbers of clinically plausible interactions generated from the main effects.

4.4 BEST SUBSETS LOGISTIC REGRESSION

An alternative to stepwise selection of variables for a model is best subset selection. This approach to model building has been available for linear regression for a number of years and makes use of the branch and bound algorithm of Furnival and Wilson (1974). Typical software implementing this method for linear regression identifies a specified number of "best" models containing one, two, three variables, and so on, up to the single model containing all p variables. Lawless and Singhal (1978, 1987a, 1987b) proposed an extension that may be used with any non-normal model. The crux of their method involves application of the Furnival-Wilson algorithm to a linear approximation of the cross-product sum-of-squares matrix that yields approximations to the maximum likelihood estimates. Selected models are then compared to the model containing all variables using a likelihood ratio test. Hosmer, Jovanovic, and Lemeshow (1989) have shown that, for logistic regression, the full generality of the Lawless and Singhal approach is not needed. Best subsets logistic regression may be performed in a straightforward manner using any program capable of best subsets linear regression. Also, some packages, including SAS, have implemented the Lawless and Singhal method in their logistic regression modules.

Applying best subsets linear regression software to perform best subsets logistic regression is most easily explained using vector and matrix notation. In this regard, we let $\mathbf{X}$ denote the $n \times (p+1)$ matrix containing the values of all p independent variables for each subject, with the first column containing 1 to represent the constant term. Here the p variables may represent the total number of variables, or those selected at the univariable stage of model building. We let $\mathbf{V}$ denote an $n \times n$ diagonal matrix with general element $v_i = \hat{\pi}_i(1-\hat{\pi}_i)$ where $\hat{\pi}_i$ is the estimated logistic probability computed using the maximum likelihood estimate $\hat{\boldsymbol{\beta}}$ and the data for the ith case, $\mathbf{x}_i$.

For the sake of clarity of presentation in this section, we repeat the expression for $\mathbf{X}$ and $\mathbf{V}$ given in Chapter 2. They are as follows:

$$\mathbf{X} = \begin{bmatrix} 1 & x_{11} & x_{12} & \cdots & x_{1p} \\ 1 & x_{21} & x_{22} & \cdots & x_{2p} \\ \vdots & \vdots & \vdots & \cdots & \vdots \\ 1 & x_{n1} & x_{n2} & \cdots & x_{np} \end{bmatrix}$$

and

$$\mathbf{V} = \begin{bmatrix} \hat{\pi}_1(1-\hat{\pi}_1) & 0 & \cdots & 0 \\ 0 & \hat{\pi}_2(1-\hat{\pi}_2) & \cdots & 0 \\ \vdots & 0 & \ddots & \vdots \\ 0 & \cdots & 0 & \hat{\pi}_n(1-\hat{\pi}_n) \end{bmatrix}.$$

As noted in Chapter 2, the maximum likelihood estimate is determined iteratively. It may be shown [see Pregibon (1981)] that $\hat{\boldsymbol{\beta}} = (\mathbf{X}'\mathbf{V}\mathbf{X})^{-1}\mathbf{X}'\mathbf{V}\mathbf{z}$, where $\mathbf{z} = \mathbf{X}\hat{\boldsymbol{\beta}} + \mathbf{V}^{-1}\mathbf{r}$ and $\mathbf{r}$ is the vector of residuals, $\mathbf{r} = (\mathbf{y} - \hat{\pi})$. This representation of $\hat{\boldsymbol{\beta}}$ provides the basis for use of linear regression software. It is easy to verify that any linear regression package, that allows weights, produces coefficient estimates identical to $\hat{\boldsymbol{\beta}}$ when used with z_i as the dependent variable and case weights, v_i, equal to the diagonal elements of $\mathbf{V}$.

To replicate the results of the maximum likelihood fit from a logistic regression package using a linear regression package, we calculate for each case, the value of a dependent variable as follows:

$$z_i = (1, x_i')\hat{\boldsymbol{\beta}} + \frac{(y_i - \hat{\pi}_i)}{\hat{\pi}_i(1 - \hat{\pi}_i)}$$

$$= \hat{\beta}_0 + \sum_{j=1}^{p} \hat{\beta}_j x_{ij} + \frac{(y_i - \hat{\pi}_i)}{\hat{\pi}_i(1 - \hat{\pi}_i)}$$

$$= \ln\left(\frac{\hat{\pi}_i}{1 - \hat{\pi}_i}\right) + \frac{(y_i - \hat{\pi}_i)}{\hat{\pi}_i(1 - \hat{\pi}_i)} \tag{4.1}$$

and a case weight

$$v_i = \hat{\pi}_i(1 - \hat{\pi}_i). \tag{4.2}$$

Note that all we need is access to the fitted values, $\hat{\pi}_i$, to compute the values of z_i and v_i. Next, we run a linear regression program using the values of z_i as the dependent variable, the values of $\mathbf{x}_i$ for our vector of independent variables, and the values of v_i for our case weights.

Proceeding further with the linear regression, it can be shown that the residuals from the fit are

$$(z_i - \hat{z}_i) = \frac{(y_i - \hat{\pi}_i)}{\hat{\pi}_i(1 - \hat{\pi}_i)}$$

and the weighted residual sum-of-squares produced by the program is

$$\sum_{i=1}^{n} v_i (z_i - \hat{z}_i)^2 = \sum_{i=1}^{n} \frac{(y_i - \hat{\pi}_i)^2}{\hat{\pi}_i(1 - \hat{\pi}_i)},$$

which is X^2, the Pearson chi-square statistic from a maximum likelihood logistic regression program. It follows that the mean residual sum-of-squares is $s^2 = X^2/(n - p - 1)$. The estimates of the standard error of the estimated coefficients produced by the linear regression program are s times the square root of the diagonal elements of the matrix $(\mathbf{X}'\mathbf{V}\mathbf{X})^{-1}$. Thus, to obtain the correct values given in equation (2.5) we need to divide the estimates of the standard error produced by the linear regression program by s, the square root of the mean square error (or standard error of the estimate).

The ability to duplicate the maximum likelihood fit in a linear regression package forms the foundation of the suggested method for performing best subsets logistic regression. In particular, Hosmer, Jovanovic, and Lemeshow (1989) show that use of any best subsets linear regression program with values of z_i in equation (4.1) for the dependent variable, case weights v_i shown in equation (4.2), and covariates $\mathbf{x}_i$, produces for any subset of q variables the approximate coefficient estimates of Lawless and Singhal (1978). Hence, we may use any best subsets linear regression program to execute the computations for best subsets logistic regression.

The subsets of variables selected for "best" models depend on the criterion chosen for "best." In best subsets linear regression three criteria have primarily been used to select variables. Two of these are based on the concept of the proportion of the total variation explained by the model. These are R^2, the ratio of the regression sum-of-squares to the total sum-of-squares, and adjusted R^2 (or AR^2), the ratio of the regression mean squares to the total mean squares. Since the adjusted R^2 is based on mean squares rather than sums-of-squares, it provides a correction for the number of variables in the model. This is important, as we must be able to compare models containing different variables and different numbers of variables. If we use R^2, the best model is always the model containing all p variables, a result that is not very helpful. An obvious extension for best subsets logistic regression is to base the R^2 measures, in a manner similar to that shown in Chapter 5, on deviance rather than Pearson chi-square. However, we do not recommend the use of the R^2 measures for best subsets logistic regression. Instead, we prefer to use the third measure used in best subsets linear regression that was developed by Mallow (1973). This is a measure of predictive squared error, denoted C_q. This measure is denoted as C_p by other authors. We use "q" instead of "p" in this text since the letter p refers to a total number of possible variables while q refers to some subset of variables.

A summary of the development of the criterion C_q in linear regression may be found in many texts on this subject, for example Ryan (1997). Hosmer, Jovanovic, and Lemeshow (1989) show that when best subsets logistic regression is performed via a best subsets linear regression package in the manner described previously in this section, Mallow's C_q has the same intuitive appeal as it does in linear regression. In particular they show that for a subset of q of the p variables

$$C_q = \frac{X^2 + \lambda^*}{X^2/(n-p-1)} + 2(q+1) - n,$$

where $X^2 = \sum\left\{(y_i - \hat{\pi}_i)^2 / \left[\hat{\pi}_i(1-\hat{\pi}_i)\right]\right\}$, the Pearson chi-square statistic for the model with p variables and λ^* is the multivariable Wald test statistic for the hypothesis that the coefficients for the $p-q$ variables not in the model are equal to zero. Under the assumption that the model fit is the correct one, the approximate expected values of X^2 and λ^* are $(n-p-1)$ and $p-q$ respectively. Substitution of these approximate expected values into the expression for C_q yields $C_q = q+1$. Hence, models with C_q near $q+1$ are candidates for a best model. The best subsets linear regression program selects as best that subset with the smallest value of C_q.

Use of the best subsets linear regression package should help select, in the same way its application in linear regression does, a core of q important covariates from the p possible covariates. At this point, we suggest that further modeling proceed in the manner described for purposeful selection of variables using a logistic regression package. Users should not be lured into accepting the variables suggested by a best subset strategy without considerable critical evaluation.

We illustrate best subsets selection with the UIS data. The variables used were those indicated in Table 1.8. History of previous IV drug use, IVHX, was coded into the same two design variables, IVHX_2 and IVHX_3, used in previous sections. A logistic regression package was used to obtain the estimated logistic probabilities for the model containing all $p=7$ variables. Following the fit of the full model the values of z and v were created using equations (4.1) and (4.2). A best subsets linear regression package was used with z as the dependent variable and v as the case weights. The possible independent variables were the 6 continuous variables plus two design variables for IVHX, for a total of 8. The best subsets linear regression program used did not have a provision for the creation of design variables from categorical scaled covariates so the design variables for IVHX were created prior to the best subsets analysis.

In Table 4.14 we present the results of the five best models selected using C_q as the criterion. In addition to the variables selected, we show the values of C_q, the values of λ^*, and the values of the likelihood ratio test, G, for the variables excluded from the model, the corresponding degrees of freedom and p-values. We note that the test statistics G and λ^* have similar values, as expected, since they test the same hypothesis and have the same asymptotic distribution.

Table 4.14 Five Best Models Identified Using Mallow's C_q. Model Covariates, Mallow's C_q, the Wald Test (λ^*), and the Likelihood Ratio Test for the Excluded Covariates, Degrees-of-Freedom and p-Value

Model	Model Covariates	C_q	λ^*	G	df	p
1	AGE, NDRGTX, IVHX_2, IVHX_3, TREAT	4.31	1.35	1.34	3	0.72
2	AGE, NDRGTX, IVXH_2, IVHX_3, TREAT, RACE	5.46	0.47	0.47	2	0.79
3	AGE, NDRGTX, IVHX_2,IVHX_3, TREAT, SITE	6.00	1.03	1.01	2	0.60
4	AGE, NDRGTX, IVHX_2, IVHX_3, TREAT, BECK	6.31	1.35	1.34	2	0.51
5	AGE, NDRGTX, IVHX_3, TREAT	6.98	6.13	6.34	4	0.18

Using only the summary statistics, we would select model 1 as the best model since it has the smallest value of C_q and the Wald and likelihood ratio tests for the excluded variables are not significant. Note that four of the five models identified as being "best" include AGE, NDRGTX, the two design variables for IVHX and TREAT. Thus these variables are important and should be in any model. The best model contains neither RACE, which is an important clinical variable, nor SITE, the study design variable. Hence, we recommend adding RACE and SITE to model 1 and proceed to the next stage of model development.

At this point the model is identical to that already presented in Section 4.2 where we focused on scale identification of continuous variables. Once we have finalized the main effects model, we could employ best subsets selection to decide on possible interactions.

Some programs, for example SAS's PROC LOGISTIC, provide a best subsets selection of covariates based on the Score test for the variables in the model. For example, the best two variable model is the one with the largest Score test among all two variable models. The output lists the covariates and Score test for a user specified number of best models of each size. The difficulty one faces when presented with this output is that the Score test increases with the number of variables in the model. Hosmer and Lemeshow (1999) show how an approximation to Mallow's C_q can be obtained from Score test output in a survival time analysis. A similar approximation can be obtained from C_q for logistic regression. First, we assume that the Pearson chi-square statistic is equal to its mean, e.g. $X^2 \approx (n-p-1)$. Next we assume that the Wald statistic for the $p-q$ excluded covariates may be approximated by the difference between the values of the Score test for all p covariates and the Score test for q covariates, namely $\lambda_q^* \approx S_p - S_q$. This results in the following approximation

$$
\begin{aligned}
C_q &= \frac{X^2 + \lambda^*}{X^2/(n-p-1)} + 2(q+1) - n \\
&\approx \frac{(n-p-1) + (S_p - S_q)}{1} + 2(q+1) - n \\
&\approx S_p - S_q + 2q - p + 1.
\end{aligned}
$$

The value of S_p is the Score test for the model containing all p covariates and is obtained from the computer output. The value of S_q is the Score test for the particular subset of q covariates and its value is also obtained from the output. The five best models identified using SAS's Score test procedure in PROC LOGISTIC are shown in Table 4.15.

The best four models in Table 4.15 are the same models obtained using the best subsets linear regression method shown in Table 4.14. We note that the approximate values of C_q in Table 4.15 are quite close to the values in Table 4.14. The fifth model is the one we eventually selected when we considered clinical and study design criteria in addition to best subsets. Thus in this example, the approximation to C_q has yielded a useful rank ordering of the models.

The advantage of the proposed method of best subsets logistic regression is that many more models can be quickly screened than was possible with the other approaches to variable identification. There is, however, one potential disadvantage with the best subsets approach: we must be able to fit the model containing all the possible covariates. In analyses that include a large number of variables this may not be possi-

Table 4.15 Five Best Models Identified Using the Score Test Approximation to Mallow's C_q, ($S_8 = 32.6798$)

Model	Model Covariates	S_q	C_q
1	AGE, NDRGTX, IVHX_2, IVHX_3, TREAT	31.1565	4.52
2	AGE, NDRGTX, IVXH_2, IVHX_3, TREAT, RACE	32.0446	5.63
3	AGE, NDRGTX, IVHX_2, IVHX_3, TREAT, SITE	31.6135	6.07
4	AGE, NDRGTX, IVHX_2, IVHX_3, TREAT, BECK	31.1569	6.52
5	AGE, NDRGTX, IVHX_2, IVHX_3, TREAT, RACE, SITE	32.6795	7.00

ble. Numerical problems can occur when we overfit a logistic regression model. If the model has many variables, we run the risk that the data are too thin to be able to estimate all the parameters. If the full model proves to be too rich, then some selective weeding out of obviously unimportant variables with univariable tests may remedy this problem. Another approach is to perform the best subsets analysis using several smaller "full" models. Numerical problems are discussed in more detail in the next section.

In summary, the ability to use weighted least squares best subsets linear regression software to identify variables for logistic regression should be kept in mind as a possible aid to variable selection. As is the case with any statistical selection method, the clinical basis of all variables should be addressed before any model is accepted as the final model.

4.5 NUMERICAL PROBLEMS

In previous chapters we have occasionally mentioned various numerical problems that can occur when fitting a logistic regression model. These problems are caused by certain structures in the data and the lack of appropriate checks in logistic regression software. The goal of this section is to illustrate these structures in certain simple situations and illustrate what can happen when the logistic regression model is fit to such data. The issue here is not one of model correctness or specification, but the effect certain data patterns have on the computation of parameter estimates.

Perhaps the simplest and thus most obvious situation is when we have a frequency of zero in a contingency table. An example of such a contingency table is given in Table 4.16. The estimated odds ratios and log odds ratios using the first level of the covariate as the reference group are given in the first two rows below the table. The point estimate of the odds ratios for level 3 versus level 1 is infinite since all subjects at level 3 responded. The results of fitting a logistic regression model to these data are given in the last two rows. The estimated coefficient in the first column is the intercept coefficient. The particular package used does not really matter as many, but not all, packages produce similar output. One program that does identify the problem is STATA. It provides an error message that $x=3$ perfectly predicts the outcome and the design variable for $x=3$ is not included in the fit of the model. Other programs may or may not provide some sort of error message indicating that convergence was not obtained or that the maximum number of iterations was used. What is rather obvious, and the tip-off that there is a problem with the model, is the large estimated coefficient for the second design variable and especially its large estimated standard error.

A common practice to avoid having an undefined point estimate is to add one-half to each of the cell counts. Adding one-half may allow us to move forward with the analysis of a single contingency table but such a simplistic remedy is rarely satisfactory with a more complex data set.

As a slightly more complex example we consider the stratified 2 by 2 tables shown in Table 4.17. The stratum-specific point estimates of the odds ratios are provided below each 2 by 2 table. The results of fitting a series of logistic regression models are provided in Table 4.18.

In the case of the data shown in Table 4.17 we do not encounter problems until we include the stratum, z, by risk factor, x, interaction terms, $x \times z\_2$ and $x \times z\_3$ in the model. The addition of the interaction terms results in a model that is equivalent to fitting a model with a single categorical variable with six levels, one for each column in Table 4.17. Thus, in a sense, the problem encountered when we include the interaction is the same one illustrated in Table 4.16. As was the case when fitting a model to the data in Table 4.16, the presence of a zero cell count is manifested by an unbelievably large estimated coefficient and estimated standard error.

The presence of a zero cell count should be detected during the univariable screening of the data. Knowing that the zero cell count is going to cause problems in the modeling stage of the analysis we could collapse the categories of the variable in a meaningful way to eliminate it, eliminate the category completely or, if the variable is at least ordinal scale, treat it as continuous.

The type of zero cell count illustrated in Table 4.17 results from spreading the data over too many cells. This problem is not likely to occur until we begin to include interactions in the model. When it does occur, we should examine the three way contingency table equivalent to the one shown in Table 4.17. The unstable results prevent us from determining whether, in fact, the interaction is important. To assess the interaction we first need to eliminate the zero cell count. One way to do this is by collapsing categories of the stratification variable. For example, in Table 4.17 we might decide that values of $z = 2$ and $z = 3$ are similar enough to pool them. The stratified analysis would then have two 2 by 2 tables the second of which results from pooling the tables

Table 4.16 A Contingency Table with a Zero Cell Count and the Results of Fitting a Logistic Regression Model to This Data

Outcome / x	1	2	3	Total
1	7	12	20	39
0	13	8	0	21
Total	20	20	20	60
$\widehat{OR}$	1	2.79	inf	
$\ln(\widehat{OR})$	0	1.03	inf	
$\hat{\beta}$	−0.62	1.03	11.7	
$\widehat{SE}$	0.47	0.65	34.9	

for $z=2$ and $z=3$. A second approach is to define a new variable equal to the combination of the stratification variable and the risk factor and to pool over levels of this variable and model it as a main effect variable. Using Table 4.17 as an example, we would have a variable with six levels corresponding to the six columns in the table. We could collapse levels five and six together. Another pooling strategy would be to pool levels three and five and four and six. This pooling strategy is equivalent to collapsing over levels of the stratification variable. The net effect is the loss of degrees-of-freedom commensurate with the amount of pooling. Twice the difference in the log-likelihood for the main effects only model and the model with the modified interaction term added provides a statistic for the significance of the coefficients for the modified interaction term.

The fitted models shown in Tables 4.16 and 4.18 resulted in large estimated coefficients and estimated standard errors. In some examples we have encountered, the magnitude of the estimated coefficient was not large enough to suspect a numerical problem; but the esti-

Table 4.17 Stratified 2 by 2 Contingency Tables with a Zero Cell Count Within One Stratum

Stratum (z)	1		2		3	
Outcome / x	1	0	1	0	1	0
1	5	2	10	2	15	1
0	5	8	2	6	0	4
Total	10	10	12	8	15	5
$\widehat{OR}$	4		15		inf	

Table 4.18 Results of Fitting Logistic Regression Models to the Data in Table 4.17

Model	1		2	
Variable	Coeff.	Std. Err.	Coeff.	Std. Err.
x	2.77	0.72	1.39	1.01
z_2	1.19	0.81	0.29	1.14
z_3	2.04	0.89	0.00	1.37
$x \times z\_2$			1.32	1.51
$x \times z\_3$			11.54	50.22
Constant	−2.32	0.77	−1.39	0.79

mated standard error always was. Hence, we believe that the best indicator of a numerical problem in logistic regression is the estimated standard error. In general, any time the estimated standard error of an estimated coefficient is large relative to the point estimate we should suspect the presence of one of the data structures described in this section.

A second type of numerical problem occurs when a collection of the covariates completely separates the outcome groups or, in the terminology of discriminant analysis, the covariates discriminate perfectly. For example, suppose that the age of every subject with the outcome present was greater than 50 and the age of all subjects with the outcome absent was less than 49. Thus, if we know the age of a subject we know with certainty the value of the outcome variable. In this situation there is no overlap in the distribution of the covariates between the two outcome groups. This type of data has been shown by Bryson and Johnson (1981) to have the property of monotone likelihood. The net result is that the maximum likelihood estimates do not exist (see Albert and Anderson (1984) and Santner and Duffy (1986)). In order to have finite maximum likelihood estimates we must have some overlap in the distribution of the covariates in the model.

A simple example illustrates the problem of complete separation and the results of fitting logistic regression models to such data. Suppose we have the following 12 pairs of covariate and outcome, (x,y): (1,0), (2,0), (3,0), (4,0), (5,0), (x_6 = 5.5, or 6.0, or 6.05, or 6.1, or 6.2, or 8.0, y_6 = 0), (6,1), (7,1), (8,1), (9,1), (10,1), (11,1). The results of fitting logistic regression models when x_6 takes on one of the values 5.5, 6.0, 6.05, 6.1, 6.2, or 8, using SAS version 6.12 are given in Table 4.19. When we use x_6 = 5.5 we have complete separation and all estimated pa-

Table 4.19 Estimated Slope ($\hat{\beta}_x$), Constant, and Estimated Standard Errors When the Data Have Complete Separation, Quasicomplete Separation, and Overlap

Estimates / x_6	5.5	6.0	6.05	6.10	6.15	6.20	8.0
$\hat{\beta}_x$	15.1	7.8	4.3	3.6	3.2	2.9	1.0
$\widehat{SE}$	19.0	35.4	6.1	4.2	3.3	2.8	0.5
$\hat{\beta}_0$	−86.7	−47.0	−26.2	−22.0	−19.5	−17.8	-6.1
$\widehat{SE}$	109.4	212.0	36.7	25.4	20.3	17.3	3.6

rameters are huge, since the maximum likelihood estimates do not exist. SAS provides a warning but at the same time provides the values of the estimates at the last iteration, leaving the ultimate decision about how to handle the output to the user. Similar behavior occurs when the value of x_6 = 6.0 is used. SAS notes this fact and again provides estimates. When overlap is at a single or a few tied values the configuration was termed by Albert and Anderson as quasicomplete separation. As the value of x_6 takes on values greater than 6 the overlap becomes greater and the estimated parameters and standard errors begin to attain more reasonable values. The sensitivity of the fit to the overlap depends on the sample size and the range of the covariate. The tip-off that something is amiss is, as in the case of the zero cell count, the very large estimated coefficients and especially the large estimated standard errors. Other programs, including STATA, do not provide output when there is complete or quasi-complete separation, e.g. $x_6 = 5.5$ or $x_6 = 6$. In the remaining cases STATA and SAS produce similar results.

The occurrence of complete separation in practice depends on the sample size, the number of subjects with the outcome present, and the number of variables included in the model. For example, suppose we have a sample of 25 subjects and only five have the outcome present. The chance that the main effects model demonstrates complete separation increases with the number of variables we include in the model. Thus, the modeling strategy that includes all variables in the model is particularly sensitive to complete separation. Albert and Anderson and Santner and Duffy provide rather complicated diagnostic procedures for determining whether a set of data displays complete or quasicomplete separation. Albert and Anderson recommend that in the absence of their diagnostic, one look at the estimated standard errors and if these tend to increase substantially with each iteration of the fit, that one sus-

Table 4.20 Data Displaying Near Collinearity Among the Independent Variables and Constant

Subject	x_1	x_2	x_3	y
1	0.225	0.231	1.026	0
2	0.487	0.489	1.022	1
3	−1.080	−1.070	1.074	0
4	−0.870	−0.870	1.091	0
5	−0.580	−0.570	1.095	0
6	−0.640	−0.640	1.010	0
7	1.614	1.619	1.087	0
8	0.352	0.355	1.095	1
9	−1.025	−1.018	1.008	0
10	0.929	0.937	1.057	1

pect the presence of complete separation. As noted in Chapter 3 the easiest way to address complete separation is to use some careful univariable analyses. The occurrence of complete separation is not likely to be of great clinical importance as it is usually a numerical coincidence rather than describing some important clinical phenomenon. It is a problem we must work around.

As is the case in linear regression, model fitting via logistic regression is also sensitive to collinearities among the independent variables in the model. Most software packages have some sort of diagnostic check, like the tolerance test employed in linear regression. Nevertheless it is possible for variables to pass these tests and have the program run, but yields output that is clearly nonsense. As a simple example, we fit logistic regression models to the data displayed in Table 4.20. In the table $x_1 \sim N(0,1)$ and the outcome variable was generated by comparing a $U(0,1)$ variate, u, to the true probability $\pi(x_1) = e^{x_1}/(1+e^{x_1})$ as follows: if $u < \pi(x_1)$ then $y = 1$, otherwise $y = 0$. The notation $N(0,1)$ indicates a random variable following the standard normal (mean = zero, variance = 1) distribution and $U(a,b)$ indicates a random variable following the uniform distribution on the interval $[a,b]$. The other variables were generated from x_1 and the constant as follows: $x_2 = x_1 + U(0,0.1)$ and $x_3 = 1 + U(0,0.01)$. Thus, x_1 and x_2 are highly correlated and x_3 is nearly collinear with the constant term. The results of fitting logistic regression models using SAS

Table 4.21 Estimated Coefficients and Standard Errors from Fitting Logistic Regression Models to the Data in Table 4.20

Var.	Coeff	Std. Err.	Coeff.	Std. Err.	Coeff.	Std. Err.	Coeff.	Std. Err.
x_1	1.4	1.0	146.4	277.0			143.0	282.2
x_2			−276.6	276.6			−141.5	281.8
x_3					2.74	21.1	−3.62	25.0
Cons.	−1.0	0.8	0.37	1.4	−1.79	20.0	3.42	26.2

version 6.12 to various subsets of the variables shown in Table 4.20 are presented in Table 4.21.

The model that includes the highly correlated variables x_1 and x_2 has both very large estimated slope coefficients and estimated standard errors. For the model containing x_3 we see that the estimated coefficients are of reasonable magnitude but the estimated standard errors are much larger than we would expect. The model containing all variables is a composite of the results of the other models. In all cases the tip-off for a problem comes from the aberrantly large estimated standard errors.

In a more complicated data set, an analysis of the associations among the covariates using a collinearity analysis similar to that performed in linear regression should be helpful in identifying the dependencies among the covariates. Belsley, Kuh, and Welsch (1980) discuss a number of methods that are implemented in many linear regression packages. One would normally not employ such an in-depth investigation of the covariates unless there was evidence of degradation in the fit similar to that shown in Table 4.21. An alternative is to use the ridge regression methods proposed by Schaefer (1986).

In general, the numerical problems of a zero cell count, complete separation, and collinearity, are manifested by extraordinarily large estimated standard errors and sometimes by a large estimated coefficient as well. New users and ones without much computer experience are especially cautioned to look at their results carefully for evidence of numerical problems. Consultation with someone more experienced may be required to ferret out and solve these numeric problems.

EXERCISES

1. Selection of the scale for continuous covariates is an important step in any modeling process. The variable systolic blood pressure at admission, SYS, in the ICU study described in Section 1.6 presents a particularly challenging example. Consider the variable vital status (STA) as the outcome variable and SYS as the covariate for a univariable logistic regression model. What is the correct scale for SYS to enter the model? As a second example, consider a univariable model with heart rate at ICU admission (HRA) as the covariate. Repeat this exercise of scale identification for SYS and HRA using a multivariable model containing these two variables plus three or four other covariates of your choice.

2. Consider the variable level of consciousness at ICU admission (LOC) as a covariate and vital status (STA) as the outcome variable. Compare the estimates of the odds ratios obtained from the cross-classification of STA by LOC and the logistic regression of STA on LOC. Use LOC = 0 as the reference group for both methods. How well did the logistic regression program deal with the zero cell? What strategy would you adopt to modeling LOC in future analyses?

3. Consider the variable vital status (STA) as the outcome variable and the remainder of the variables in the ICU data set as potential covariates. Use each of the variable selection methods discussed in this chapter to find a "best" model. Document thoroughly the rationale for each step in each process you follow. Compare and contrast the models resulting from the different approaches to variable selection. Note that in all cases the analysis should address not only identification of main effects but also appropriate scale for continuous covariates and potential interactions. Display the results of your final model in a table. Include in the table point and 95% CI estimates of all relevant odds ratios. Document the rationale for choosing the final model.

4. Repeat Exercises 1 and 3 for the Low Birthweight data and the Prostatic Cancer data.

CHAPTER 5

Assessing the Fit of the Model

5.1 INTRODUCTION

We begin our discussion of methods for assessing the fit of an estimated logistic regression model with the assumption that we are at least preliminarily satisfied with our efforts at the model building stage. By this we mean that, to the best of our knowledge, the model contains those variables (main effects as well as interactions) that should be in the model and that variables have been entered in the correct functional form. Now we would like to know how effectively the model we have describes the outcome variable. This is referred to as its *goodness-of-fit*.

If we intend to assess the goodness-of-fit of the model, then we should have some specific ideas about what it means to say that a model fits. Suppose we denote the observed sample values of the outcome variable in vector form as $\mathbf{y}$ where $\mathbf{y}'=(y_1,y_2,y_3,...,y_n)$. We denote the values predicted by the model, or *fitted values*, as $\hat{\mathbf{y}}$ where $\hat{\mathbf{y}}'=(\hat{y}_1,\hat{y}_2,\hat{y}_3,...,\hat{y}_n)$. We conclude that the model fits if (1) summary measures of the distance between $\mathbf{y}$ and $\hat{\mathbf{y}}$ are small and (2) the contribution of each pair $(y_i,\hat{y}_i)$, $i=1,2,3,...,n$ to these summary measures is unsystematic and is small relative to the error structure of the model. Thus, a complete assessment of the fitted model involves both the calculation of summary measures of the distance between $\mathbf{y}$ and $\hat{\mathbf{y}}$, and a thorough examination of the individual components of these measures.

When the model building stage has been completed, a series of logical steps may be used to assess the fit of the model. The components of the proposed approach are (1) computation and evaluation of overall measures of fit, (2) examination of the individual components of the summary statistics, often graphically, and (3) examination of other measures of the difference or distance between the components of $\mathbf{y}$ and $\hat{\mathbf{y}}$.

5.2 SUMMARY MEASURES OF GOODNESS-OF-FIT

We begin with the summary measures of goodness-of-fit, as they are routinely provided as output with any fitted model and give an overall indication of the fit of the model. Summary statistics, by nature, may not provide information about the individual model components. A small value for one of these statistics does not rule out the possibility of some substantial and thus interesting deviation from fit for a few subjects. On the other hand, a large value for one of these statistics is a clear indication of a substantial problem with the model.

Before discussing specific goodness-of-fit statistics, we must first consider the effect the fitted model has on the degrees of freedom available for the assessment of model performance. We use the term *covariate pattern* to describe a single set of values for the covariates in a model. For example, in a data set containing values of age, race, sex and weight for each subject, the combination of these factors may result in as many different covariate patterns as there are subjects. On the other hand, if the model contains only race and sex, each coded at two levels, there are only four possible covariate patterns. We note that during model development it is not necessary to be concerned about the number of covariate patterns. The degrees-of-freedom for tests are based on the difference in the number of parameters in competing models, not on the number of covariate patterns. However, the number of covariate patterns may be an issue when the fit of a model is assessed.

Goodness-of-fit is assessed over the constellation of fitted values determined by the covariates in the model, not the total collection of covariates. For instance, suppose that our fitted model contains p independent variables, $\mathbf{x}' = (x_1, x_2, x_3, ..., x_p)$, and let J denote the number of distinct values of $\mathbf{x}$ observed. If some subjects have the same value of $\mathbf{x}$ then $J < n$. We denote the number of subjects with $\mathbf{x} = \mathbf{x}_j$ by m_j, $j = 1, 2, 3, ..., J$. It follows that $\sum m_j = n$. Let y_j denote the number of positive responses, $y = 1$, among the m_j subjects with $\mathbf{x} = \mathbf{x}_j$. It follows that $\sum y_j = n_1$, the total number of subjects with $y = 1$. The distribution of the goodness-of-fit statistics is obtained by letting n become large. If the number of covariate patterns also increases with n then each value of m_j tends to be small. Distributional results obtained under the condition that only n becomes large are said to be based on *n-asymptotics*. If we fix $J < n$ and let n become large then each value of m_j also tends to become large. Distributional results based on each m_j becoming large are said to be based on *m-asymptotics*. The difference between these asymptotics and the need to distinguish between them should become clearer as we discuss summary statistics in greater detail.

Initially, we assume that $J \approx n$, as we expect whenever there is at least one continuous covariate in the model. This is the case most frequently encountered in practice. It also presents the greatest challenge in developing distributions of goodness-of-fit statistics.

5.2.1 Pearson Chi-Square Statistic and Deviance

In linear regression, summary measures of fit, as well as diagnostics for casewise effect on the fit, are functions of a residual defined as the difference between the observed and fitted value $(y-\hat{y})$. In logistic regression there are several possible ways to measure the difference between the observed and fitted values. To emphasize the fact that the fitted values in logistic regression are calculated for each covariate pattern and depend on the estimated probability for that covariate pattern, we denote the fitted value for the jth covariate pattern as $\hat{y}_j$ where

$$\hat{y}_j = m_j\hat{\pi}_j = m_j \frac{e^{\hat{g}(\mathbf{x}_j)}}{1+e^{\hat{g}(\mathbf{x}_j)}} ,$$

where $\hat{g}(\mathbf{x}_j)$ is the estimated logit.

We begin by considering two measures of the difference between the observed and the fitted values: the Pearson residual and the deviance residual. For a particular covariate pattern the Pearson residual is defined as follows:

$$r(y_j,\hat{\pi}_j) = \frac{(y_j - m_j\hat{\pi}_j)}{\sqrt{m_j\hat{\pi}_j(1-\hat{\pi}_j)}}. \tag{5.1}$$

The summary statistic based on these residuals is the Pearson chi-square statistic

$$X^2 = \sum_{j=1}^{J} r\left(y_j, \hat{\pi}_j\right)^2 . \tag{5.2}$$

The deviance residual is defined as

$$d\left(y_j, \hat{\pi}_j\right) = \pm\left\{2\left[y_j \ln\left(\frac{y_j}{m_j \hat{\pi}_j}\right) + \left(m_j - y_j\right)\ln\left(\frac{\left(m_j - y_j\right)}{m_j\left(1-\hat{\pi}_j\right)}\right)\right]\right\}^{1/2}, \tag{5.3}$$

where the sign, + or –, is the same as the sign of $\left(y_j - m_j\hat{\pi}_j\right)$. For covariate patterns with $y_j = 0$ the deviance residual is

$$d\left(y_j, \hat{\pi}_j\right) = -\sqrt{2m_j\left|\ln\left(1-\hat{\pi}_j\right)\right|}$$

and the deviance residual when $y_j = m_j$, is

$$d(y_j, \hat{\pi}_j) = \sqrt{2m_j\left|\ln\left(\hat{\pi}_j\right)\right|} .$$

The summary statistic based on the deviance residuals is the deviance

$$D = \sum_{j=1}^{J} d\left(y_j, \hat{\pi}_j\right)^2 . \tag{5.4}$$

In a setting where $J = n$, this is the same quantity shown in equation (1.10).

The distribution of the statistics X^2 and D under the assumption that the fitted model is correct in all aspects is supposed to be chi-square with degrees-of-freedom equal to $J-(p+1)$. For the deviance this statement follows from the fact that D is the likelihood ratio test statistic of a saturated model with J parameters versus the fitted model with $p+1$ parameters. Similar theory provides the null distribution of X^2. The problem is that when $J \approx n$, the distribution is obtained under n-asymptotics, and hence the number of parameters is increasing at the same rate as the sample size. Thus, p-values calculated for these two statistics when $J \approx n$, using the $\chi^2(J-p-1)$ distribution, are incorrect.

One way to avoid the above noted difficulties with the distributions of X^2 and D when $J \approx n$ is to group the data in such a way that m-asymptotics can be used. To understand the rationale behind the various grouping strategies that have been proposed, it is helpful to think of X^2 as the Pearson and D as the log-likelihood chi-square statistics that result from a $2 \times J$ table. The rows of the table correspond to the two values of the outcome variable, $y = 1, 0$. The J columns correspond to the J possible covariate patterns. The estimate of the expected value under the hypothesis that the logistic model in question is the correct model for the cell corresponding to the $y = 1$ row and j^{th} column is $m_j \hat{\pi}_j$. It follows that the estimate of the expected value for the cell corresponding to the $y = 0$ row and jth column is $m_j(1 - \hat{\pi}_j)$. The statistics X^2 and D are calculated in the usual manner from this table.

Thinking of the statistics as arising from the $2 \times J$ table gives some intuitive insight as to why we cannot expect them to follow the $\chi^2(J - p - 1)$ distribution. When chi-square tests are computed from a contingency table the p-values are correct under the null hypothesis when the estimated expected values are "large" in each cell. This condition holds under m-asymptotics. Although this is an oversimplification of the situation, it is essentially correct. In the $2 \times J$ table described above the expected values are always quite small since the number of columns increases as n increases. To avoid this problem we may collapse the columns into a fixed number of groups, g, and then calculate observed and expected frequencies. By fixing the number of columns, the estimated expected frequencies become large as n becomes large. Thus, m-asymptotics hold. The theory required to derive the distribution of the statistics is not quite so straightforward but the intuitive appeal of thinking in this manner is most helpful. The relevant distribution theory presented in a series of papers by Moore (1971), and Moore and Spruill (1975), considers what happens to chi-square goodness-of-fit tests when the boundaries forming the cells are functions of random variables.

5.2.2 The Hosmer-Lemeshow Tests

Hosmer and Lemeshow (1980) and Lemeshow and Hosmer (1982) proposed grouping based on the values of the estimated probabilities. Suppose for sake of discussion, that $J = n$. In this case we think of the n columns as corresponding to the n values of the estimated probabilities, with the first column corresponding to the smallest value, and the nth column to the largest value. Two grouping strategies were proposed as follows: (1) collapse the table based on percentiles of the estimated probabilities and (2) collapse the table based on fixed values of the estimated probability.

With the first method, use of $g = 10$ groups results in the first group containing the $n'_1 = n/10$ subjects having the smallest estimated probabilities, and the last group containing the $n'_{10} = n/10$ subjects having the largest estimated probabilities. With the second method, use of $g = 10$ groups results in cutpoints defined at the values $k/10$, $k = 1, 2, ..., 9$, and the groups contain all subjects with estimated probabilities between adjacent cutpoints. For example, the first group contains all subjects whose estimated probability is less than or equal to 0.1, while the tenth group contains those subjects whose estimated probability is greater than 0.9. For the $y = 1$ row, estimates of the expected values are obtained by summing the estimated probabilities over all subjects in a group. For the $y = 0$ row, the estimated expected value is obtained by summing, over all subjects in the group, one minus the estimated probability. For either grouping strategy, the Hosmer-Lemeshow goodness-of-fit statistic, $\hat{C}$, is obtained by calculating the Pearson chi-square statistic from the $g \times 2$ table of observed and estimated expected frequencies. A formula defining the calculation of $\hat{C}$ is as follows:

$$\hat{C} = \sum_{k=1}^{g} \frac{\left(o_k - n'_k \bar{\pi}_k\right)^2}{n'_k \bar{\pi}_k \left(1 - \bar{\pi}_k\right)}, \tag{5.5}$$

where n'_k is the total number of subjects in the kth group, c_k denotes the number of covariate patterns in the kth decile,

$$o_k = \sum_{j=1}^{c_k} y_j$$

is the number of responses among the c_k covariate patterns, and

$$\bar{\pi}_k = \sum_{j=1}^{c_k} \frac{m_j \hat{\pi}_j}{n'_k}$$

is the average estimated probability.

Using an extensive set of simulations, Hosmer and Lemeshow (1980) demonstrated that, when $J = n$ and the fitted logistic regression model is the correct model, the distribution of the statistic $\hat{C}$ is well approximated by the chi-square distribution with $g - 2$ degrees of freedom, $\chi^2(g-2)$. While not specifically examined, it is likely that $\chi^2(g-2)$ also approximates the distribution when $J \approx n$.

An alternative to the denominator shown in equation (5.5) is obtained if we consider o_k to be the sum of independent nonidentically distributed random variables. This suggests that we should standardize the squared difference between the observed and estimated expected frequency by

$$\sum_{j=1}^{c_k} m_j \hat{\pi}_j (1 - \hat{\pi}_j).$$

It is easy to show that

$$\sum_{j=1}^{c_k} m_j \hat{\pi}_j (1 - \hat{\pi}_j) = n'_k \bar{\pi}_k (1 - \bar{\pi}_k) - \sum_{j=1}^{c_k} m_j (\hat{\pi}_j - \bar{\pi}_k)^2 .$$

In a series of simulations Xu (1996) showed that use of

$$\sum_{j=1}^{c_k} m_j \hat{\pi}_j (1 - \hat{\pi}_j)$$

results in a trivial increase in the value of the test statistic. Thus, in practice we calculate $\hat{C}$ using equation (5.5).

Additional research by Hosmer, Lemeshow, and Klar (1988) has shown that the grouping method based on percentiles of the estimated probabilities is preferable to the one based on fixed cutpoints in the sense of better adherence to the $\chi^2(g-2)$ distribution, especially when many of the estimated probabilities are small (i.e., less than 0.2). Thus, unless stated otherwise, $\hat{C}$ is based on the percentile-type of grouping, usually with $g = 10$ groups. These groups are often referred to as the "deciles of risk." This term comes from health sciences research where the outcome $y = 1$ often represents the occurrence of some disease. Most if not all logistic regression software packages provide the capability to obtain $\hat{C}$ and its p-value, usually based on 10 groups. In addition many packages provide the option to obtain the 10×2 table listing the observed and estimated expected frequencies in each decile.

The results of applying the decile of risk grouping strategy to the estimated probabilities computed from the model for UIS study in Table 4.9 are shown in Table 5.1. For example, the observed frequency in the drug free group, $(DFREE=1)$, for the fifth decile of risk is 16. This value is obtained from the sum of the observed outcomes for the 58 subjects in this group. In a similar fashion the corresponding estimated expected frequency for this decile is 12.7, which is the sum of the 58 estimated probabilities for these subjects. The observed frequency for the return to drug use group, $(DFREE=0)$, is $58-16=42$, and the estimated expected frequency is $58-12.7=45.3$.

The value of the Hosmer-Lemeshow goodness-of-fit statistic computed from the frequencies in Table 5.1 is $\hat{C}=4.39$ and the corresponding p-value computed from the chi-square distribution with 8 degrees of freedom is 0.820. This indicates that the model seems to fit quite well. A comparison of the observed and expected frequencies in each of the 20 cells in Table 5.1 shows close agreement within each decile of risk.

Because the distribution of $\hat{C}$ depends on m-asymptotics, the appropriateness of the p-value depends on the validity of the assumption that the estimated expected frequencies are large. Examining Table 5.1 we see that only one of the estimated expected frequencies is less than five. In general, our point of view is a bit more liberal than those who maintain that with tables of this size (about 20 cells), all expected frequencies must be

Table 5.1 Observed (Obs) and Estimated Expected (Exp) Frequencies Within Each Decile of Risk, Defined by Fitted Value (Prob.) for DFREE = 1 and DFREE = 0 Using the Fitted Logistic Regression Model in Table 4.9

		DFREE = 1		DFREE = 0		
Decile	Prob.	Obs	Exp	Obs	Exp	Total
1	0.094	4	4.1	54	53.9	58
2	0.126	5	6.2	52	50.8	57
3	0.163	8	8.5	50	49.5	58
4	0.204	11	10.4	46	46.6	57
5	0.234	16	12.7	42	45.3	58
6	0.279	11	14.5	46	42.5	57
7	0.324	18	17.5	40	40.5	58
8	0.376	24	19.8	33	37.2	57
9	0.459	23	23.9	35	34.1	58
10	0.728	27	29.3	30	27.7	57

greater than 5. In this case, we feel that there is reason to believe that the calculation of the p-value is accurate enough to support the hypothesis that the model fits. If one is concerned about the magnitude of the expected frequencies, selected adjacent rows of the table may be combined to increase the size of the expected frequencies while, at the same time, reducing the number of degrees-of-freedom.

A few additional comments about the calculation of $\hat{C}$ are needed. When the number of covariate patterns is less than n, we have the possibility that one or more of the empirical deciles will occur at a pattern with $m_j > 1$. If this happens then the value of $\hat{C}$ will depend, to some extent, on how these ties are assigned to deciles. The results presented in Table 5.1 were obtained using STATA's lfit command where ties are assigned to the same decile in such as way as to make the column totals as close to $n/10$ as possible. Other statistical packages may use different strategies to handle ties. For example, fitting the same model in SAS version 6.12 yielded the same results shown in Table 4.9 but with $\hat{C} = 2.873$ and the corresponding p-value is 0.9421. The use of different methods to handle ties by different packages is not likely to be an issue unless the number of covariate patterns is so small that assigning all tied values to one decile results in a huge imbalance in decile size, or worse, considerably fewer than 10 groups. In this case the computed value of $\hat{C}$ may be quite different from one package to the next. In addition, when too few groups are used to calculate $\hat{C}$, we run the risk that we will not have the sensitivity needed to distinguish observed from expected frequencies. It has been our experience that when $\hat{C}$ is calculated from fewer than 6 groups it will almost always indicate that the model fits.

The advantage of a summary goodness-of-fit statistic like $\hat{C}$ is that it provides a single, easily interpretable value that can be used to assess fit. The great disadvantage is that in the process of grouping we may miss an important deviation from fit due to a small number of individual data points. Hence we advocate that, before finally accepting that a model fits, an analysis of the individual residuals and relevant diagnostic statistics be performed. These methods are presented in the next section.

Our experience is that a table such as the one presented in Table 5.1 contains valuable descriptive information for assessing the adequacy of the fitted model over the deciles of risk. Comparison of observed to expected frequencies within each cell may indicate regions where the model does not perform satisfactorily.

Other grouping strategies have been proposed which lead to statistics similar to $\hat{C}$. Tsiatis (1980) suggested a goodness-of-fit statistic based on an explicit partition of the covariate space into g regions. A categorical variable with g levels is introduced into the model corresponding to the g groups. The goodness-of-fit test is the Score test of the coefficients for the new grouping variable. Tsiatis showed that the Score test for this variable is based on a comparison of the observed frequency to estimated expected frequency within each of the g groups. The test has $g-1$ degrees of freedom. This test can be easily carried out in the EGRET and SAS packages. An alternative in packages not having the capability to perform the Score test is to use the maximum partial likelihood test for the coefficients for the $g-1$ design variables. When it is difficult or unclear how to partition the covariate space into meaningful groups, then an alternative to explicit partitioning is to use deciles of risk. Application of the maximum likelihood test to assess the fit of the model in Table 4.9 using the deciles of risk shown in Table 5.1 yields a value of 4.89 which, with 9 degrees of freedom, gives a p-value of 0.843. Hence, this test also supports the fit of the model. One disadvantage of using the maximum partial likelihood or Score test is that actual values of the observed and estimated expected frequencies need not be obtained. These quantities may be useful, when there is evidence of lack of fit, in indicating those deciles where it is occurring.

The limitation of using the Pearson chi-square statistic with $J-(p+1)$ degrees of freedom has generated quite a bit of work on goodness-of-fit tests in recent years. Osius and Rojek (1992) extend work by McCullagh (1985a, 1985b, and 1986) and derive an easily computed large sample normal approximation to the distribution of the Pearson chi-square statistic. Su and Wei (1991) propose a test based on cumulative sums of residuals whose p-value must be determined by complicated and time consuming simulations. Le Cessie and van Houwelingen (1991 and 1995) propose tests based on sums of squares of smoothed residuals whose p-values may be evaluated using either a normal approximation or an easily computed scaled chi-square distribution. Stukel (1988) proposes a two degree of freedom test to ascertain whether a generalized logistic model is better than a standard model fit to the data. Her test is similar to, but more easily computed than, the test proposed by Brown (1982), although the Brown test is automatically computed in BMDP program LR. Hosmer, Hosmer, le Cessie, and Lemeshow (1997) the distributional properties of these tests examine via simulations. They recommend that overall assessment of fit be examined using a combination of tests: the Hosmer-Lemeshow decile of risks test, the Osius and Rojek normal approximation to the distribution of the Pearson chi-square statistic, and Stukel's test.

The large sample normal approximation to the distribution of the Pearson chi-square statistic derived by Osius and Rojek (1992) may be easily computed in any package that has the option to save the fitted values from the logistic regression model and do a weighted linear regression. The essential steps in the procedure when we have J covariate patterns are as follows:

1. Save the fitted values from the model, denoted as $\hat{\pi}_j, j = 1,2,3,\ldots,J$.

2. Create the variable $v_j = m_j\hat{\pi}_j(1-\hat{\pi}_j), j = 1,2,3,\ldots,J$.

3. Create the variable $c_j = \dfrac{(1-2\hat{\pi}_j)}{v_j}, j = 1,2,3,\ldots,J$.

4. Compute the Pearson chi-square statistic shown in (5.2), namely,

$$X^2 = \sum_{j=1}^{J} \frac{(y_j - m_j\hat{\pi}_j)^2}{v_j}.$$

5. Perform a weighted linear regression of c, defined in step 3, on $\mathbf{x}$, the model covariates, using weights v, defined in step 2. Note that the sample size for this regression is J, the number of covariate patterns. Let RSS denote the residual sum-of-squares from this regression. Some packages, for example STATA, scale the weights to sum to 1.0. In this case the reported residual sum-of-squares must be multiplied by the mean of the weights to obtain the correct RSS.

6. Compute the correction factor for the variance, denoted for convenience as A, as follows:

$$A = 2\left(J - \sum_{j=1}^{J} \frac{1}{m_j}\right).$$

7. Compute the standardized statistic

$$z = \frac{[X^2 - (J-p-1)]}{\sqrt{A + RSS}}.$$

8. Compute a two-tailed p-value using the standard normal distribution.

Application of the eight-step procedure using the model in Table 4.9 yields $X^2 = 511.781$, $RSS = 189.658$, $A = 49.667$ and

$$z = \frac{511.781 - (521 - 10 - 1)}{\sqrt{49.667 + 189.658}} = 0.115.$$

The two-tailed p-value is $p = 0.908$. Again, we cannot reject the null hypothesis that the models fits.

To carry out the above analysis it is necessary to form an aggregated data set. This is easy to do in some software packages and impossible in others. In these latter packages we suggest using a second package to create the aggregated data set and then returning to the logistic regression package with this new data set. The essential steps in any package are: (1) Define as aggregation variables the main effects in the model. This defines the covariate patterns. (2) Calculate the sum of the outcome variable and the number of terms in the sum over the aggregation variables. This produces y_j and m_j for each covariate pattern. (3) Output a new data set containing the values of the aggregation variables, covariate patterns, and the two calculated variables, y_j and m_j.

Weesie (1998) has written a STATA program implementing a method proposed by Windmeijer (1990) for computing the significance of the Pearson chi-square statistic using the standard normal distribution. The approach is similar to the above eight-step procedure but is only appropriate in settings when there are n covariate patterns. Thus it is less general than the above method.

Windmeijer (1990) points out that both the Pearson chi-square and the estimator of its variance used to form z in step 7 are quite sensitive to large or small estimated probabilities. Both values are inflated. He suggests that subjects with very small or large fitted values, near 0 or 1, be excluded when using the Pearson chi-square statistic. The default exclusion criteria in Weesie's STATA program are $\hat{\pi} < 1.0 \times 10^{-5}$ or $\hat{\pi} > \left(1 - 1.0 \times 10^{-5}\right)$. In general we think this is good advice but urge considerable caution and complete honesty in reporting what was done so as to avoid possible criticism that the data have been tinkered with in order to obtain a good fitting model.

Stukel (1988) proposes a two degree-of-freedom test that determines whether two parameters in a generalized logistic model are equal to zero. Briefly, the two additional parameters allow the tails of the logistic regression model (i.e., the small and large probabilities) to be either heavier/longer or lighter/shorter than the standard logistic regression model. This test is not a goodness-of-fit test since it does not compare observed and fitted values. However it does provide a test of the basic logistic regression model assumption and in that sense we feel it is a useful adjunct to the Hosmer-Lemeshow and Osius-Rojek goodness-of-fit tests. The test has not been implemented in any package; but it can be easily obtained from the following procedure:

1. Save the fitted values from the model, denoted as $\hat{\pi}_j, j = 1, 2, 3, \ldots, J$.

2. Compute the estimated logit

$$\hat{g}_j = \ln\left(\frac{\hat{\pi}_j}{1-\hat{\pi}_j}\right) = \mathbf{x}_j'\hat{\boldsymbol{\beta}},\ j = 1, 2, 3 \ldots, J.$$

3. Compute two new covariates $z_{1j} = 0.5 \times \hat{g}_j^2 \times \mathrm{I}(\hat{\pi}_j \geq 0.5)$ and $z_{2j} = -0.5 \times \hat{g}_j^2 \times \mathrm{I}(\hat{\pi}_j < 0.5)$, $j = 1, 2, 3, \ldots, J$, where $\mathrm{I}(\arg) = 1$ if arg is true and zero if arg is false. Note that in a setting when all the fitted values are either less than or greater than 0.5 only one variable is created.

4. Perform the Score test for the addition z_1 and/or z_2 to the model. If a package does not perform the Score test then the partial likelihood ratio test can be used.

Application of the four-step procedure to the fitted model in Table 4.9 yields a value for the partial likelihood ratio test of 3.95 which, with two degrees of freedom, yields $p = 0.139$. Again we cannot reject the hypothesis that the logistic regression model is the correct model.

As we have mentioned at various points in this section, a complete assessment of fit is a multi-faceted investigation involving summary tests and measures as well as diagnostic statistics. This is especially important to keep in mind when using overall goodness-of-fit tests. The desired outcome for most investigators is the decision not to reject the null hypothesis that the model fits. With this decision one is subject to the possibility of the Type II error and hence the power of the test becomes an issue. The simulation results reported in Hosmer, Hosmer, le Cessie and Lemeshow (1997) indicate that none of the overall goodness-of-fit tests is especially powerful for small to moderate sample sizes $n < 400$. One should keep this firmly in mind when using goodness-of-fit tests.

Before we discuss diagnostic statistics we present a few other measures of model performance that are often useful supplements to the overall tests of fit just discussed.

5.2.3 Classification Tables

An intuitively appealing way to summarize the results of a fitted logistic regression model is via a classification table. This table is the result of cross-classifying the outcome variable, y, with a dichotomous variable whose values are derived from the estimated logistic probabilities.

To obtain the derived dichotomous variable we must define a cutpoint, c, and compare each estimated probability to c. If the estimated probability exceeds c then we let the derived variable be equal to 1; otherwise it is equal to 0. The most commonly used value for c is 0.5. The appeal of this type of approach to model assessment comes from the close relationship of logistic regression to discriminant analysis when the distribution of the covariates is multivariate normal within the two outcome groups. However, it is not limited to this model (e.g., see Efron (1975)).

In this approach, estimated probabilities are used to predict group membership. Presumably, if the model predicts group membership accurately according to some criterion, then this is thought to provide evidence that the model fits. Unfortunately, this may or may not be the case. For example, it is easy to construct a situation where the logistic regression model is in fact the correct model and thus fits, but classification is poor. Suppose that $P(Y=1)=\theta_1$ and that $X \sim N(0,1)$ in the group with $Y=0$ and $X \sim N(\mu,1)$ in the group with $Y = 1$. In this discriminant analysis model the slope coefficient for the logistic regression model is (see equation (1.23)) $\beta_1 = \mu$ and the intercept is (see equation (1.22))

$$\beta_0 = \ln\left[\frac{\theta_1}{(1-\theta_1)}\right] - \frac{\mu^2}{2} .$$

The probability of misclassification, PMC, may be shown to be

Table 5.2 Classification Table Based on the Logistic Regression Model in Table 4.9 Using a Cutpoint of 0.5

Classified	Observed		Total
	DFREE = 1 Drug Free	DFREE = 0 Returned to Drug Use	
DFREE = 1	16	11	27
DFREE = 0	131	417	548
Total	147	428	575

Sensitivity = 16/147=10.9%; Specificity=417/428=97.4%

$$\text{PMC} = \theta_1 \Phi\left\{\frac{1}{\beta_1}\ln\left[\frac{(1-\theta_1)}{\theta_1}\right] - \frac{\beta_1}{2}\right\} + (1-\theta_1)\Phi\left\{\frac{1}{\beta_1}\ln\left[\frac{\theta_1}{(1-\theta_1)}\right] - \frac{\beta_1}{2}\right\},$$

where Φ is the cumulative distribution function of the N(0,1) distribution. Thus, the expected error rate is a function of the magnitude of the slope, not necessarily of the fit of the model. Accurate or inaccurate classification does not address our criteria for goodness-of-fit: that the distances between observed and expected values be unsystematic, and within the variation of the model. However, the classification table may be a useful adjunct to other measures based more directly on residuals.

The results of classifying the observations of the UIS using the fitted model given in Table 4.9 are presented in Table 5.2. The classification table shown in Table 5.2 is fairly typical of those seen in many logistic regression applications. The overall rate of correct classification is estimated as $75.3\% = 100[(16+417)/575]\%$, with 97.4% (417/428) of the drug free group (specificity) and only 10.9% (16/147) of the returned to drug use group (sensitivity) being correctly classified. Classification is sensitive to the relative sizes of the two component groups and always favors classification into the larger group, a fact that is also independent of the fit of the model. This may be seen by considering the expression for PMC as a function of θ_1. The disadvantage of using PMC as a criterion is that it reduces a probabilistic model where outcome is measured on a continuum, to a dichotomous model where predicted outcome is binary. For practical purposes there is little difference between the values of $\hat{\pi} = 0.48$ and $\hat{\pi} =$

0.52, yet use of a 0.5 cutpoint would establish these two individuals as markedly different.

An important reason why measures derived from a 2×2 classification table (such as sensitivity and specificity) should not be used as measures of model performance is that they depend heavily on the distribution of the probabilities in the sample. Thus, if two models are being compared, differences between them with respect to sensitivity and specificity may depend entirely on "patient mix" rather than on the superiority of one model over another.

In the discussion that follows we must keep in mind the meaning of probability which is that, among n subjects, each having the same probability of the outcome of interest, $\hat{\pi}$, the number who are expected to develop the outcome is $n\hat{\pi}$ and the number expected not to develop the outcome is $n(1-\hat{\pi})$. (This logic formed the basis of the discussion in Section 5.2.2 on goodness-of-fit testing.) Suppose that 0.50 was the cutpoint being used for classification purposes and suppose that 100 subjects had a probability $\hat{\pi} = 0.51$. All of these subjects would be predicted to have the outcome but, assuming the model is well calibrated, 51 of the subjects would actually develop the outcome whereas 49 would be expected not to develop the outcome. Thus 49 of the 100 patients would be misclassified.

Consider again the 2×2 classification table from the UIS presented in Table 5.2. An examination of the estimated probabilities of return to drug use in the two classification groups reveals that among the 27 subjects predicted to be drug free, probabilities ranged from 0.503 to 0.728, with a mean of 0.553. Among the 548 subjects predicted to return to drug use, probabilities ranged from 0.029 to 0.498, with a mean of 0.241. Clearly, because so many of the subjects in this study have probabilities close to the cutpoint we expect a considerable amount of misclassification. In Table 5.2 we see that 417 of the 548 subjects predicted to return to drug use actually did returned to drug use whereas 11 of the 27 subjects predicted to be drug free were misclassified. Thus, of the total 147 subjects who were actually drug free, only 16 of them were correctly predicted (i.e., sensitivity = 16/147 = 10.9%).

Suppose now that we keep the prediction unchanged for each subject but alter the distribution of probabilities among the subjects predicted to return to drug use and among those predicted not to return to drug use. The rule we use is:

$$\text{if} \quad \hat{\pi} < 0.50, \text{ then let } \hat{\pi} = 0.05$$
$$\text{and if} \quad \hat{\pi} \geq 0.50, \text{ then let } \hat{\pi} = 0.95.$$

Table 5.3 Classification Table Based on the Logistic Regression Model in Table 4.9 Using a Cutpoint of 0.5, but All Probabilities $\hat{\pi} < 0.50$ Are Replaced with $\hat{\pi} = 0.05$ and All Probabilities $\hat{\pi} \geq 0.50$ Are Replaced with $\hat{\pi} = 0.95$

Classified	Observed		Total
	DFREE = 1 Drug Free	DFREE = 0 Returned to Drug Use	
DFREE = 1	26	1	27
DFREE = 0	27	521	548
Total	53	522	575

Sensitivity = 26/53=49.1%; Specificity=521/522=99.8%

Clearly, this modification would reflect a population that was very polarized with respect to their likelihood of remaining drug free. If the model was well *calibrated* (i.e., probabilities reflecting the true outcome experience in the data), then only 5% of those predicted to return to drug use would actually be misclassified and, similarly, only 5% of those predicted not to return to drug use would be misclassified. The resulting 2×2 table would be as presented in Table 5.3. Note that both the sensitivity and specificity are considerably greater than they were for the actual population seen in Table 5.2, where there was a wide range of probabilities. The reason for the sensitivity being so low even in this highly polarized population is that there were relatively few subjects whose probabilities of remaining drug free were above 0.50.

Now consider a second hypothetical population where

$$\text{if} \quad \hat{\pi} < 0.50, \text{ then let } \hat{\pi} = 0.45$$
$$\text{and if} \quad \hat{\pi} \geq 0.50, \text{ then let } \hat{\pi} = 0.55.$$

This homogenous population is one where a great deal of misclassification would be expected. Assuming the probabilities accurately reflect the outcome experience in these data, the 2×2 table would be as presented in Table 5.4. Note that the sensitivity and specificity are much worse than was the case with the actual, heterogeneous, population.

For these reasons, one cannot compare models on the basis of measures derived from 2×2 classification tables since these measures are completely confounded by the distribution of probabilities in the samples upon which they are based. The same model, evaluated in two populations,

Table 5.4 Classification Table Based on the Logistic Regression Model in Table 4.9 Using a Cutpoint of 0.5, but All Probabilities $\hat{\pi} < 0.50$ Are Replaced with $\hat{\pi} = 0.45$ and All Probabilities $\hat{\pi} \geq 0.50$ are Replaced with $\hat{\pi} = 0.55$

Classified	Observed		Total
	DFREE = 1 Drug Free	DFREE = 0 Returned to Drug Use	
DFREE = 1	15	12	27
DFREE = 0	301	247	548
Total	316	259	575

Sensitivity = 15/316=4.7%; Specificity=247/259=95.4%

could give very different impressions of performance if sensitivity or specificity was used as the measure of performance.

In summary, the classification table is most appropriate when classification is a stated goal of the analysis; otherwise it should only supplement more rigorous methods of assessment of fit.

5.2.4 Area Under the ROC Curve

Sensitivity and specificity rely on a single cutpoint to classify a test result as positive. A more complete description of classification accuracy is given by the area under the ROC (Receiver Operating Characteristic) curve. This curve, originating from signal detection theory, shows how the receiver operates the existence of signal in the presence of noise. It plots the probability of detecting true signal (sensitivity) and false signal (1 – specificity) for an entire range of possible cutpoints.

The area under the ROC curve, which ranges from zero to one, provides a measure of the model's ability to *discriminate* between those subjects who experience the outcome of interest versus those who do not. As an example, consider the model for estimating the probability that a subject will remain drug free as given in Table 4.9. Suppose that we were interested in *predicting* the outcome for each patient. One rule we might try is the one shown in Table 5.2, where we predict the subject will remain drug free if $\Pr(y=1) \geq 0.50$, and predict the subject will return to drug use if $\Pr(y=1) < 0.50$. There are some statistical benefits associated with using 0.5 but we could consider what happens when we use other cutpoints. For

Table 5.5 Classification Table Based on the Logistic Regression Model in Table 4.9 Using a Cutpoint of 0.6

Classified	Observed		Total
	DFREE = 1 Drug Free	DFREE = 0 Returned to Drug Use	
DFREE = 1	5	0	5
DFREE = 0	142	428	570
Total	147	428	575

Sensitivity = 5/147=3.4%; Specificity=428/428=100%

example, suppose that we used a cutpoint of 0.6 instead. This would result in the classification table shown in Table 5.5, where the sensitivity is only 3.4% but the specificity is 100%. In fact, the same can be done for any possible choice of cutpoint. Table 5.6 summarizes the results of choosing all possible cutpoints between 0.05 and 0.60 in increments of 0.05.

If our objective was to choose an optimal cutpoint for the purposes of classification, one might select a cutpoint that maximizes both sensitivity and specificity. This choice is facilitated through a graph such as the one shown in Figure 5.1 where we see that an "optimal" choice for a cutpoint might be 0.26 as that is approximately where the sensitivity and specificity curves cross.

Table 5.6 Summary of Sensitivity, Specificity, and 1-Specificity for Classification Tables Based on the Logistic Regression Model in Table 4.9 using a Cutpoint of 0.05 to 0.60 in Increments of 0.05

Cutpoint	Sensitivity	Specificity	1–Specificity
0.05	99.32%	2.57%	97.43%
0.10	95.92%	15.19%	84.81%
0.15	90.48%	31.78%	68.22%
0.20	81.63%	46.26%	53.74%
0.25	65.99%	61.21%	38.79%
0.30	57.14%	72.20%	27.80%
0.35	40.14%	82.01%	17.99%
0.40	29.25%	87.38%	12.62%
0.45	18.37%	92.06%	7.94%
0.50	10.88%	97.43%	2.57%
0.55	5.44%	99.30%	0.70%
0.60	3.40%	100.00%	0.00%

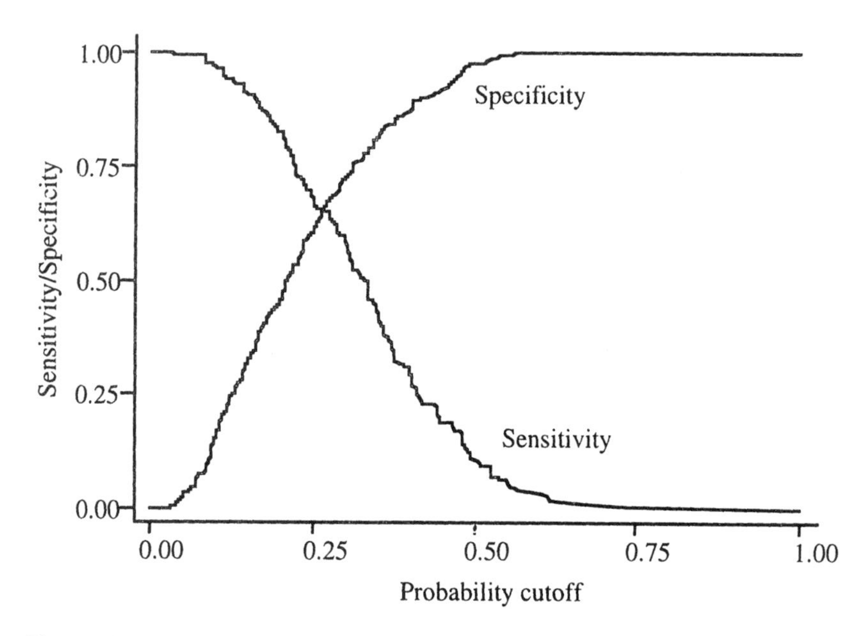

Figure 5.1 Plot of sensitivity and specificity versus all possible cutpoints in the UIS.

A plot of sensitivity versus 1 – specificity over all possible cutpoints is shown in Figure 5.2. The curve generated by these points is called the *ROC Curve* and the area under the curve provides a measure of discrimination which is the likelihood that a subject who remains drug free will have a higher $\Pr(y=1)$ than a subject who returns to drug use. The area under the ROC Curve in Figure 5.2 is 0.6989.

As a general rule:

If ROC = 0.5:	this suggests no discrimination (i.e., we might as well flip a coin)
If $0.7 \le ROC < 0.8$:	this is considered acceptable discrimination
If $0.8 \le ROC < 0.9$:	this is considered excellent discrimination
If $ROC \ge 0.9$:	this is considered outstanding discrimination.

In practice it is extremely unusual to observe areas under the ROC Curve greater than 0.9. In fact, as we noted in Chapter 4, Section 5, when there is complete separation it is impossible to estimate the coefficients of a logistic regression model, yet nearly complete separation would be required for the area under the ROC Curve to be >90%.

We note here that a poorly fitting model (i.e., poorly calibrated as assessed by the goodness-of-fit measures presented in section 5.2.2) may still

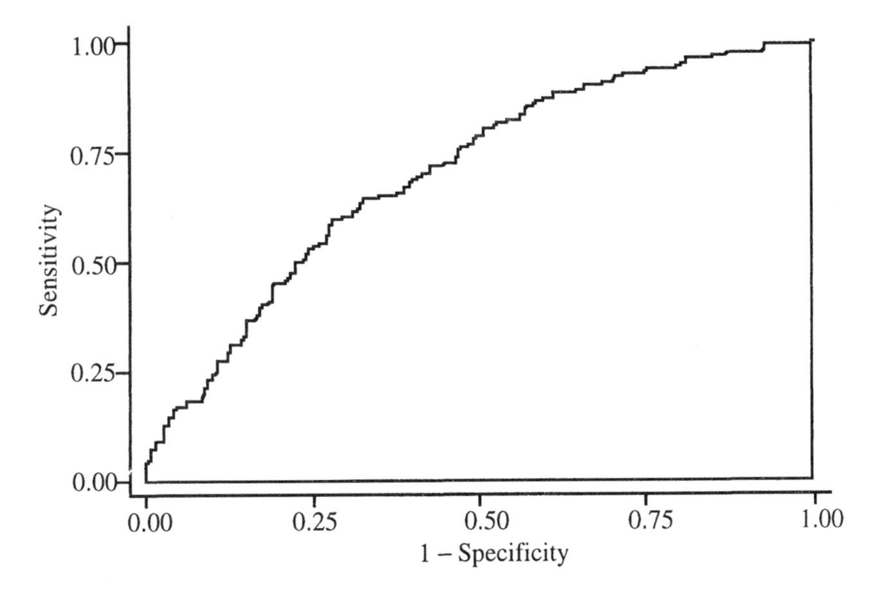

Figure 5.2 Plot of sensitivity versus 1–specificity for all possible cutpoints in the UIS. The resulting curve is called the ROC Curve.

have good discrimination. For example, if one added 0.25 to every probability in a good fitting logistic model with good ability to discriminate, the new model would be poorly calibrated whereas the discrimination would not be affected at all. We believe that model performance should be assessed by considering *both* calibration and discrimination.

Another perhaps more intuitive way to understand the meaning of the area under the ROC Curve is as follows: recall that we let n_1 denote the number of subjects with $y=1$ and n_0 denote the number of subjects with $y=0$. We then create $n_1 \times n_0$ pairs: each subject with $y=1$ is paired with each subject with $y=0$. Of these $n_1 \times n_0$ pairs, we determine the proportion of the time that the subject with $y=1$ had the higher of the two probabilities. This proportion may be shown to be equal to the area under the ROC Curve. For example, in the UIS, there were 575 subjects. Of these, 147 remained drug free while 428 did not. A total of $147 \times 428 = 62{,}916$ comparisons are made and we count the number of times that the probability of remaining drug free is higher for the subject that did remain drug free than for the subject who did not. (When the probability is the same for both subjects we add 1/2 to the count.) For these data the count of the number of times that the subject with $y=1$ had a higher probability than the subject with $y=0$ was 43,972.5. (The reader may recognize that this count is the Mann-Whitney U statistic for these data.) The ratio $43,972.5/62,916=.6989$ is the area under the ROC curve, the same value as we obtained before.

5.2.5 Other Summary Measures

For sake of completeness we present a short discussion of R^2 measures that have been proposed for use with logistic regression models. In general, these measures are based on various comparisons of the predicted values from the fitted model to those from model(0), the no data or intercept only model and, as a result, do not assess goodness-of-fit. We think that a true measure of fit is one based strictly on a comparison of observed to predicted values from the fitted model. However there may be settings where the R^2 measures can provide useful statistics for comparing competing models fit to the same set of data. Mittlböck and Schemper (1996) study the properties of 12 different measures using the criteria: (1) the measure has an easily understood interpretation (2) the squared measure can attain a lower bound of 0 and an upper bound of 1 and (3) the measure is consistent with the character of logistic regression (i.e., not being changed by a linear transformation of model covariates). They recommend two for routine use: the squared Pearson correlation coefficient of observed outcome with the predicted probability and a linear regression-like sum-of-squares R^2. All other measures, including some popular likelihood-based R^2 statistics are judged to be inadequate on at least one of their criteria.

In a setting with n covariate patterns the squared Pearson correlation coefficient is

$$\mathrm{r}^2=\frac{\left[\sum_{i=1}^{n}(y_i-\bar{y})(\hat{\pi}_i-\bar{\pi})\right]^2}{\left[\sum_{i=1}^{n}(y_i-\bar{y})^2\right]\times\left[\sum_{i=1}^{n}(\hat{\pi}_i-\bar{\pi})^2\right]}, \tag{5.6}$$

where $\bar{y}=\bar{\pi}=n_1/n$. The linear regression-like measure is

$$R_{ss}^2 = 1 - \frac{\sum_{i=1}^{n}(y_i - \hat{\pi}_i)^2}{\sum_{i=1}^{n}(y_i - \bar{y})^2}. \quad (5.7)$$

The case of $J < n$ covariate patterns was not considered by Mittlböck and Schemper (1996). However, the extensions of the two measures to this setting are

$$r_c^2 = \frac{\left[\sum_{j=1}^{J}(y_j - m_j\bar{y})(m_j\hat{\pi}_j - m_j\bar{\pi})\right]^2}{\left[\sum_{j=1}^{J}(y_j - m_j\bar{y})^2\right] \times \left[\sum_{j=1}^{J}(m_j\hat{\pi}_j - m_j\bar{\pi})^2\right]} \quad (5.8)$$

and

$$R_{ssc}^2 = 1 - \frac{\sum_{j=1}^{J}(y_j - m_j\hat{\pi}_j)^2}{\sum_{j=1}^{J}(y_j - m_j\bar{y})^2}. \quad (5.9)$$

Using the fitted model in Table 4.9 and evaluating the squared Pearson correlation coefficient defined in (5.6), we obtain $r^2 = 0.0946$. The value of the linear regression like sum-of-squares measure from (5.7) is

$$R_{ss}^2 = 1 - (99.061/109.4168) = 0.0946.$$

The fitted model has 521 covariate patterns. Evaluating the covariate pattern version of the Pearson correlation coefficient in (5.8) yields $r_c^2 = 0.3564$. The increase from the value of 0.0946 in the $J = n$ case is due to increased range of y_j (0-2) versus y_i (0-1) in the values being correlated. The sum-of-squares measure is $R_{ssc}^2 = 1 - (94.261/104.696)$ $= 0.0997$.

We obtain another version of R_{ss}^2 when we use log-likelihoods in place of sums-of-squares. Mittlböck and Schemper (1996) do not recommend it for routine use, as it is not as intuitively easy to explain. However the measure is calculated in a number of packages under various names (e.g., pseudo R^2 in STATA). If we let L_0 and L_P denote the log-likelihoods for models containing only the intercept and the model containing the intercept plus the p covariates respectively, then the log likelihood-based R^2 is

$$R_L^2 = \frac{L_0 - L_p}{L_0} = 1 - \frac{L_p}{L_0} . \tag{5.10}$$

The maximum value for R_L^2 is obtained when we fit the saturated model. If $J = n$ then $L_s = 0$ and we see that R_L^2 is equal to 1.0. However, if $J < n$ then the maximum is less than 1.0. A modification of the statistic that can attain 1.0 in the $J < n$ case is

$$R_{LS}^2 = \frac{L_0 - L_p}{L_0 - L_S} . \tag{5.11}$$

The value of the log-likelihood from the saturated model, L_S, may be easily obtained from the deviance for the model with p covariates and its log-likelihood is computed as

$$L_S = L_P + 0.5D.$$

Hence, it would seem prudent to calculate L_S whenever $J < n$ and to use R_{LS}^2.

As an example, we evaluate (5.10) using the fitted model in Table 4.9 and, assuming $J = 575$, we obtain

$$R_L^2 = 1 - \frac{-298.981}{-326.864} = 0.0853 .$$

In order to evaluate (5.11) we need the value of L_S using $J = 521$ covariate patterns. The value of the deviance from (5.2) is $D = 530.74$ and from the above expression we obtain

$$L_S = (-298.981) + 0.5 \times (530.74) = -33.611$$

and

$$R_{LS}^2 = \frac{[(-326.864) - (-298.981)]}{[(-326.864) - (-33.611)]} = 0.0951.$$

All the various R^2 values for this example are low when compared to R^2 values typically encountered with good linear regression models. Unfortunately low R^2 values in logistic regression are the norm and this presents a problem when reporting their values to an audience accustomed to seeing linear regression values. As we demonstrate throughout this chapter, the fitted model in Table 4.9 is a good model (based on goodness-of-fit and discrimination). Thus we do not recommend routine publishing of R^2 values with results from fitted logistic regression models. However, they may be helpful in the model building stage as a statistic to evaluate competing models.

5.3 Logistic Regression Diagnostics

The summary statistics based on the Pearson chi-square residuals described in the previous section provide a single number that summarizes the agreement between observed and fitted values. The advantage (as well as the disadvantage) of these statistics is that a single number is used to summarize considerable information. Therefore, before concluding that the model "fits", it is crucial that other measures be examined to see if fit is supported over the entire set of covariate patterns. This is accomplished through a series of specialized measures falling under the general heading of *regression diagnostics*. We assume that the reader has had some experience with diagnostics for linear regression. For a brief introduction to linear regression diagnostics see Kleinbaum, Kupper, Muller and Nizam (1998). A more detailed presentation may be found in Cook and Weisberg (1982) and Belsley, Kuh, and Welsch (1980). Pregibon (1981) provided the theoretical work that extended linear regression diagnostics to logistic regression. Since that key paper, work has focused on refining the use of logistic regression diagnostics in assessing goodness-of-fit. We begin by briefly describing logistic regression diagnostics. In this development we assume that the fitted model contains p covariates and that they form J covariate patterns. Deriving the diagnostic statistics requires a higher mathematical level than most of the other material in this text. However, an understanding of the mathematical development is not required for the effective application of the diagnostics in practice. Thus, less sophisticated mathematical readers may wish to skip to Chapter 5, Section 4 where the discussion of the calculations and uses of the diagnostics begins.

The key quantities for logistic regression diagnostics, as in linear regression, are the components of the "residual sum-of-squares." In linear regression a key assumption is that the error variance does not depend on the conditional mean, $E(Y_j|\mathbf{x}_j)$. However, in logistic regression we have binomial errors and, as a result, the error variance is a function of the conditional mean:

$$\begin{aligned}\text{var}(Y_j|\mathbf{x}_j) &= m_j E(Y_j|\mathbf{x}_j)\times\left[1-E(Y_j|\mathbf{x}_j)\right]\\ &= m_j\pi(\mathbf{x}_j)\left[1-\pi(\mathbf{x}_j)\right].\end{aligned}$$

Thus, we begin with residuals as defined in (5.1) and (5.3) which have been "divided" by estimates of their standard errors; this may not be entirely obvious in the case of the deviance residual. Let r_j and d_j denote the values of the expressions given in equation (5.1) and (5.3), respectively, for covariate pattern $\mathbf{x}_j$. Since each residual has been divided by an approximate estimate of its standard error, we expect that if the logistic regression model is correct these quantities have a mean approximately equal to zero and a variance approximately equal to 1. We discuss their distribution shortly.

In addition to the residuals for each covariate pattern, other quantities central to the formation and interpretation of linear regression diagnostics are the "hat" matrix and the leverage values derived from it. In linear regression the hat matrix is the matrix that provides the fitted values as the projection of the outcome variable into the covariate space. Let $\mathbf{X}$ denote the $J\times(p+1)$ matrix containing the values for all J covariate patterns formed from the observed values of the p covariates, with the first column being one to reflect the presence of an intercept in the model. The matrix $\mathbf{X}$ is often called the design matrix. In linear regression the hat matrix is $\mathbf{H}=\mathbf{X}(\mathbf{X}'\mathbf{X})^{-1}\mathbf{X}'$; for example, $\hat{\mathbf{y}}=\mathbf{H}\,\mathbf{y}$. The linear regression residuals, $(\mathbf{y}-\hat{\mathbf{y}})$, expressed in terms of the hat matrix are $(\mathbf{I}-\mathbf{H})\mathbf{y}$ where $\mathbf{I}$ is the $J\times J$ identity matrix. Using weighted least squares linear regression as a model, Pregibon (1981) derived a linear approximation to the fitted values, which yields a hat matrix for logistic regression. This matrix is

$$\mathbf{H}=\mathbf{V}^{1/2}\mathbf{X}(\mathbf{X}'\mathbf{V}\mathbf{X})^{-1}\mathbf{X}'\mathbf{V}^{1/2}, \tag{5.12}$$

where $\mathbf{V}$ is a $J\times J$ diagonal matrix with general element

$$v_j = m_j\hat{\pi}(\mathbf{x}_j)\left[1-\hat{\pi}(\mathbf{x}_j)\right].$$

In linear regression the diagonal elements of the hat matrix are called the *leverage values* and are proportional to the distance from $\mathbf{x}_j$ to the mean of the data. This concept of distance to the mean is important in linear regression, as points that are far from the mean may have considerable influence on the values of the estimated parameters. The extension of the concept of leverage to logistic regression requires additional discussion and clarification.

Let the quantity h_j denote the jth diagonal element of the matrix $\mathbf{H}$ defined in equation (5.12). It may be shown that

$$h_j = m_j\hat{\pi}(\mathbf{x}_j)\left[1-\hat{\pi}(\mathbf{x}_j)\right]\mathbf{x}'_j(\mathbf{X}'\mathbf{V}\mathbf{X})^{-1}\mathbf{x}'_j = v_j \times b_j\,, \qquad (5.13)$$

where

$$b_j = \mathbf{x}'_j(\mathbf{X}'\mathbf{V}\mathbf{X})^{-1}\mathbf{x}'_j$$

and $\mathbf{x}'_j = (1, x_{1j}, x_{2j}, \ldots x_{pj})$ is the vector of covariate values defining the jth covariate pattern. The sum of the diagonal elements of $\mathbf{H}$ is, as is the case in linear regression, $\sum h_j = (p+1)$, the number of parameters in the model. In linear regression the dimension of the hat matrix is usually $n \times n$ and thus ignores any common covariate patterns in the data. With this formulation, any diagonal element in the hat matrix has an upper bound of $1/k$ where k is the number of subjects with the same covariate pattern. If we formulate the hat matrix for logistic regression as an $n \times n$ matrix then each diagonal element is bounded from above by $1/m_j$, where m_j is the total number of subjects with the same covariate pattern. When the hat matrix is based upon data grouped by covariate patterns, the upper bound for any diagonal element is 1.

It is important to know whether the statistical package being used calculates the diagnostic statistics by covariate pattern. For example, STATA's logistic command uses individual subject data to fit models. Following estimation it computes all diagnostic statistics by covariate pattern but retains the size of the original data set. Thus all subjects in a particular covariate pattern have the same covariate values, fitted value and diagnostic statistics, but each subject has an individual outcome. On the other hand, SAS's logistic procedure computes diagnostic statistics based on the data structure in its model statement. If one assumes that there are n covariate patterns (and the outcome is either 0 or 1) then diagnostic statistics are based on individual subjects. However, if the data have been previously collapsed or grouped into covariate patterns and binomial trials input (y_j/m_j) is used, then diagnostic statistics are by covariate pattern. In general, we recommend that diagnostic statistics be computed taking into account covariate patterns. This is especially important when the number of covariate patterns, J, is much smaller than n, or if some values of m_j are larger than 5. For example, in the final model for the UIS data shown in Table 4.9 we have $J = 521$ and $n = 575$. In this situation we definitely should compute the diagnostic statistics by covariate pattern. If, on the other hand, we had a model with $J = 570$ and we were using SAS, we might not go to the trouble to aggregate the data by covariate patterns.

When the number of covariate patterns is much smaller than n there is the risk that we may fail to identify influential and/or poorly fit covariate patterns. Consider a covariate pattern with m_j subjects, $y_j = 0$ and estimated logistic probability $\hat{\pi}_j$. The Pearson residual defined in equation (5.1), computed individually for each subject with this covariate pattern, is

$$r_i = \frac{\left(0 - \hat{\pi}_j\right)}{\sqrt{\hat{\pi}_j\left(1 - \hat{\pi}_j\right)}}$$

$$= -\sqrt{\frac{\hat{\pi}_j}{\left(1 - \hat{\pi}_j\right)}},$$

while the Pearson residual based on all subjects with this covariate pattern is

$$r_i = \frac{\left(0 - m_j\hat{\pi}_j\right)}{\sqrt{m_j\hat{\pi}_j\left(1 - \hat{\pi}_j\right)}}$$

$$= -\sqrt{m_j}\sqrt{\frac{\hat{\pi}_j}{\left(1 - \hat{\pi}_j\right)}}$$

which increases negatively as m_j increases. If $m_j = 1$ and $\hat{\pi}_j = 0.5$, then $r_j = -1$ which is not a large residual. On the other hand, if there were $m_j = 16$ subjects with this covariate pattern, then $r_j = -4.0$ which is quite large. If we performed the analysis in STATA then the Pearson residual would be -4.0 for each of the 16 subjects in the covariate pattern. If we performed the analysis in SAS with a sample of size n then the Pearson

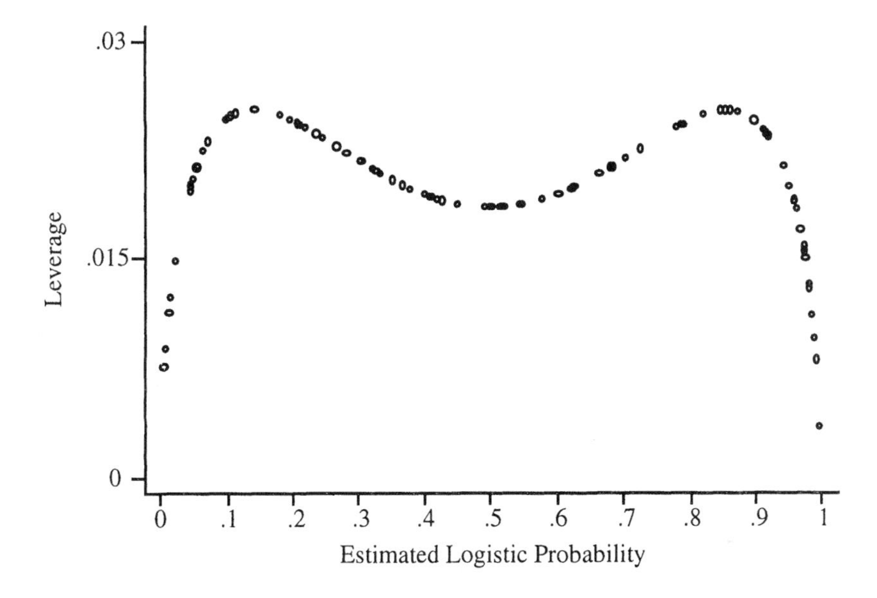

Figure 5.3 Plot of leverage (h) versus the estimated logistic probability ($\hat{\pi}$) for a hypothetical univariable logistic regression model.

residual would be −1.0 for all 16 subjects. Thus the diagnostic statistics are different even though both packages produce the same fitted model.

A major point that must be kept in mind when interpreting the magnitude of the leverage is the effect that v_j has on h_j in equation (5.13). Pregibon (1981) notes that the fit determines the estimated coefficients and, since the estimated coefficients determine the estimated probabilities, points with large values of h_j are extreme in the covariate space and thus lie far from the mean. Lesaffre (1986, p.117) refutes this point, where he shows that the term v_j in the expression for h_j cannot be ignored. The following example demonstrates that, up to a point, both Pregibon and Lesaffre are correct.

Figure 5.3 presents a plot of the leverage values versus the estimated probabilities for a sample of 100 observations from a logistic model with $g(x) = 0.8x$ and $x \sim N(0,9)$. Recall that the notation $N(0,9)$ describes a variable following a normal distribution with mean zero and variance 9.

We see that the leverage increases as the estimated probability gets further from 0.5 (x gets further from its mean, nominally zero) until the estimated probabilities become less than 0.1 or greater than 0.9. At that point the leverage decreases and rapidly approaches zero. This example shows that the most extreme points in the covariate space may have the smallest leverage. This is the exact opposite of the situation in linear regression, where the leverage is a monotonic increasing function of the distance of a covariate pattern from the mean. The practical consequence of this is that to interpret a particular value of the leverage in logistic regression correctly, we need to know whether the estimated probability is small (<0.1) or large (>0.9). If the estimated probability lies between 0.1 and 0.9 then the leverage gives a value that may be thought of as distance. When the estimated probability lies outside the interval 0.1 to 0.9, then the value of the leverage may not measure distance in the sense that further from the mean implies a larger value.

A quantity that does increase with the distance from the mean is $b_j = \mathbf{x}'_j(\mathbf{X}'\mathbf{V}\mathbf{X})^{-1}\mathbf{x}'_j$. Thus, if we are only interested in distance then we should focus on b_j. A plot of the b_j versus the estimated probability for the example is shown in Figure 5.4. In this Figure we see that b_j provides a measure of distance in the covariate space and, as a result, is more like the leverage values in linear regression. However, since the most useful

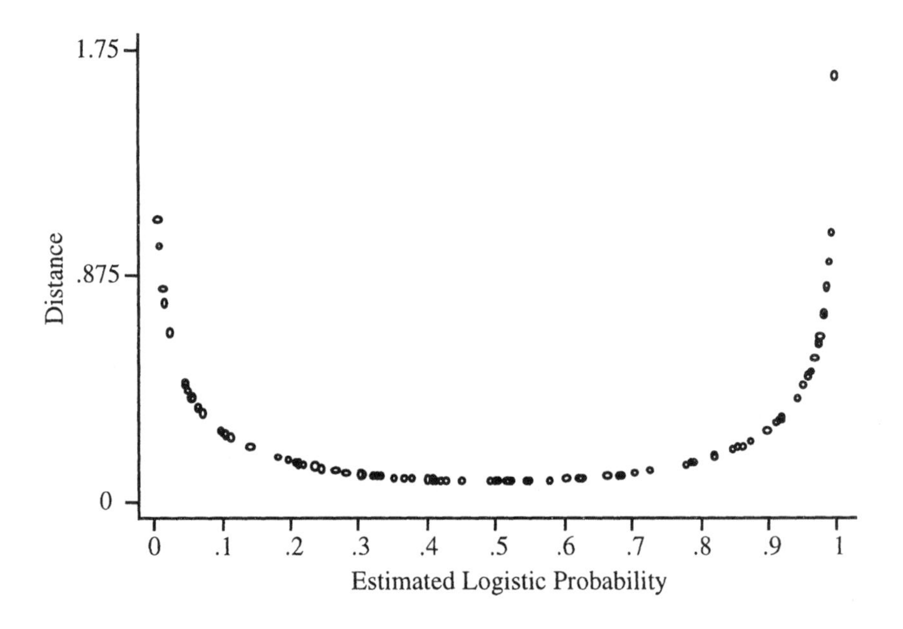

Figure 5.4 Plot of the distance portion of leverage (b) versus the estimated logistic probability ($\hat{\pi}$) for a hypothetical univariable logistic regression model.

diagnostic statistics for logistic regression are functions of the full leverage, h_j, the distance portion, b_j, is not discussed further here.

If we use the Pregibon (1981) linear regression-like approximation for the residual for the jth covariate pattern, $\left[y_j - m_j\hat{\pi}(\mathbf{x}_j)\right] \approx (1-h_j)y_j$, then the variance of the residual is

$$m_j\hat{\pi}(\mathbf{x}_j)\left[1-\hat{\pi}(\mathbf{x}_j)\right](1-h_j),$$

which suggests that the Pearson residuals do not have variance equal to 1 unless they are further standardized. Recall that we denote by r_j the Pearson residual given in equation (5.1). The standardized Pearson residual for covariate pattern $\mathbf{x}_j$ is

$$r_{sj} = \frac{r_j}{\sqrt{1-h_j}} . \tag{5.14}$$

Another useful diagnostic statistic is one that examines the effect that deleting all subjects with a particular covariate pattern has on the value of the estimated coefficients and the overall summary measures of fit, X^2 and D. The change in the value of the estimated coefficients is analogous to the measure proposed by Cook (1977, 1979) for linear regression. It is obtained as the standardized difference between $\hat{\boldsymbol{\beta}}$ and $\hat{\boldsymbol{\beta}}_{(-j)}$, where these represent the maximum likelihood estimates computed using all J covariate patterns and excluding the m_j subjects with pattern $\mathbf{x}_j$ respectively, and standardizing via the estimated covariance matrix of $\hat{\boldsymbol{\beta}}$. Pregibon (1981) showed, to a linear approximation, that this quantity for logistic regression is

$$\begin{aligned}\Delta\hat{\boldsymbol{\beta}}_j &= \left(\hat{\boldsymbol{\beta}}-\hat{\boldsymbol{\beta}}_{(-j)}\right)'(\mathbf{X'VX})\left(\hat{\boldsymbol{\beta}}-\hat{\boldsymbol{\beta}}_{(-j)}\right)\\ &= \frac{r_j^2 h_j}{(1-h_j)^2}\\ &= \frac{r_{sj}^2 h_j}{(1-h_j)} .\end{aligned} \tag{5.15}$$

Using similar linear approximations it can be shown that the decrease in the value of the Pearson chi-square statistic due to deletion of the subjects with covariate pattern $\mathbf{x}_j$ is

$$\Delta X_j^2 = \frac{r_j^2}{(1-h_j)} \\ = r_{sj}^2 . \tag{5.16}$$

A similar quantity may be obtained for the change in the deviance,

$$\Delta D_j = d_j^2 + \frac{r_j^2 h_j}{(1-h_j)} .$$

If we replace r_j^2 by d_j^2 this yields the approximation

$$\Delta D_j = \frac{d_j^2}{(1-h_j)} , \tag{5.17}$$

which is similar in form to the expression in equation (5.16).

These diagnostic statistics are conceptually quite appealing, as they allow us to identify those covariate patterns that are poorly fit (large values of ΔX_j^2 and/or ΔD_j), and those that have a great deal of influence on the values of the estimated parameters (large values of $\Delta\hat{\boldsymbol{\beta}}_j$). After identifying these influential patterns (subjects), we can begin to address the role they play in the analysis.

Before proceeding to the use of the diagnostics in an example, we make a few summary comments on what we might expect their application to tell us. Consider first the measure of fit, ΔX_j^2. This measure is smallest when y_j and $m_j\hat{\pi}(\mathbf{x}_j)$ are close. This is most likely to happen when y_j = 0 and $\hat{\pi}(\mathbf{x}_j) < 0.1$ or $y_j = m_j$ and $\hat{\pi}(\mathbf{x}_j) > 0.9$. Similarly ΔX_j^2 is largest when y_j is furthest from $m_j\hat{\pi}(\mathbf{x}_j)$. This is most likely to occur if we have a value of $y_j = 0$ and $\hat{\pi}(\mathbf{x}_j) > 0.9$, or with $y_j = m_j$ and $\hat{\pi}(\mathbf{x}_j) < 0.1$. These same covariate patterns are not likely to have a large $\Delta\hat{\boldsymbol{\beta}}_j$ since, when $\hat{\pi}(\mathbf{x}_j) < 0.1$ or $\hat{\pi}(\mathbf{x}_j) > 0.9$, $\Delta\hat{\boldsymbol{\beta}}_j \approx \Delta X_j^2 h_j$, and h_j is approaching zero. The influence diagnostic, $\Delta\hat{\boldsymbol{\beta}}_j$, is large when both ΔX_j^2 and h_j are at least moderate. This is most likely to occur when $0.1 < \hat{\pi}(\mathbf{x}_j) < 0.3$, or $0.7 < \hat{\pi}(\mathbf{x}_j) < 0.9$. As we know from Figure 5.3, these are the intervals where the leverage, h_j, is largest. In the region where $0.3 < \hat{\pi}(\mathbf{x}_j) < 0.7$ the chances are not as great that either ΔX_j^2 or h_j is large. Table 5.7 summarizes these observations. This table reports what might be expected, not what may actually happen in any particular example. Therefore, it should only be used as a guide to further understanding and interpretation of the diagnostic statistics.

In linear regression essentially two approaches are used to interpret the value of the diagnostics often in conjunction with each other. The first is graphical. The second employs the distribution theory of the linear regression model to develop the distribution of the diagnostics under the assumption that the fitted model is correct. In the graphical approach, large values of diagnostics either appear as spikes or reside in the extreme corners of plots. A value of the diagnostic statistic for a point appearing to lie away from the balance of the points is judged to be extreme if it exceeds some percentile of the relevant distribution. This may sound a little too hypothesis-testing oriented but, under the assumptions of linear regression with normal errors, there is a known statistical distribution whose percentiles provide some guidance as to what constitutes a large value. Presumably, if the model is correct and fits then no values should be exceptionally large, and the plots should appear as expected under the distribution of the diagnostic.

In logistic regression we have to rely primarily on visual assessment, as the distribution of the diagnostics under the hypothesis that the model fits is known only in certain limited settings. For instance, consider the Pearson residual, r_j. It is often stated that the distribution of this quantity is approximately $N(0,1)$ when the model is correct. This statement is only true when m_j is sufficiently large to justify that the normal distribution provides an adequate approximation to the binomial distribution, a condition obtained under m-asymptotics. For example, if $m_j = 1$ then r_j has only two possible values and can hardly be expected to be normally distributed. Jennings (1986b) has stated this point clearly and with all the necessary technical details. All of the diagnostics are evaluated by covariate pattern; hence any approximations to their distributions based on the normal distribution, under binomial errors, depend on the number of subjects with that pattern. When a fitted model contains some continuous

Table 5.7 Likely Values of Each of the Diagnostic Statistics ΔX^2, $\Delta\hat{\beta}$, and h Within Each of Five Regions Defined by the Value of the Estimated Logistic Probability ($\hat{\pi}$)

	Diagnostic Statistic		
$\hat{\pi}$	ΔX^2	$\Delta\hat{\beta}$	h
< 0.1	Large or Small	Small	Small
0.1—0.3	Moderate	Large	Large
0.3—0.7	Moderate to Small	Moderate	Moderate to Small
0.7—0.9	Moderate	Large	Large
> 0.9	Large or Small	Small	Small

covariates then the number of covariate patterns, J, is of the same order as n, and m-asymptotic results cannot be relied upon. Thus, in practice, an assessment of "large" is, of necessity, a judgment call based on experience and the particular set of data being analyzed. Using the $N(0,1)$, or equivalently, the $\chi^2(1)$ distribution for squared quantities may provide some guidance as to what large is. However, we urge that these percentiles be used with extreme caution. There is no substitute for experience in the effective use of diagnostic statistics.

We have defined seven diagnostic statistics which may be divided into three categories: (1) the basic building blocks, which are of interest in themselves, but also are used to form other diagnostics, (r_j, d_j, h_j); (2) derived measures of the effect of each covariate pattern on the fit of the model, $(r_{sj}, \Delta X_j^2, \Delta D_j)$; and (3) a derived measure of the effect of each covariate pattern on the value of the estimated parameters, $(\Delta\hat{\beta}_j)$. Most logistic regression software packages provide the capability to obtain at least one of the measures within each group.

A number of different types of plots have been suggested for use, each directed at a particular aspect of fit. Some are formed from the seven diagnostics while others require additional computation. For example, see the methods based on grouping and smoothing in Landwehr, Pregibon, and Shoemaker (1984) and Fowlkes (1987). It is impractical to consider all possible suggested plots, so we restrict attention to a few of the more easily obtained ones that are meaningful in logistic regression analysis. We consider them to be the core of an analysis of diagnostics. These consist of the following:

(1) Plot ΔX_j^2 versus $\hat{\pi}_j$

(2) Plot ΔD_j versus $\hat{\pi}_j$

(3) Plot $\Delta\hat{\boldsymbol{\beta}}_j$ versus $\hat{\pi}_j$.

Other plots that are sometimes useful include:

(4) Plot ΔX_j^2 versus h_j

(5) Plot ΔD_j versus h_j

(6) Plot $\Delta\hat{\boldsymbol{\beta}}_j$ versus h_j,

as these allow direct assessment of the contribution of leverage to the value of the diagnostic statistic. One additional plot that we have found especially useful is a plot of ΔX_j^2 versus $\hat{\pi}_j$ where the size of the plotting symbol is proportional to the size of $\Delta\hat{\boldsymbol{\beta}}_j$. This plot is used in the examples that follow.

To illustrate the use of the diagnostic statistics and their related plots, we consider the final model for UIS data given in Table 4.9. Recall that the summary statistics indicated that the model fits. Thus, we do not expect an analysis of diagnostics to show large numbers of covariate patterns

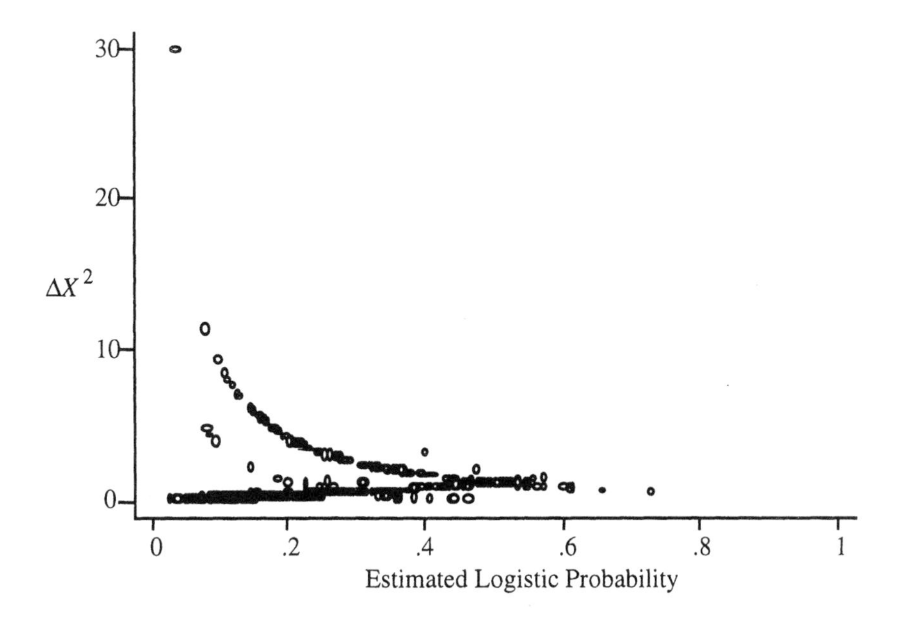

Figure 5.5 Plot of ΔX^2 versus the estimated probability from the fitted model in Table 4.9, UIS $J = 521$ covariate patterns.

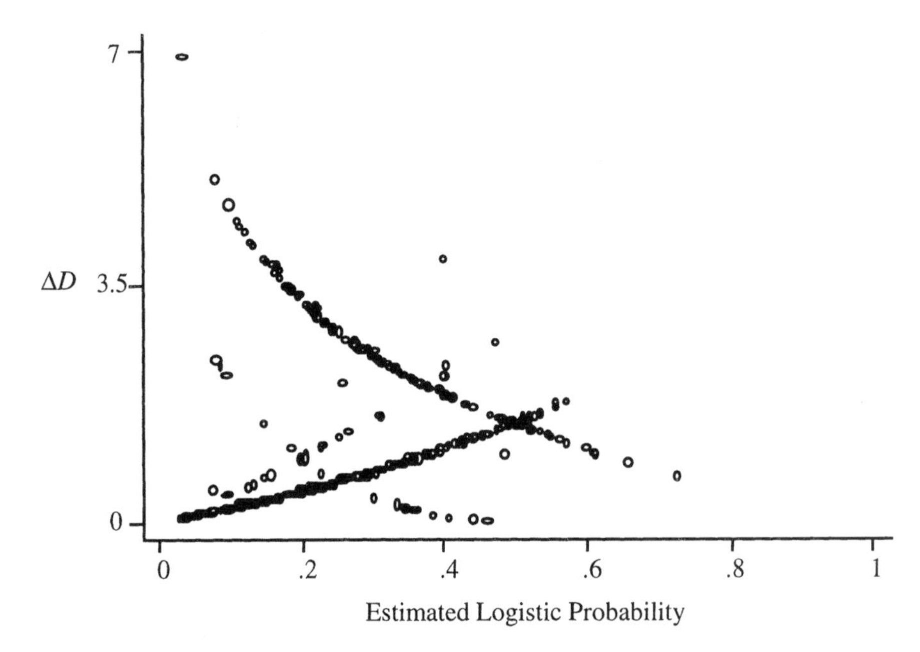

Figure 5.6 Plot of ΔD versus the estimated probability from the fitted model in Table 4.9, UIS $J = 521$ covariate patterns.

being fit poorly. We might uncover a few covariate patterns which do not fit, or which have considerable influence on the estimated parameters. The key plots are given in Figures 5.5 to Figure 5.8. We discuss each plot in turn.

The diagnostics ΔX^2 and ΔD plotted versus the estimated logistic probabilities are shown in Figure 5.5 and Figure 5.6, respectively. We prefer to use these plots instead of plots of r_j and d_j versus $\hat{\pi}_j$. The reasons for this choice are as follows: (1) When $J \approx n$, most positive residuals correspond to covariate patterns where $y_j = m_j$ (e.g., 1) and negative residuals to those with $y_j = 0$. Hence, the sign of the residual is not useful. (2) Large residuals, regardless of sign, correspond to poorly fit points. Squaring these residuals further emphasizes the lack of fit and removes the issue of sign. (3) The shape of the plot allows us to determine which patterns have $y_j = 0$ and which have $y_j = m_j$.

The shapes of the plots in Figures 5.5 and 5.6 are similar and show quadratic like curves. The points on the curves going from the top left to

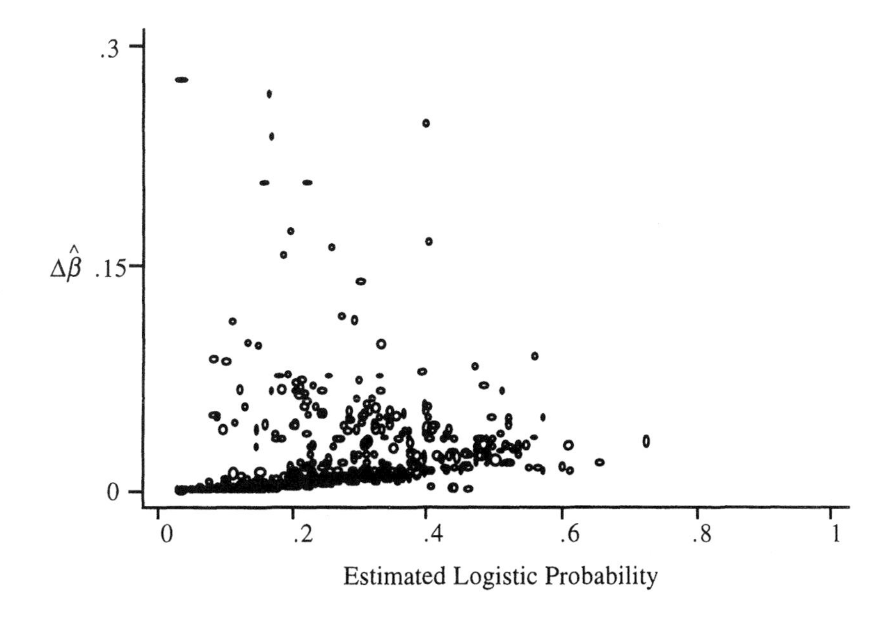

Figure 5.7 Plot of $\Delta\hat{\boldsymbol{\beta}}$ versus the estimated probability from the fitted model in Table 4.9, UIS $J = 521$ covariate patterns.

bottom right corner correspond to covariate patterns with $y_j = m_j$. The ordinate for these points is proportional to $\left(1-\hat{\pi}_j\right)^2$ since $m_j = 1$ for most covariate patterns. The points on the other curves, going from the bottom left to top right corner, correspond to covariate patterns with $y_j = 0$. The ordinate for these points is proportional to $\left(0-\hat{\pi}_j\right)^2$. Covariate patterns that are poorly fit will generally be represented by points falling in the top left or top right corners of the plots. We look for points that fall some distance from the balance of the data plotted. Assessment of this distance is partly based on numeric value and partly based on visual impression.

In Figure 5.5 we see 1 point (i.e., covariate pattern) that is extremely poorly fit in the top left corner of the plot, $\Delta X^2 \approx 30$. There is one other point that lies a bit away from the others with $\Delta X^2 \approx 12$. These same two points are easily seen in Figure 5.6.

The range of ΔX^2 is much greater than ΔD. This is a property of Pearson versus deviance residuals. Whenever possible we prefer to use plots of both ΔX^2 and ΔD versus $\hat{\pi}$.

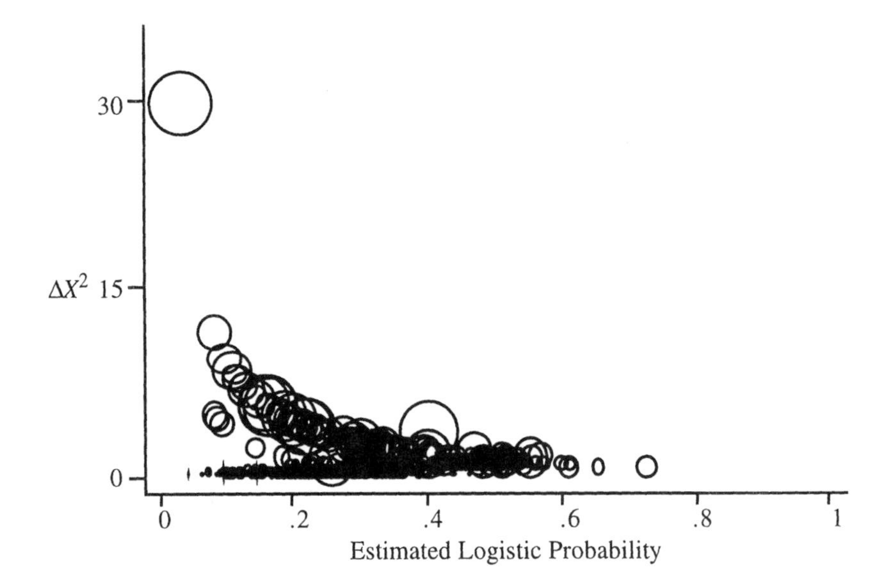

Figure 5.8 Plot of ΔX^2 versus the estimated probability from the fitted model in Table 4.9 with size of the plotting symbol proportional to $\Delta\hat{\boldsymbol{\beta}}$, UIS $J = 521$ covariate patterns.

Aside from the two points noted, the plots show that the model fits reasonably well. Most of the values of ΔX^2 and ΔD are less than, or at least not much larger than, 4. We use 4 as a crude approximation to the upper ninety-fifth percentile of the distribution of ΔX^2 and ΔD as, under m-asymptotics, these quantities would be distributed approximately as $\chi^2(1)$ with $\chi^2_{0.95}(1) = 3.84$.

The influence diagnostic $\Delta\hat{\boldsymbol{\beta}}$ is plotted versus $\hat{\pi}$ in Figure 5.7. We see four points that lie somewhat away from the rest of the data. The values themselves are not especially large, as all are less than 0.3. In our experience the influence diagnostic must be larger than 1.0 for an individual covariate pattern to have an effect on the estimated coefficients. However there are always exceptions and it is good practice to note outlying values of $\Delta\hat{\boldsymbol{\beta}}$, regardless of the actual magnitude.

We noted in Table 5.7 that the largest values of $\Delta\hat{\boldsymbol{\beta}}$ are most likely to occur when both ΔX^2 and leverage are at least moderately large. However large values can also occur when either component is large. This is the case in Figure 5.7 where the covariate pattern with the largest influence diagnostic is the one with the largest value of ΔX^2. The other points are in the region of estimated probabilities where both ΔX^2 and ΔD can be moderately large.

In Figure 5.8 we plot ΔX^2 versus $\hat{\pi}$ with the size of the symbol proportional to $\Delta\hat{\beta}$. This plot allows us to ascertain the contributions of residual and leverage to $\Delta\hat{\beta}$. The large circle in the top left corner corresponds to the largest value of ΔX^2. Another large circle can be partially seen at $\hat{\pi} \approx 0.4$. This point has a small value of ΔX^2 but is in the region where we expect to find maximum leverage.

One problem with the influence diagnostic $\Delta\hat{\beta}$ is that it is a summary measure of change over all coefficients in the model simultaneously. For this reason it is important to examine the changes in the individual coefficients due to specific covariate patterns identified as influential.

Examination of Figures 5.5 to Figure 5.8 identifies four covariate patterns with outlying values on one or more of the diagnostics statistics. These include the pattern with large values of ΔX^2 and ΔD, and three more with outlying values of $\Delta\hat{\beta}$. Information on these patterns is presented in Table 5.8. The quantity *P#* in Table 5.8 refers to the covariate pattern number. This number is somewhat arbitrary, as its value depends on how the data were aggregated. It should be noted that *P#* is not the original study identification code.

The results in Table 5.8 provide examples of what can be learned about a fitted model through diagnostic statistics. Consider covariate pattern 31. If this covariate pattern is deleted from the data set, we expect to see a substantial decrease in X^2 and a somewhat smaller decrease in D. However, as shown in Table 5.9, when we delete this covariate pattern and refit the model the actual observed decrease in the Pearson chi-square statistic is about 7. This change is much less than the value suggested by the diagnostic statistic. On the other hand the deviance decreases by about 22 even though the value of the diagnostic statistic suggests the change should be about 7. It has been our experience that ΔX^2 and ΔD tend to be modestly positively correlated with actual observed changes when covariate patterns are deleted. Thus, we recommend that one fit the model with covariate pattern(s) deleted to obtain the actual change(s)/effect(s). Even though covariate pattern 31 has the largest value of $\Delta\hat{\beta}$, its numeric value is not large enough for us to expect to see major changes in the estimated coefficients. As shown in Table 5.9, the maximum change in any coefficient in the model is less than 10 percent. In summary, covariate pattern 31 is typical of one type of pattern that has a large value of ΔX^2 and ΔD.

Namely, the fitted model predicts that it is unlikely for the subject to respond when in fact they do (i.e. $\hat{\pi}$ is small and $y=1$). While the opposite type of poor fit ($\hat{\pi}$ is large and $y=0$) is not present in the UIS model we have seen it occur in many other analyses. The exercises at the end of this chapter contain a number of different problems designed to highlight various aspects of the performance of the diagnostic statistics.

The other three covariate patterns described in Table 5.8 have outlying values of $\Delta\hat{\beta}$ relative to the rest of the values of this statistic. The values of the change in fit diagnostics are modest and the leverage values are small. As shown in Table 5.9 there are no substantial changes in model fit or estimated parameters when we delete each pattern.

The last column of Table 5.9 presents the changes when the model was fit deleting all four covariate patterns (a total of 5 subjects). The collective effect is substantial. Numerous estimated coefficients change by more than 20 percent and the fit measures, X^2 and D, also show substantial decreases. For these reasons one might consider removing these five subjects from the analysis. We consulted with our colleagues and they found the covariate values for these five subjects to be quite reasonable and therefore felt that the subjects should not be deleted. With this decision made, we can move to the final step, presentation and interpretation of the

Table 5.8 Covariate Values, Observed Outcome (y_j), Number (m_j), Estimated Logistic Probability ($\hat{\pi}$), and the Value of the Four Diagnostic Statistics $\Delta\hat{\beta}$, ΔX^2, ΔD, and Leverage (h) for the Four Most Extreme Covariate Patterns ($P\#$)

P#	31	477	105	468
AGE	24	41	26	40
NDRGTX	20	0	0	0
IHVX	2	3	1	3
RACE	0	1	1	1
TREAT	0	0	0	0
SITE	1	0	0	0
y_j	1	1	2	1
m_j	1	1	2	1
$\hat{\pi}$	0.033	0.163	0.403	0.168
$\Delta\hat{\beta}$	0.277	0.267	0.246	0.236
ΔX^2	29.925	5.403	3.191	5.192
ΔD	6.909	3.812	3.916	3.735
h	0.009	0.047	0.072	0.044

model using the fitted model from Table 4.9. However, before doing so, we use the results seen here as the basis of a short discussion on the reasons for the changes seen in Table 5.9 and a more general discussion of the role of diagnostic statistics in analysis.

We note that the values of the goodness-of-fit statistic based on deciles of risk, $\hat{C}$, in Table 5.9 are all small and the smallest is for the original model. In practice, one cannot use $\hat{C}$ to select a "best" fitting model from a collection of models that all fit. The statistic shows that each of the six models shown in Table 5.9 seems to provide an overall fit to the data.

The net effect of the deletion of the five subjects is an increase in the coefficients involving age, the number of previous drug treatments and treatment duration while the coefficient for race decreases. The reason for the effect on race is that 4 of the five subjects had RACE = Other and remained drug free. Thus once they are removed there is a less pronounced difference between the two racial groups. All five subjects were on the shorter duration treatment and all were drug free at 12 months. Thus re-

Table 5.9 Estimated Coefficients from All Data, the Percent Change when the Covariate Pattern Is Deleted, and Values of Goodness-of-Fit Statistics for Each Model

		Covariate Pattern Deleted				
Variable	All Data	31	477	105	468	All Four
AGE	0.117	8.9	5.3	−2.8	4.8	18.0
NDRGFP1	1.669	9.6	6.3	−1.6	5.7	22.4
NDRGFP2	0.434	9.4	4.1	2.1	3.8	21.1
INHX_2	−0.635	8.8	0.4	0.3	0.7	10.6
IVHX_3	−0.705	0.5	5.6	<−0.1	5.8	12.9
RACE	0.684	1.3	−5.8	−8.5	−5.9	−20.3
TREAT	0.435	5.2	3.6	7.3	3.6	20.8
SITE	0.516	−5.6	<0.1	2.9	0.1	−2.3
AGE×NDRGFP1	−0.015	9.9	14.6	−8.3	13.1	33.6
RACE×SITE	−1.429	−0.5	−3.6	−4.2	−3.6	−12.5
Constant	−6.844	7.7	3.3	−1.3	3.0	14.0
Goodness-of-Fit						
D	511.78	489.94	511.57	508.70	511.61	482.63
X^2	530.74	523.62	526.85	526.88	526.94	511.11
$\hat{C}$	4.39	5.55	6.36	6.69	6.36	6.86

moving these subjects removes some of the "positive" effect of the shorter treatment and leads to a more pronounced difference between the two levels. The reason for the observed behavior in age and the number of previous drug treatments is not clear. There is no clear pattern in the ages or number of previous treatments. In addition these variables interact and the number of previous drug treatments is highly non-linear in the model.

The model for the UIS data is an example where the model fits well, and use of diagnostics identified only a few covariate patterns where the model did not fit, and/or the patterns were influential. Suppose instead that we have a model where the summary statistics indicate that there is substantial deviation from fit. In this situation, we have evidence that for more than a few covariate patterns, y_j differs from $m_j\hat{\pi}_j$. One or more of three things has likely happened: (1) the logistic model does not provide a good approximation to the correct relationship between the conditional mean, $E(Y|\mathbf{x}_j)$, and $\mathbf{x}_j$, (2) we have not measured and/or not included an important covariate into the model, or (3) at least one of the covariates in the model has not been entered in the correct scale. We discuss each of these in turn.

The logistic regression model is remarkably flexible. Unless we are dealing with a set of data where most of the probabilities are very small or very large, or where the fit is extremely poor in an identifiable systematic manner, it is unlikely that any alternative model will provide a better fit. Cox (1970) demonstrates that the logistic and other, similar symmetric models are virtually identical in the region from 0.2 to 0.8. If one suspects, based on clinical or other reasons (such as graphical presentations, or Stukel's test, described in Section 5.2.2) that the logistic model is the wrong one, then careful thought should be given to the choice of the alternative model. Particular attention should be given to issues of interpretation. Are the coefficients clinically interpretable? The approach that tries all other possible models and selects the "best fitting" one is not recommended, as no thought is given to the clinical implications of the selected model. In some situations, inadequacy of a fitted logistic model can be corrected by returning to model building and rechecking variable selection and scale identification. Model fitting is an iterative procedure. We rarely obtain a final model on the first pass through the data. However, we must keep in mind the distinction between getting a model to fit and having the theoretically correct model.

Some interesting theoretical work has been done by White (1982, 1989) and Hjort (1988, 1999) on the use of maximum likelihood estimation with a misspecified model. These authors show that the fitted logistic regression model is the one that minimizes the Kullbeck-Leibler information distance between the theoretically correct model and the logistic model. In this sense the fitted logistic regression model is a best approximation to the true model. Recently, Maldonado and Greenland (1993) examine the interpretation of model coefficients in this setting and conclude that if one follows a thorough model building paradigm, similar to one presented in Chapter 4 and this chapter, then the estimated coefficients can provide useful estimates of effect even when the model is somewhat misspecified. Along these same lines Lin, Pstay and Kronmal (1998) present a method to quantify the sensitivity of estimates of effect to unmeasured confounders. It is not clear to us how practical the method may prove to be, as it requires one to specify the distribution of the model covariates within each level of the outcome variable.

White (1982, 1989) provides a test for the hypothesis that the fitted model is the theoretically correct one. The test is elegant but is difficult to compute in practice and its power has not been adequately studied. Hence, we recommend that assessment of the adequacy of the fitted logistic model be performed using the methods suggested in this chapter. When there is evidence that the logistic model does not fit the data an alternative model should be selected on the basis of clinical considerations.

When performing an analysis, we hope that the study was designed carefully so that data on all major covariates were collected. However, it is possible that the clinical factors associated with the outcome variable are not well known and in this case a key variable may not be present in the observed data. The potential biases and pitfalls of this oversight are enormous. Little can be done if this is the case, except to go back and collect these data. This approach of retroactive data collection is also impractical in most research situations.

Lack of fit may also occur if the variability in the outcome variable exceeds what would be predicted by the model and binomial variation. Much of the work on this problem is motivated by toxicological experiments where a dependence in the observations is present due to the outcome being measured on littermates having the same parentage, see Haseman and Hogan (1975), Haseman and Kupper (1979), Legler and Ryan (1997), Ryan (1992), and Williams (1975). This source of lack of fit is often called extrabinomial variation. Another setting where this problem occurs is where the dependence is due to a general clustering of groups of responses (e.g., when a treatment is randomly assigned to a group of subjects such as a school or patients of a physician). The clustering can also be due to repeated observations on subjects over time. This is an active area of methodological research and several software packages now incorporate the capability to fit appropriately modified logistic regression models. Because of its practical importance we consider methods for the analysis of clustered binary data in some detail in Section 8.3.

In summary, one should not proceed to presenting the results from a fitted model until the fit of model has been thoroughly assessed using both summary measures and diagnostic statistics.

5.4 ASSESSMENT OF FIT VIA EXTERNAL VALIDATION

In some situations it may be possible to exclude a subsample of our observations, develop a model based on the remaining subjects, and then test the model in the originally excluded subjects. In other situations it may be possible to obtain a new sample of data to assess the goodness-of-fit of a previously developed model. This type of assessment is often called model validation, and may be especially important when the fitted model is used to predict outcome for future subjects. The reason for considering this type of assessment of model performance is that the fitted model always performs in an optimistic manner on the developmental data set. Harrell, Lee, and Mark (1996) discuss this within a general model building context. The use of validation data amounts to an assessment of goodness-of-fit where the fitted model is considered to be theoretically known, and no estimation is performed. Some of the diagnostics discussed in Section 5.3 ($\Delta X^2, \Delta D, \Delta\hat{\beta}$) mimic this idea by computing, for each covariate pattern, a quantity based on the exclusion of the particular covariate pattern. With a new data set a more thorough assessment is possible.

The methods for assessment of fit in the validation sample parallel those described in Sections 5.2 and 5.3 for the developmental sample. The major difference is that the values of the coefficients in the model are regarded as fixed constants rather than estimated values.

Suppose that the validation sample consists of n_v observations $(y_i, \mathbf{x}_i)$, $i = 1, 2, ..., n_v$, which may be grouped into J_v covariate patterns. In keeping with previous notation, let y_j denote the number of positive responses among the m_j subjects with covariate pattern $\mathbf{x} = \mathbf{x}_j$ for $j = 1, 2, ..., J_v$. The logistic probability for the jth covariate pattern is π_j, the value of the previously estimated logistic model using the covariate pattern, $\mathbf{x}_j$, from the validation sample. These quantities become the basis for the computation of the summary measures of fit, X^2, D, and C, from the validation sample. Each of these is considered in turn.

The computation of the Pearson chi-square statistic follows directly from equation (5.2), with obvious substitution of quantities from the validation sample. In this case X^2 is computed as the sum of J_v independent terms. If each $m_j\pi_j$ is large enough to use the normal approximation to the binomial distribution, then X^2 is distributed as $\chi^2(J_v)$ under the hypothesis that the model is correct. We expect that in practice the observed numbers of subjects within each covariate pattern is small, with most $m_j = 1$. Hence, we cannot employ m-asymptotics. In this case we can use results presented in Osius and Rojek (1992) to obtain a statistic that follows the standard normal distribution under the hypothesis that the model is correct and J_v is sufficiently large. The procedure is similar to the one presented in Section 5.2. Specifically one computes the standardized statistic

$$z = \frac{X^2 - J_v}{\sigma_v},$$

where

$$\sigma_v^2 = 2J_v + \sum_{j=1}^{J_v} \frac{1}{m_j\pi_j(1-\pi_j)} - 6\sum_{j=1}^{J_v} \frac{1}{m_j}.$$

The test uses a two-tailed p-value based on z.

The same line of reasoning discussed in Section 5.2.2 to develop the Hosmer-Lemeshow test may be used to obtain an equivalent statistic for the validation sample. Assume that we wish to use 10 groups composed of the deciles of risk. Any other grouping strategy could be used with obvious modifications in the calculations. Let n_k denote the approximately $n_v/10$ subjects in the kth decile of risk. Let $o_k = \sum y_j$ be the number of positive responses among the covariate patterns falling in the kth decile of risk. The estimate of the expected value of o_k under the assumption that the model is correct is $e_k = \sum m_j\pi_j$, where the sum is over the covariate patterns in the decile of risk. The Hosmer-Lemeshow statistic is obtained as the Pearson chi-square statistic computed from the observed and expected frequencies

$$C_v = \sum_{k=1}^{g} \frac{(o_k - e_k)^2}{n_k\bar{\pi}_k(1-\bar{\pi}_k)}, \tag{5.18}$$

where $\bar{\pi}_k = \sum m_j \pi_j / n_k$. The subscript, v, has been added to C to emphasize that the statistic has been calculated from a validation sample. Under the hypothesis that the model is correct, and the assumption that each e_k is sufficiently large for each term in C_v to be distributed as $\chi^2(1)$, it follows that C_v is distributed as $\chi^2(10)$. In general, if we use g groups then the distribution is $\chi^2(g)$. In addition to calculating a p-value to assess overall fit, we recommend that each term in C_v be examined to assess the fit within each decile of risk. The comments given in Section 5.2.2 regarding modification of the denominator of the test statistic, $\hat{C}$, in equation (5.5) also apply to C_v in equation (5.18).

The classification table is the remaining summary statistic that we are likely to use with the validation sample and then only in instances where classification is an important use of the model. The classification table is obtained in exactly the same manner as shown in Section 5.2.3, with the modification that probabilities are no longer thought of as being estimated. The resulting table may then be used to compute statistics such as sensitivity, specificity, positive and negative predictive power. Interpretation of these quantities depends on the particular situation.

5.5 INTERPRETATION AND PRESENTATION OF THE RESULTS FROM A FITTED LOGISTIC REGRESSION MODEL

Once we are satisfied that the fit of the model is adequate, we are ready to use the model to address the inferential goals of the particular study. In our experience this almost always involves using the estimates of model coefficients to obtain estimates of odds ratios. We use the model presented in Table 4.9, whose fit was checked earlier in this chapter, as an example. For convenience Table 5.10 presents more detailed results from the fitted model.

While the model in Table 5.10 is much more complicated than the typical model one finds in a subject matter journal, it is an excellent example for teaching purposes. It contains a dichotomous main effect covariate (TREAT), a polychotomous main effect covariate (IVHX), two dichotomous covariates and their interaction (RACE and SITE), a linear continuous covariate, a non-linear continuous covariate and their interaction (AGE and NDRGTX modeled via NDRGFP1 and NDRGFP2). We begin with the nominal scale covariates that appear only as main effects.

Table 5.10 Estimated Coefficients, Standard Errors, *z*-Scores, Two-Tailed *p*-Values and 95% Confidence Intervals for the Final Logistic Regression Model for the UIS (*n* = 575)

Variable	Coeff.	Std. Err.	*z*	P>|z|	95 % CI
AGE	0.117	0.0289	4.04	<0.001	0.060, 0.173
NDRGFP1	1.669	0.4072	4.10	<0.001	0.871, 2.467
NDRGFP2	0.434	0.1169	3.71	<0.001	0.205, 0.663
IVHX_2	−0.635	0.2987	−2.13	0.034	−1.220, −0.049
IVHX_3	−0.705	0.2616	−2.70	0.007	−1.217, −0.192
RACE	0.684	0.2641	2.59	0.010	0.166, 1.202
TREAT	0.435	0.2038	2.14	0.033	0.035, 0.834
SITE	0.516	0.2549	2.03	0.043	0.017, 1.016
AGE×NDRGFP1	−0.015	0.0060	−2.53	0.011	−0.027, −0.391
RACE×SITE	−1.429	0.5298	−2.70	0.007	−2.468, −0.391
Constant	−6.844	1.2193	−5.61	<0.001	−9.234, −4.454

As shown in Chapter 3 we obtain estimates of the odds ratios and their confidence intervals for dichotomous covariates (coded zero or one) and polychotomous covariates, with zero or one reference cell design variables, by exponentiating their respective coefficients and the end points of their respective confidence intervals. The odds ratios and confidence intervals for TREAT and IVHX obtained from the results in Table 5.10 are presented in Table 5.11.

In the first column of Table 5.11 we indicate the covariate and each of its levels. The reference level is the one with an odds ratio equal to 1.0. Some readers may question the need to include all levels, preferring instead to indicate the reference level by exclusion. Either approach is acceptable; however, we feel the explicit method shown in Table 5.11 is clearer and thus makes the discussion easier to follow.

The estimate of the odds ratio for treatment is 1.54. The correct interpretation is that the odds of remaining drug free for 12 months for a subject on the longer duration treatment is estimated to be 1.54 times larger than the odds for a similar (with respect to the other covariates in the model) subject on the shorter duration treatment. In many, if not most, subject matter journals this interpretation would be stated more concisely, but incorrectly, as subjects on the longer treatment are 1.54 times more likely to remain drug free for 12 months. The second interpretation relies on the "odds ratio approximates relative risk" argument. We go into this in more detail in Chapter 6 where we discuss case-control studies, but it is suffi-

Table 5.11 Estimated Odds Ratios and 95% Confidence Intervals for Treatment and History of IV Drug Use in the UIS (n = 575)

Variable	Value	Odds Ratio	95 % CI
Treatment			
	Short	1.00	
	Long	1.54	1.04, 2.30
IV Drug Use			
	Never	1.00	
	Previous	0.53	0.30, 0.95
	Recent	0.49	0.30, 0.83

cient at this point to indicate that this is only true when the outcome is "rare". As a rule of thumb, this argument is likely to be true when the outcome occurs less than 10 percent of the time. In our example this means that the logistic probability of remaining drug free should be small. This is not true since overall 25.6 percent of the subjects remained drug free for 12 months and the range of fitted values is from 0.02 to 0.78. In addition, 21 percent of subjects in the shorter duration treatment group remained drug free for 12 months. Zhang and Yu (1998) examine the extent to which the odds ratio over-estimates the relative risk when the outcome is not rare. Their results show that the over-estimation can be quite pronounced for odds ratios greater than 2.5 or less than 0.5. How important their results are in practice depends on how the estimated odds ratio is going to be used. For example, in our model there is a statistically significant (p = 0.033) benefit to the longer treatment and, since both interpretations of the estimated odds ratio provide a reasonable statement of this fact, either one could be used in a paper presenting the results of the study. On the other hand, if it is vitally important to have an accurate estimate of the increase in the likelihood of remaining drug free then one should present the odds ratio using the correct interpretation and attempt to correct the over-estimation. One can obtain a crude correction to the odds ratio from the figure in Zhang and Yu (1998). For example, an odds ratio of 1.5 with an "incidence among the unexposed" of 21 percent corrects to a relative risk of about 1.3. In our experience the estimated odds ratios from the vast majority of fitted logistic regression models are used to present "broad-strokes" estimates of effect and not precise estimates of the increase in the likelihood of the event. Thus, in the remainder of this section, we use the more concise "relative risk" type interpretation of the odds ratio.

The confidence interval estimate in Table 5.11 suggests that odds for the longer treatment could be as little as 1.04 or as much as 2.3 times as great as the odds for the shorter treatment.

The estimates of the odds ratios for history of IV drug use in Table 5.11 are both less than 1.0. In our experience new users of logistic regression modeling have a particularly difficult time in this case. If the odds ratio is less than one then the covariate is often referred to as "protective" for the outcome. This statement comes from the fact that in many examples the $y=1$ outcome is, in a practical sense, the less desirable of the two outcomes. If an odds ratio is less than one then the $y=1$ outcome is less likely to occur. In our example the situation is reversed since $y=1$ corresponds to the desirable outcome of remaining drug free. Consider a subject with a history of previous IV drug use, the estimate of the odds ratio in Table 5.11 is 0.53 with a 95 percent confidence interval of (0.30, 0.95). The interpretation is that a subject with a history of previous IV drug use is approximately 0.53 times as likely to remain drug free for 12 months as a similar subject with no history of previous drug use and the odds could be as much as 0.3 times or as little as 0.95 times smaller with 95 percent confidence. The odds ratio for recent IV drug use is 0.49 with a 95 percent confidence interval (0.30, 0.83). Thus a subject with a recent history of drug use is also about one-half as likely to remain drug free for 12 months as a subject with no history.

Before leaving our discussion of IVHX we note that the estimates of the coefficients in Table 5.10, and thus the odds ratios in Table 5.11, are quite similar for both previous and recent users. This suggests that we could consider modeling IVHX with a dichotomous covariate coded "never-ever". The principle of parsimony in modeling (i.e., use as few parameters as possible) suggests this might be a better model. We specifically kept the full three-level coding to demonstrate modeling methods with a multi-category covariate. We leave it as an exercise for the reader to examine whether we can pool the previous and recent categories and use the resulting dichotomous covariate.

We now turn to interpreting the results for RACE and SITE, which interact in our model. The covariate SITE is included to control for the location of the program and is of less importance to the subject matter team than possible racial differences. Thus we present in Table 5.12 the estimates of the odds ratios for non-white versus white within the two sites. The details of estimating odds ratios in the presence of this type of interaction are discussed in Section 3.7 and are not repeated here. However, we

Table 5.12 Estimated Odds Ratios and 95% Confidence Intervals for Race within Site in the UIS (n = 575)

Site	Race	Odds Ratio	95 % CI
Site A			
	White	1.00	
	Other	1.98	1.18, 3.33
Site B			
	White	1.00	
	Other	0.47	0.19, 1.18

encourage the reader to verify that the results presented in Table 5.12 come from appropriately specified logit differences from Table 5.10.

The interpretation of the results in Table 5.12 are: (1) A non-white subject at Site A is almost 2 times more likely to remain drug free for 12 months than a similar white subject. The confidence interval suggests that the difference could be as little as 1.2 times more likely or as much as 3.3 times more likely. (2) At site B, a non-white subject is estimated to be 0.47 times as likely to remain drug free for 12 months than a similar white subject. However, the confidence interval for the odds ratio at Site B includes 1.0 so the difference between the two racial groups is not significant at this site. As mentioned, we are less interested in comparing sites within levels of race, but we leave this comparison as an exercise.

We next estimate odds ratios for age and the number of previous drug treatments. The model is quite complicated in these two covariates but the process is same as that used to estimate the odds ratios for race within site in Table 5.12. Namely, we specify the covariate of interest and the levels for the odds ratio and the value(s) of the interaction variable. For example we may be interested in calculating the odds ratio for a 5-year increase in age for subjects with 0, 1, 3 and 6 previous drug treatments holding all other covariates constant. We follow the basic steps of evaluating the two logits, taking their difference and exponentiating the result. Since we are holding IVHX, RACE, TREAT and SITE constant we do not include their respective coefficients and values in the expressions for the logits. Thus the abbreviated basic form of the estimated logit in AGE and fractional polynomial transformation of NDRGTX is:

$$\hat{g}(AGE, NDRGTX) = \hat{\beta}_0 + \hat{\beta}_1(AGE) + \hat{\beta}_2(NDRGFP1) + \hat{\beta}_3(NDRGFP2) + \hat{\beta}_4(AGE \times NDRGFP1) \quad (5.19)$$

and, at AGE + 5, it is

$$\hat{g}(AGE+5, NDRGTX) = \hat{\beta}_0 + \hat{\beta}_1(AGE+5) + \hat{\beta}_2(NDRGFP1) + \hat{\beta}_3(NDRGFP2) + \hat{\beta}_4((AGE+5) \times NDRGFP1), \quad (5.20)$$

where $\hat{\beta}_1, \hat{\beta}_2, \hat{\beta}_3$, and $\hat{\beta}_4$ represent the respective estimated coefficients in Table 5.10. Recall that the two fractional polynomial covariates are

$$NDRGFP1 = \left(\frac{NDRGTX+1}{10}\right)^{-1}$$

and

$$NDRGFP2 = NDRGFP1 \times \ln\left(\frac{NDRGTX+1}{10}\right).$$

It follows that the estimated logit difference is

$$\hat{g}(AGE+5, NDRGTX) - \hat{g}(AGE, NDRGTX) = \hat{\beta}_1 \times 5 + \hat{\beta}_4 \times 5 \times NDRGFP1. \quad (5.21)$$

The two logits may be complicated but their difference is a simple linear function of NDRGFP1 and the change in age. We estimate the odds ratio by evaluating (5.21) at values of NDRGFP1 at the specified values of NDRGTX and exponentiating the results. The estimator of the variance of the logit difference in (5.21) is

$$\widehat{\text{Var}}\left(\hat{\beta}_1 \times 5 + \hat{\beta}_4 \times 5 \times NDRGFP1\right) = 5^2\widehat{\text{Var}}\left(\hat{\beta}_1\right) + (5 \times NDRGFP1)^2\widehat{\text{Var}}\left(\hat{\beta}_4\right) + 2 \times 5 \times 5 \times \text{NDRGFP1} \times \widehat{\text{Cov}}\left(\hat{\beta}_1, \hat{\beta}_4\right). \quad (5.22)$$

To simplify the notation, we denote the square root of the quantity from (5.22) as

$$\widehat{\text{SE}}(AGE+5, AGE) = \left\{\widehat{\text{Var}}\left(\hat{\beta}_1 \times 5 + \hat{\beta}_4 \times 5 \times NDRGFP1\right)\right\}^{0.5}. \quad (5.23)$$

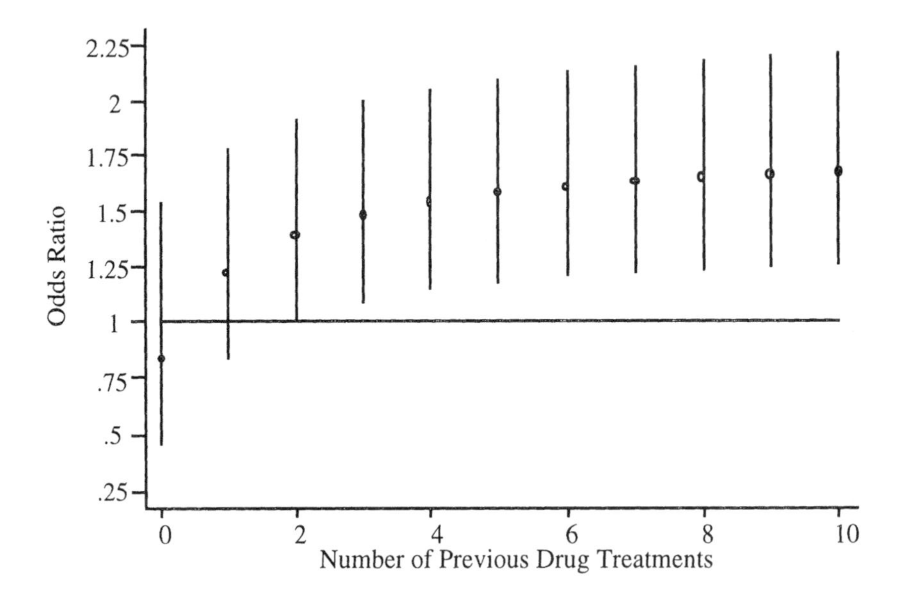

Figure 5.9 Estimated odds ratio and 95 percent confidence limits for a five-year increase in age based on the model in Table 5.10.

The endpoints of the $100\times(1-\alpha)\%$ confidence interval estimator for the logit difference are

$$\left[\hat{\beta}_1\times 5+\hat{\beta}_4\times 5\times NDRGFP1\right]\pm z_{1-\alpha/2}\widehat{\text{SE}}(AGE+5,AGE)\ . \qquad (5.24)$$

We obtain the endpoints of the $100\times(1-\alpha)\%$ confidence interval estimator for the odds ratio by exponentiating the endpoints from (5.24). At this point we have the option of presenting a table of estimated odds ratios and confidence intervals at various values of NDRGTX, or we can present the results graphically over a range of values. In this case we feel a graphical presentation is more effective than a table of results.

Figure 5.9 presents point and interval estimates of the odds ratio for a five-year increase in AGE for up to 10 previous drug treatments. The point-wise 95 percent limits are indicated by the vertical bars. The graph indicates that an older (by 5 years) subject with two or more previous drug treatments is significantly and progressively more likely to remain drug free than a younger subject. The estimates of the odds ratio gradually increase to about 1.7 at 10 previous treatments. (*Note*: The lower bound of the confidence interval estimator at two drug treatments is 1.01.) Since the confidence intervals include 1.0 for fewer than two previous drug treatments we conclude that a five-year increase in age is not significant at these levels. Only slight additional increases in the odds ratios for a five-year AGE increase are seen at more than 10 previous drug treatments and, as such, we chose not to include them in the graph. For example, at 10 previous treatments the estimate of the odds ratio in Figure 5.9 is 1.67 and the 95 percent confidence interval is (1.26, 2.21). At the maximum of 40 previous treatments we obtain, by the same methods, an estimate of the odds ratio of 1.76 with a 95 percent confidence interval (1.32, 2.33). We feel that the graph presented in Figure 5.9 portrays the effect of older age at the values of previous drug treatments much more clearly than is possible in a tabulation of this same information.

To complete the presentation of the results we need to describe the effect of the number of previous drug treatments controlling for age. Since the model is non-linear in the number of previous drug treatments the effect must be described at each value. We begin by assessing the effect of an increase of one previous drug treatment. The abbreviated logit at an increase of one drug treatment from (5.19) is

$$\begin{aligned}\hat{g}(AGE, NDRGTX+1) = {} & \hat{\beta}_0 + \hat{\beta}_1(AGE) + \hat{\beta}_2(NDRGFP11) \\ & + \hat{\beta}_3(NDRGFP21) + \hat{\beta}_4((AGE)\times NDRGFP11), \quad (5.25)\end{aligned}$$

where

$$NDRGFP11 = \left(\frac{(NDRGTX+1)+1}{10}\right)^{-1} = \left(\frac{NDRGTX+2}{10}\right)^{-1}$$

and

$$NDRGFP21 = NDRGFP11 \times \ln\left(\frac{NDRGTX+2}{10}\right).$$

The logit difference of (5.25) and (5.19) is

$$\begin{aligned}\hat{g}(AGE, NDRGTX+1) - \hat{g}(AGE, NDRGTX) = {} & \\ & \hat{\beta}_2(NDRGFP11 - NDRGFP1) \\ & + \hat{\beta}_3(NDRGFP21 - NDRGFP2) \\ & + \hat{\beta}_4(AGE)\times(NDRGFP11 - NDRGFP1)\,. \quad (5.26)\end{aligned}$$

In order to simplify the expressions we let

$$A = NDRGFP11 - NDRGFP1 = \left(\frac{NDRGTX+2}{10}\right)^{-1} - \left(\frac{NDRGTX+1}{10}\right)^{-1}$$

and

$$B = NDRGFP21 - NDRGFP2 = \left\{NDRGFP11 \times \ln\left(\frac{NDRGTX+2}{10}\right)\right\} - \left\{NDRGFP1 \times \ln\left(\frac{NDRGTX+1}{10}\right)\right\}.$$

It follows that the logit difference in (5.26) is

$$\hat{g}(AGE, NDRGTX+1) - \hat{g}(AGE, NDRGTX) = \hat{\beta}_2 A + \hat{\beta}_3 B + \hat{\beta}_4 AGE \times A. \qquad (5.27)$$

We obtain estimates of odds ratios by evaluating (5.27) at particular values of AGE and NDRGTX. The estimates depend, in a fairly complex and not easily envisioned manner, on both AGE and NDRGTX. Thus we choose a graphical presentation but, before presenting the graph, we describe how to obtain the confidence interval estimator.

The estimator of the variance of the logit difference in (5.27) is

$$\begin{aligned}\widehat{\text{Var}}\left[\hat{\beta}_2 A + \hat{\beta}_3 B + \hat{\beta}_4 AGE \times A\right] = \; & A^2 \widehat{\text{Var}}\left(\hat{\beta}_2\right) + B^2 \widehat{\text{Var}}\left(\hat{\beta}_3\right) \\ & + (AGE \times A)^2 \widehat{\text{Var}}\left(\hat{\beta}_4\right) + 2AB\widehat{\text{Cov}}\left(\hat{\beta}_2, \hat{\beta}_3\right) \\ & + 2A(AGE \times A)\widehat{\text{Cov}}\left(\hat{\beta}_2, \hat{\beta}_4\right) \\ & + 2B(AGE \times A)\widehat{\text{Cov}}\left(\hat{\beta}_3, \hat{\beta}_4\right),\end{aligned}$$

and the estimator of the standard error is

$$\widehat{\text{SE}}(NDRGTX+1, NDRGTX) = \left\{\widehat{\text{Var}}\left[\hat{\beta}_2 A + \hat{\beta}_3 B + \hat{\beta}_4 AGE \times A\right]\right\}^{0.5}.$$

The endpoints of the $100(1-\alpha)\%$ confidence interval estimator are

$$\left[\hat{g}(AGE, NDRGTX+1) - \hat{g}(AGE, NDRGTX)\right] \\ \pm z_{1-\alpha/2}\widehat{\text{SE}}(NDRGTX+1, NDRGTX)\,. \qquad (5.28)$$

We obtain the end points for the confidence interval estimator for the odds ratio by exponentiating the end points in (5.28).

Due to the complicated nature of the relationship between age and the number of previous drug treatments we feel that a graph is the best way to see the effect of increased numbers of drug treatments. Figure 5.10 presents the estimated odds ratios and associated 95 percent confidence intervals for an increase of one previous drug treatment at ages 20, 25, 30 and 35 (top left to bottom right in the figure). A horizontal line is drawn at 1.0 in each plot to provide a quick way to see whether the odds ratio is significantly different from 1.0.

In each plot the odds ratio at NDRGTX = 0 is for one treatment versus zero treatments. The confidence interval covers 1.0 in all but the plot for AGE = 35. In this case a 35 year old subject with one previous treatment is significantly more likely to remain drug free for 12 months than a

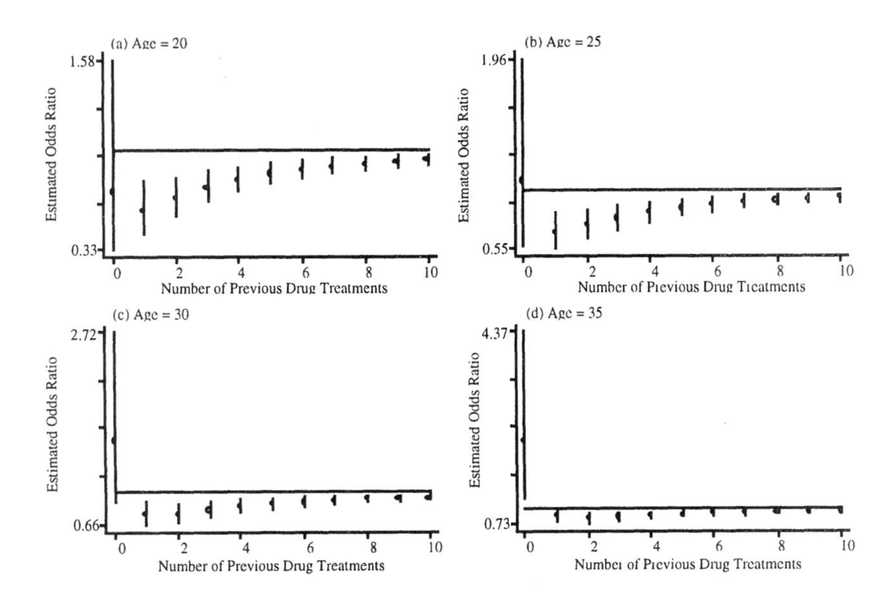

Figure 5.10 Estimated odds ratios and 95 percent confidence limits for an increase of one drug treatment from the plotted value of NDRGTX for a subject of age (a) 20, (b) 25, (c) 30 and (d) 35.

35 year old with no previous drug treatments. The remaining odds ratios in each plot compare subjects of a given age having 2 versus 1, 3 versus 2, up to 11 versus 10 previous treatments. In each plot the lines denoting the confidence intervals do not contain 1.0, hence the odds ratios are significantly less than one. The odds ratios approach 1.0 as the referent number of drug treatments increases. The overall interpretation is that with each increment of one drug treatment subjects are significantly more likely to remain drug free for 12 months. The increment in risk decreases as the referent number of drug treatments increases. That is, the risk of 11 versus 10 (a 10 percent increase) is not as great as the risk of 2 versus 1 (a 50 percent increase). In three of the four plots the smallest odds ratio for remaining drug free is for 2 versus 1 previous treatment. Thus the odds of remaining drug free is greatest for subjects with one previous treatment. This suggests that we calculate odds ratios comparing each number of previous drug treatments to a referent value of 1 treatment.

The individual logit difference and its confidence interval for the difference in the number of previous drug treatments at a particular AGE are obtained from (5.27) and (5.28) by defining A and B to be the difference in the fractional polynomial variables for a general value of NDRGTX and NDRGTX = 1 as follows:

$$\begin{aligned} A &= NDRGFP1 - \left(\frac{1+1}{10}\right)^{-1} \\ &= NDRGFP1 - 5 \end{aligned}$$

and

$$\begin{aligned} B &= NDRGFP2 - \left(\frac{1+1}{10}\right)^{-1} \ln\left(\frac{1+1}{10}\right) \\ &= NDRGFP2 - 5\ln\left(\frac{1}{5}\right) \\ &= NDRGFP2 + 5\ln(5)\,. \end{aligned}$$

We obtain the odds ratios and their confidence intervals by exponentiating the results for the logit differences from (5.27) and (5.28) with A and B defined above. These results are shown in Figure 5.11.

The odds ratios and confidence intervals plotted at 0 in the four plots in Figure 5.11 are the inverse of the odds ratios plotted in Figure 5.10. The odds ratios plotted at 2 are the same as those plotted at 1 in Figure 5.10. The others may be shown to be products of the odds ratios in Figure 5.10.

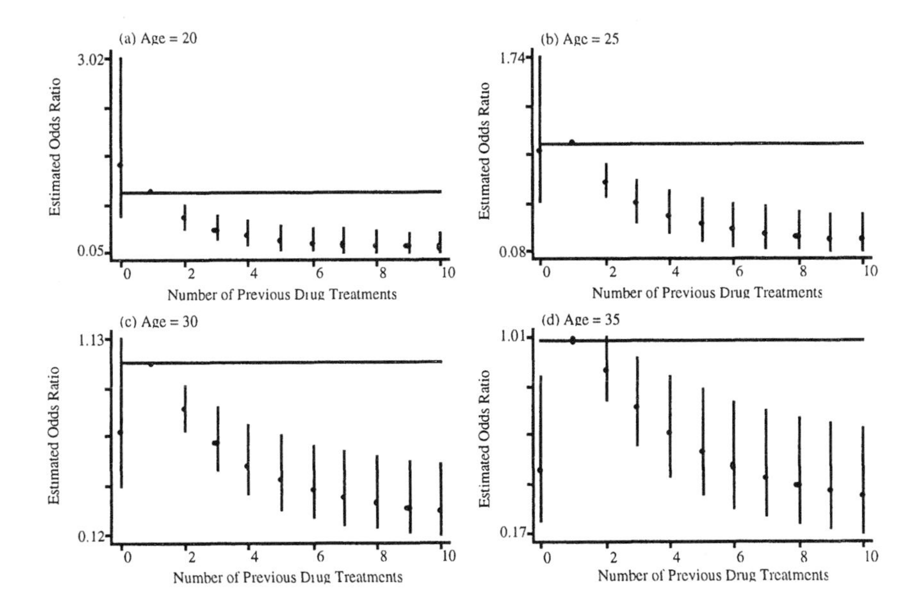

Figure 5.11 Estimated odds ratios and 95 percent confidence limits comparing zero, two, three up to 10 previous drug treatments to one previous treatment for a subject of age (a) 20, (b) 25, (c) 30 and (d) 35.

The general picture that emerges from Figure 5.11 is that at any age subjects with progressively more previous drug treatments are significantly less likely to remain drug free for 12 months when compared to a subject with one previous treatment. For older subjects the odds ratio at 0 is also less than one and is significant at age 35.

In general we conclude that, at most ages, a subject with one previous drug treatment is most likely to remain drug free for 12 months, that the odds ratio becomes progressively smaller (further from 1.0) with increasing numbers of previous drug treatments. The figures give an impression of approaching an age-specific lower bound.

Before we leave this section we make a few comments on model building. Comparatively speaking the model in Table 5.10 is more complicated than virtually every logistic regression model we have encountered in the health sciences literature. The typical model in the literature has a few continuous covariates modeled linearly, a few design variables and, in rare instances, an interaction. There seems to be some reluctance on the part of subject matter scientists to consider more complicated models. We think the reason is a lack of confidence in being able to determine when more complicated non-linear terms are needed and, if they are included, insecurity on how to interpret the results. What we hope we have accomplished in Chapters 4 and 5 is to provide a set of methods that can serve a basic paradigm for model building, model evaluation and model presentation that will allow the reader to feel confident that he/she has developed the best possible model within the constraints of the data and to feel secure in his/her ability to interpret the results of the model, regardless of how complicated it may appear to be. In particular, we hope that through the discussion in this section the reader has developed a firm grasp of the fundamental principal that the estimate of an odds ratio comes from exponentiating a logit difference.

EXERCISES

1. As is the case in linear regression, effective use of diagnostic statistics depends on our ability to interpret and understand the values of the statistics. The purpose of this problem is to provide a few structured examples to examine the effect on the fitted logistic regression model and diagnostic statistics when data are moved away from the model (i.e., poorer fit), and also toward the model (i.e., better fit). Table 5.13 lists values of the independent variable, x, and seven different columns of the outcome variable, y, labeled "Model." All models fit in this problem use the given values of x for the covariate. Different models are fit using the seven different columns for the outcome variable. The data for the column labeled "Model 0" are constructed to represent a "typical" realization when the logistic regression model is correct. In the columns labeled " Model 1" to "Model 3" we have changed some of the y values away from the original model. Namely some cases with small values of x have had y changed from 0 to 1 and others with large values of x have had the y values changed from 1 to 0. For models labeled "Model −1" and "Model −2" we have moved the y values in the direction of the model. That is, we have changed y from 1 to 0 for some small values of x and have changed y from 0 to 1 for some large values of x. Fit the six logistic regression models for the data in columns "Model −2" to "Model 3." Compute for each fitted model the values of the leverage, h, the change in chi-square, ΔX^2, and the influence diagnostic, $\Delta\hat{\beta}$. Plot each of these versus the fitted values, predicted logistic probabilities. Compare the plots over the various models. Do the statistics pick out poorly fit and influential cases? How do

Table 5.13 Hypothetical Data to Illustrate the Use of Diagnostic Statistics to Detect Poorly Fit and Influential Subjects and Complete Separation

	Model						
x	–i	–2	–1	0	1	2	3
–5.65	0	0	0	0	0	0	0
–4.75	0	0	0	0	0	1	1
–3.89	0	0	0	0	0	0	0
–3.12	0	0	0	0	0	0	0
–2.93	0	0	0	0	0	0	0
–2.87	0	0	0	0	0	0	0
–1.85	0	0	0	0	1	1	1
–1.25	0	1	1	1	1	1	1
–0.97	0	0	0	0	0	0	0
–0.19	1	1	1	1	1	1	1
–0.15	1	1	1	1	1	1	1
0.69	1	1	1	1	1	1	1
1.07	1	1	1	1	1	1	1
1.18	1	1	1	1	1	1	1
1.45	1	1	0	0	0	0	0
2.33	1	1	1	0	0	0	0
3.57	1	1	1	1	1	1	1
4.41	1	1	1	1	1	1	1
4.57	1	1	1	1	1	1	0
5.85	1	1	1	1	1	1	1

the estimated coefficients change relative to Model 0? Fit "Model –i." What happens and why? Refer to the discussion in Section 4.5 on complete separation.

2. In the exercises in Chapter 4, Problem 3, multivariable models for the ICU study were formed. Assess the fit of the model(s) that you feel was (were) best among those considered. This assessment should include an overall assessment of fit and use of the diagnostic statistics. Does the model fit? Are there any particular subjects, or covariate patterns, which seem to be poorly fit or overly influential? If so, how would you propose to deal with them?

3. Repeat Exercise 2 for the models developed for the low birth weight data and the prostate cancer data in Chapter 4, Exercise 4.

4. Fit the final model for the UIS shown in Table 5.10 and obtain the estimated covariance matrix of the estimated coefficients.

5. Use the method of logit differences to estimate the odds ratio for site A versus B within racial groups.

6. As a more complicated exercise in using the method of logit differences used in Section 5.5, estimate the odds ratio for a 25 year old white subject with 1 previous drug treatment, no history of IV drug use and on the longer treatment at site A compared to a 30 year old nonwhite subject with 5 previous drug treatments, a recent IV drug user and on the shorter treatment at site B. We make no claim that this is a clinically useful comparison. The purpose of the exercise is to illustrate the general applicability of the method of logit differences.

7. Obtain 95 percent confidence intervals for the estimates in Exercises 5 and 6.

8. We noted in Section 5.5 that an argument could be made for combining the previous and recent levels of IV drug use into one group yielding a dichotomous "never-ever" covariate. Evaluate the model using this covariate in place of history of IV drug use coded at three levels. Is the dichotomous covariate significant? Do the coefficients for the other covariates in the model change? How does the fit of the model compare with the fit of the original model (Table 5.10)?

CHAPTER 6

Application of Logistic Regression with Different Sampling Models

6.1 INTRODUCTION

Up to this point we have assumed that our data have come from a simple random sample. Considerable progress has been made in recent years to extend the use of the logistic regression model to other types of sampling. In this chapter we begin with a review of the classic cohort study. Next we consider the case-control study and the stratified case-control study. We conclude with a section that deals with fitting models when data come from a complex sample survey. The goals are to briefly describe some of the mathematics involved in fitting the model, to indicate how the model can be fit using available software and to discuss the interpretation of the estimated parameters. References to the literature for more detailed treatment of these topics are provided.

Throughout this chapter we assume that the outcome variable is dichotomous, coded as 0 or 1, and that its conditional probability given a vector of covariates is the logistic regression model. In addition, we assume that the number of covariate patterns is equal to the sample size. Modifications to allow for replication at covariate patterns are a notational detail, not a conceptual problem.

6.2 COHORT STUDIES

Several variations of the cohort (or prospective) study are in common use. In the simplest design, a simple random sample of subjects is chosen and the values of the covariates are determined. These subjects are then followed for a fixed period of time and the outcome variable is measured. This type of sample is identical to what is often referred to as the *regression sampling model*, in which we assume that the values of the covariates are fixed and measured without error and the outcome is measured conditionally on the observed values of the covariates. Under these assumptions and independence of the observations, the likelihood function for a sample of size n is

$$L_1(\boldsymbol{\beta}) = \prod_{i=1}^{n} \mathrm{P}(Y_i = y_i | \mathbf{x}_i). \tag{6.1}$$

When the observed values of y and the logistic regression model are substituted into the expression for the conditional probability, $L_1(\boldsymbol{\beta})$ simplifies to the likelihood function in equation (1.3).

A modification of this situation is a randomized trial where subjects are first chosen via a simple random sample and then allocated independently and with known probabilities into "treatment" groups. Subjects are followed over time and the outcome variable is measured for each subject. If the responses are such that a normal errors model is appropriate we would be naturally led to consider a normal theory analysis of covariance model which would contain appropriate design variables for treatment, relevant covariates, and any interactions between treatment and covariates deemed necessary. The extension of the likelihood function in equation (6.1) to incorporate treatment and covariate information when the outcome is dichotomous is obtained by including these variables in the logistic regression model.

Another modification is for the design to incorporate a stratification variable such as location or clinic. In this situation the likelihood function is the product of the stratum-specific likelihood functions, each of which is similar in form to $L_1(\boldsymbol{\beta})$. We would perhaps add terms to the model to account for stratum-specific responses. These might include a design variable for stratum and interactions between this design variable and other covariates.

In each of these designs we use the likelihood function $L_1(\boldsymbol{\beta})$ as a basis for determining the maximum likelihood estimates of the unknown parameters in the vector $\boldsymbol{\beta}$. Tests and confidence intervals for the parameters follow from well-developed theory for maximum likelihood estimation [see Cox and Hinkley (1974)]. The estimated parameters may be used in the logistic regression model to estimate the conditional probability of response for each subject. The fact that the estimated logistic probability provides a model-based estimate of the probability of response permits the development of methods for assessment of goodness-of-fit such as those discussed in Chapter 5. Chambless and Boyle (1985) have extended $L_1(\boldsymbol{\beta})$ to the case where the data come from a stratified simple random sample.

In some prospective studies the outcome variable of interest is the time to the occurrence of some event. In these studies the time to event is now frequently modeled using the proportional hazards model [see Hosmer and Lemeshow (1999)]. In these situations a method of analysis which is sometimes used is to ignore the actual failure time and model the occurrence or nonoccurrence of the event via logistic regression. This method of analysis became a popular way to analyze time to event data when easily used logistic regression software became available in the major software packages. However, now that software to fit the Cox or proportional hazards model is just as available and just as easy to use, we no longer recommend that logistic regression analysis be used to approximate a time to event analysis.

6.3 CASE-CONTROL STUDIES

One of the major reasons the logistic regression model has seen such wide use, especially in epidemiologic research, is the ease of obtaining adjusted odds ratios from the estimated slope coefficients when sampling is performed conditional on the outcome variables, as in a case-control study. Breslow (1996) has written an excellent review paper. Besides tracing the development of the case-control study he describes the statistical issues and controversies surrounding some famous studies such as the first Surgeon General's report on smoking and health (Surgeon General (1964)). He presents some of the newer innovative applications involving nesting and matching as well as some of the current limitations of this study design. We encourage any reader not familiar with this powerful and frequently employed study design to read this paper. We only consider the use of logistic regression in the simplest case-control designs in this section. More advanced applications may be found in Breslow (1996) and cited references.

As noted by Breslow (1996), Cornfield (1951) is generally given credit for first observing that the odds ratio is invariant under study design (cohort or case-control). However, it was not until the work of Farewell (1979) and Prentice and Pyke (1979) that the mathematical details justifying the common practice of analyzing case-control data as if they were cohort data were worked out.

In contrast to cohort studies, the binary outcome variable in a case-control study is fixed by stratification. The dependent variables in this setting are one or more primary covariates, exposure variables in $\mathbf{x}$. In this type of study design, samples of fixed size are chosen from the two strata defined by the outcome variable. The values of the primary exposure variables and the relevant covariates are then measured for each subject selected. The covariates are assumed to include all relevant exposure, confounding, and interaction terms. The likelihood function is the product of the stratum-specific likelihood functions depend on the probability that the subject was selected for the sample, and the probability distribution of the covariates.

It is not difficult algebraically to manipulate the case-control likelihood function to obtain a logistic regression model in which the dependent variable is the outcome variable of interest to the investigator. The key steps in this development are two applications of Bayes theorem. Since the likelihood function is based on subjects selected, we need to define a variable that records the selection status for each subject in the population. Let the variable s denote the selection ($s = 1$) or non-selection ($s = 0$) of a subject. The full likelihood for a sample of size n_1 cases ($y = 1$) and n_0 controls ($y = 0$) is

$$\prod_{i=1}^{n_1} \mathrm{P}(\mathbf{x}_i | y_i = 1, s_i = 1) \prod_{i=1}^{n_0} \mathrm{P}(\mathbf{x}_i | y_i = 0, s_i = 1). \tag{6.2}$$

For an individual term in the likelihood function shown in equation (6.2) the first application of Bayes theorem yields

$$\mathrm{P}(\mathbf{x} | y, s = 1) = \frac{\mathrm{P}(y | \mathbf{x}, s = 1)\ \mathrm{P}(\mathbf{x} | s = 1)}{\mathrm{P}(y | s = 1)}. \tag{6.3}$$

The second application of Bayes theorem is to the first term in the numerator of equation (6.3). This yields, when $y = 1$,

$$\mathrm{P}(y = 1 | \mathbf{x}, s = 1) = \frac{\mathrm{P}(y = 1 | \mathbf{x})\ \mathrm{P}(s = 1 | \mathbf{x}, y = 1)}{\mathrm{P}(y = 0 | \mathbf{x})\ \mathrm{P}(s = 1 | \mathbf{x}, y = 0) + \mathrm{P}(y = 1 | \mathbf{x})\ \mathrm{P}(s = 1 | \mathbf{x}, y = 1)}. \tag{6.4}$$

Assume that the selection of cases and controls is independent of the covariates with respective probabilities τ_1 and τ_0; then

$$\tau_1 = P(s=1|y=1,\mathbf{x}) = P(s=1|y=1),$$

and

$$\tau_0 = P(s=1|y=0,\mathbf{x}) = P(s=1|y=0).$$

Substitution of τ_1, τ_0 and the logistic regression model, $\pi(\mathbf{x})$, for $P(y=1|\mathbf{x})$, into equation (6.4) yields

$$P(y=1|\mathbf{x},s=1) = \frac{\tau_1\pi(\mathbf{x})}{\tau_0[1-\pi(\mathbf{x})]+\tau_1\pi(\mathbf{x})}. \tag{6.5}$$

If we divide the numerator and denominator of the expression on the right side of equation (6.5) by $\tau_0[1-\pi(\mathbf{x})]$, the result is a logistic regression model with intercept term $\beta_0^* = \ln(\tau_1/\tau_0)+\beta_0$. To simplify the notation, let $\pi^*(\mathbf{x})$ denote the right side of equation (6.5). Since we assume that sampling is carried out independent of covariate values, $P(\mathbf{x}|s=1) = P(\mathbf{x})$, where $P(\mathbf{x})$ denotes the probability distribution of the covariates. The general term in the likelihood shown in equation (6.3) then becomes, for $y=1$,

$$P(\mathbf{x}|y=1,s=1) = \frac{\pi^*(\mathbf{x})P(\mathbf{x})}{P(y=1|s=1)}. \tag{6.6}$$

A similar term for $y = 0$ is obtained by replacing $\pi^*(\mathbf{x})$ by $[1-\pi^*(\mathbf{x})]$ in the numerator and $P(y=1|s=1)$ by $P(y=0|s=1)$ in the denominator of equation (6.6). If we let

$$L^*(\boldsymbol{\beta}) = \prod_{i=1}^{n} \pi^*(\mathbf{x}_i)^{y_i}[1-\pi^*(\mathbf{x}_i)]^{1-y_i},$$

the likelihood function shown in equation (6.2) becomes

$$L^*(\boldsymbol{\beta})\prod_{i=1}^{n}\left[\frac{P(\mathbf{x}_i)}{P(y_i|s_i=1)}\right]. \tag{6.7}$$

The first term in equation (6.7), $L^*(\boldsymbol{\beta})$, is the likelihood obtained when we pretend the case-control data were collected in a cohort study, with the outcome of interest modeled as the dependent variable. If we assume that the probability distribution of $\mathbf{x}$, $P(\mathbf{x})$, contains no information about the coefficients in the logistic regression model, then maximization of the full likelihood with respect to the parameters in the logistic model, $\pi^*(\mathbf{x})$, is only subject to the restriction that $\mathrm{P}(y_i = 1|s_i = 1) = n_1/n$ and $\mathrm{P}(y_i = 0|s_i = 1) = n_0/n$. The likelihood equation obtained by differentiating with respect to the parameter β_0^* assures that this condition is satisfied. Thus, maximization of the full likelihood with respect to the parameters in $\pi^*(\mathbf{x})$ need only consider that portion of the likelihood which looks like a cohort study. The implication of this is that *analysis of data from case-control studies via logistic regression may proceed in the same way and using the same computer programs as cohort studies.* Nevertheless, inferences about the intercept parameter β_0 are not possible without knowledge of the sampling fractions within cases and controls, τ_0 and τ_1.

The assumption that the marginal distribution of $\mathbf{x}$ contains no information about the parameters in the logistic regression model requires additional discussion, as it is not true in one historically important situation, the normal theory discriminant function model. This model was discussed briefly in Chapters 1 and 2. When the assumptions for the normal discriminant function model hold, the maximum likelihood estimators of the coefficients for the logistic regression model obtained from conditional likelihoods such as those in equations (6.2) and (6.7) are less efficient than the discriminant function estimator shown in equation (2.11) [see Efron (1975)]. However, the assumptions for the normal theory discriminant function model are rarely, if ever, attained in practice. Application of the normal discriminant function when its assumptions do not hold may result in substantial bias, especially when some of the covariates are dichotomous variables. As a general rule, estimation should be based on equations (6.2) and (6.7), unless there is considerable evidence in favor of the normal theory discriminant function model.

Prentice and Pyke (1979) have shown that the maximum likelihood estimators obtained by pretending that the case-control data resulted from a cohort sample have the usual properties associated with maximum likelihood estimators. Specifically, they are asymptotically normally distributed, with covariance matrix obtained from the inverse of the information matrix. Thus, percentiles from the $N(0,1)$ distribution may be used in conjunction with estimated standard errors produced from standard logistic regression software to form Wald statistics and confidence interval estimates. The theory of likelihood ratio tests may be employed to compare models via the difference in the deviance of the two models, assuming of course that the models are nested. Scott and Wild (1986) have shown that inferences based on this approach are sensitive to incorrect specifications of the logit function. They show that failure to include necessary higher order terms in the logit produces a model with estimated standard errors that are too small. These results are special cases of more general results obtained by White (1982).

Modification of the likelihood function to incorporate additional levels of stratification beyond case-control status follows in the same manner as described for cohort data (i.e., inclusion of relevant design variables and interaction terms). Thus, model building and inferences from fitted models for case-control data may proceed using the methods developed for cohort data, as described in Chapters 4 and 5. However, this approach is not valid for matched or highly stratified data. Appropriate methods for the analysis of the latter are presented in detail in Chapter 7.

Fears and Brown (1986) proposed a method for the analysis of stratified case-control data that arise from a two-stage sample. Breslow and Cain (1988) and Scott and Wild (1991) provide further discussion and refinement of the method. This approach requires that we know the sampling rates for the first stage and the total number of subjects in each stratum. This information is used to define the relative sampling rates for cases and controls within each stratum. The ratio of these is included in the model in the form of an additional known constant added to the stratum-specific logit. Specifically, suppose we let n_j be the total number of subjects with $y=j$ observed out of a possible N_j and let the kth stratum-specific quantities be n_{jk} and N_{jk}, $j=0,1$ and $k=1,2,\ldots K$. The relative stratum-specific sampling rates are $w_{1k}=(n_{1k}/N_{1k})/(n_1/N_1)$ and $w_{0k}=(n_{0k}/N_{0k})/(n_0/N_0)$. The Fears and Brown model uses stratum-specific logits of

$$g_k(\mathbf{x})=\ln(w_{1k}/w_{0k})+\beta_0+\boldsymbol{\beta}'\mathbf{x},$$

$k=1,2,\ldots K$. This model may be handled with standard logistic regression software by defining a new variable, typically referred to as an offset, which takes on the value $\ln(w_{1k}/w_{0k})$ and forcing it into the model with a coefficient equal to 1.0.

Breslow and Cain (1988) show that the estimator proposed by Brown and Fears is asymptotically normally distributed and derive an estimator of the covariance matrix. Breslow and Zhao (1988) and Scott and Wild (1991) point out that the estimated standard errors produced when standard logistic regression software is used to implement the Brown and Fears method overestimates the true standard errors. They provide expressions for a covariance matrix that yields consistent estimates of the variances and covariances of the estimated regression coefficients. The matrix is complicated to compute, as it requires a special purpose program or a high degree of skill in using a package allowing matrix calculations such as SAS, STATA or S-Plus. For these reasons we do not present the variance estimator in detail. We note that Breslow and Zhao use a slightly different offset, $\ln\left[\left(n_{1k}/N_{1k}\right)/\left(n_{0k}/N_{0k}\right)\right]$, which yields the same estimates of the regression coefficients but a different intercept.

Before leaving our discussion of logistic regression in the case-control setting, we briefly consider the application of the chi-square goodness-of-fit tests for the logistic regression model presented in Section 5.2. The essential feature of these tests is that for a particular covariate pattern, the number of subjects with the response of interest among m sampled is distributed binomially with parameters m and response probability given by the hypothesized logistic regression model. Recall that for cohort data, the likelihood function was parameterized directly in terms of the logistic probability. For case-control data, the function $\pi^*(\mathbf{x})$ is the probability $P(y=1|\mathbf{x},s=1)$. For a particular covariate pattern, conditioning on the number of subjects m observed to have a given covariate pattern is equivalent to conditioning on the event, $(\mathbf{x},s=1)$. Thus, for case-control studies in which the logistic regression model assumption is correct, the conditional distribution of the number of subjects responding among the m observed to have a particular covariate pattern is binomial with parameters m and $\pi^*(\mathbf{x})$. Hence, the results developed in Chapter 5 based on m-asymptotics also apply.

It is often the case that data from case-control studies do not arise from simple random samples within each stratum. For example, the design may call for the inclusion of all subjects with $y=1$ and a sample of subjects with $y=0$. For these designs there is an obvious dependency among the observations. If this dependency is not too great, or if we appeal to a super-population model [see Prentice (1986)], then employing a theory that ignores it should not bias the results significantly.

6.4 FITTING LOGISTIC REGRESSION MODELS TO DATA FROM COMPLEX SAMPLE SURVEYS

Some of the more recent improvements in logistic regression statistical software include routines to perform analyses with data obtained from complex sample surveys. These routines may be found in STATA, SUDAAN (1997) and other less well-known special purpose packages. Our goal in this section is to provide a brief introduction to these methods and to illustrate them with an example data set. The reader who needs more detail is encouraged to see Korn and Graubard (1990), Roberts, Rao and Kumar (1987), Skinner, Holt and Smith (1989) and Thomas and Rao (1987).

The essential idea, as discussed in Roberts, Rao and Kumar (1987), is to set up a function that approximates the likelihood function in the finite sampled population with a likelihood function formed from the observed sample and known sampling weights. Suppose we assume that the population may be broken into $k=1,2,\ldots,K$ strata, $j=1,2,\ldots,M_k$ primary sampling units in each stratum and $i=1,2,\ldots,N_{kj}$ elements in the kjth primary sampling unit. Suppose our observed data consist of n_{kj} elements from m_k primary sampling units from stratum k. Denote the total number of observations as $n=\sum_{k=1}^{K}\sum_{j=1}^{m_k} n_{kj}$. Denote the known sampling weight for the kjith observation as w_{kji}, the vector of covariates as $\mathbf{x}_{kji}$ and the dichotomous outcome as y_{kji}. The approximate log-likelihood function is

$$\sum_{k=1}^{K}\sum_{j=1}^{m_k}\sum_{i=1}^{n_{kj}}\left[w_{kji}\times y_{kji}\right]\times\ln\left[\pi\left(\mathbf{x}_{kji}\right)\right]+\left[w_{kji}\times\left(1-y_{kji}\right)\right]\times\ln\left[1-\pi\left(\mathbf{x}_{kji}\right)\right]. \quad (6.8)$$

Differentiating this equation with respect to the unknown regression coefficients yields the vector of $p+1$ score equations

$$\mathbf{X}'\mathbf{W}(\mathbf{y}-\boldsymbol{\pi})=\mathbf{0}, \quad (6.9)$$

where $\mathbf{X}$ is the $n\times(p+1)$ matrix of covariate values, $\mathbf{W}$ is an $n\times n$ diagonal matrix containing the weights, $\mathbf{y}$ is the $n\times 1$ vector of observed outcomes and $\boldsymbol{\pi}=\left(\pi\left(\mathbf{x}_{111}\right),\ldots,\pi\left(\mathbf{x}_{Km_K n_{Kj}}\right)\right)'$ is the $n\times 1$ vector of logistic probabilities. In theory, any logistic regression package that allows weights could be used to obtain the solutions to equation (6.9). The problem comes in obtaining the correct estimator of the covariance matrix of the estimator of the coefficients. Naive use of a standard logistic regression package with weight matrix $\mathbf{W}$ would yield estimates on the matrix $(\mathbf{X'DX})^{-1}$ where $\mathbf{D}=\mathbf{WV}$ is an $n\times n$ diagonal matrix with general element $w_{kji}\times\hat{\pi}(\mathbf{x}_{kji})[1-\hat{\pi}(\mathbf{x}_{kji})]$. The correct estimator is

$$\hat{\text{Var}}(\hat{\boldsymbol{\beta}})=(\mathbf{X'DX})^{-1}\mathbf{S}(\mathbf{X'DX})^{-1}, \tag{6.10}$$

where $\mathbf{S}$ is a pooled within-stratum estimator of the covariance matrix of the left side of equation (6.9). Denote a general element in the vector in (6.9) as $\mathbf{z}'_{kji}=\mathbf{x}'_{kji}w_{kji}(y_{kji}-\pi(\mathbf{x}_{kji}))$, the sum over the n_{kj} sampled units in the jth primary sampling unit in the kth stratum as $\mathbf{z}_{kj}=\sum_{i=1}^{n_{kj}}\mathbf{z}_{kji}$ and their stratum-specific mean as $\bar{\mathbf{z}}_k=\frac{1}{m_k}\sum_{j=1}^{m_k}\mathbf{z}_{kj}$. The within-stratum estimator for the kth stratum is

$$\mathbf{S}_k=\frac{m_k}{m_k-1}\sum_{j=1}^{m_k}(\mathbf{z}_{kj}-\bar{\mathbf{z}}_k)(\mathbf{z}_{kj}-\bar{\mathbf{z}}_k)'.$$

The pooled estimator is $\mathbf{S}=\sum_{k=1}^{K}(1-f_k)\mathbf{S}_k$. The quantity $(1-f_k)$ is called the finite population correction factor where $f_k=m_k/M_k$ is the ratio of the number of observed primary sampling units to the total number of primary sampling units in stratum k. In settings where M_k is unknown it is common practice to assume it is large enough that f_k is quite small and the correction factor is equal to one.

The likelihood function in (6.8) is only an approximation to the true likelihood. Thus, we would expect that inferences about model parameters should be based on univariable and multivariable Wald statistics comparing estimated coefficients to an estimate of their variance computed from specific elements of (6.10) in the same manner as described in Chapter 2. However, simulations in Korn and Graubard (1990) as well as Thomas and Rao (1987) show that when data come from a complex sample survey from a finite population, use of a modified Wald statistic and the F-distribution (details to follow) yield tests with better adherence to the stated alpha-level. STATA and SUDAAN report results from these modified Wald tests. The problem is that none of the simulations referred to actually examines logistic regression models fit using continuous and categorical covariates with estimates obtained from (6.9) and variances from (6.10). Korn and Graubard appear to use a linear regression with normal errors model and refer to theoretical results in Anderson (1984) that depend on rather stringent assumptions of multivariate normality. Thomas and Rao examine models with a dichotomous or polychotomous outcome and a few categorical covariates. Another problem, in our opinion, is the fact that software pacakges, for example STATA, use the t-distribution to assess significance of Wald statistics for individual coefficients. Given the paucity of appropriate simulations and theory we are not convinced that there is sufficient evidence to support the use of the modified Wald statistic with the F-distribution with logistic regression models. One possible justification is that the use of the modified Wald statistic with the F-distribution is conservative in that significance levels using this approach are, in general, larger than those obtained from treating the Wald statistics as being multivariate normal for sufficiently large samples (as is assumed in previous chapters). We present results based on both tests in the example.

The relationship between the Wald test and the modified Wald test is as follows. Let W denote the Wald statistic for testing that all p slope coefficients in a fitted model are equal to zero, i.e.,

$$W = \hat{\boldsymbol{\beta}}' \left[\widehat{\text{Var}}\left(\hat{\boldsymbol{\beta}}\right)_{p \times p} \right]^{-1} \hat{\boldsymbol{\beta}}, \tag{6.11}$$

where $\hat{\boldsymbol{\beta}}$ denotes the vector of p slope coefficients and $\widehat{\text{Var}}\left(\hat{\boldsymbol{\beta}}\right)_{p \times p}$ is the $p \times p$ sub-matrix obtained from the full $(p+1) \times (p+1)$ matrix in equation (6.10). That is, one leaves out the row and column for the constant term. The p-value is computed using a chi-square distribution with p degrees of freedom as p-value $= \text{P}\left[\chi^2(p) \geq W\right]$.

The adjusted Wald statistic is

$$F = \frac{(s - p + 1)}{sp} W, \tag{6.12}$$

where $s = \left(\sum_{k=1}^{K} m_k\right) - K$ is the total number of sampled primary sampling units minus the number of strata. The p-value is computed using an F-distribution with p and $(s-p+1)$ degrees of freedom as p-value $= \text{P}\left[\text{F}(p, s-p+1) \geq F\right]$.

As an example of a study involving stratification, clustering and unique sampling weights, we used data from the National Health and Nutrition Examination Survey (NHANES III), conducted by the National Center for Health Statistics (NCHS) between 1988 and 1994. This was the third in a series of data collection programs carried out by the NCHS in order to obtain health and nutrition data on the population of the United States. (NHANES I took place between 1971-74, NHANES II between 1976-80.) Data were collected via physical examinations and clinical and laboratory tests. Prevalence data were collected for specific diseases and health conditions. In this survey a multistage probability sample of 39,695 subjects was selected representing more than 250 million people living in the United States. For purposes of this example we consider only adults 20 years of age or older. There were a total of 17,030 subjects in this subset who represented 177.2 million adults living in the U.S. at that time. In the NHANES III design, 49 pseudo-strata were created with 2 pseudo-psu's identified within each such stratum (see NCHS (1996)). These features must be adhered to in the appropriate analysis of the data.

A code sheet describing the variables used in this example is presented in Table 6.1. For this example, a dichotomous outcome variable, HBP, was generated representing whether the subject had high blood pressure (defined by an average systolic blood pressure PEPMNK1R > 140 mmHg).

It should be noted that the NHANES III, like just about any other large scale survey, suffers from the fact that complete data are not available for every subject. This problem is exacerbated in complex sample surveys since every subject carries along a unique statistical weight representing the number of individuals in the population he or she represents. Hence, if that subject is missing a measurement on just one of the variables involved in a multivariable problem, then that subject will be eliminated from the analysis and the sum of the statistical weights of the subjects remaining will not equal the size of the population for which inference is to be made.

This problem has been addressed extensively by survey statisticians and solutions to the problem range from redistributing the statistical weights of the dropped subjects among the subjects remaining to imputing every missing value so that the weights will be preserved. Another, perhaps simplistic, approach is simply to run the analyses with the

Table 6.1 Variables in the NHANES III Data Set

Variable	Description	Codes / Values	Name
1	Respondent Identification Number		SEQN
2	Pseudo-PSU	1,2	SDPPSU6
3	Pseudo-stratum	01 – 49	SDPSTRA6
4	Statistical weight	225.93 – 139744.9	WTPFHX6
5	Age	(in years)	HSAGEIR
6	Sex	0 = Female 1 = Male	HSSEX
7	Race	1 = White 2 = Black 3 = Other	DMARACER
8	Body Weight	(in pounds)	BMPWTLBS
9	Standing Height	(inches)	BMPHTIN
10	Average Systolic BP	(mm Hg)	PEPMNK1R
11	Average Diastolic BP	(mm Hg)	PEPMNK5R
12	Has respondent smoked > 100 cigarettes in life	1 = Yes 2 = No	HAR1
13	Does repondent smoke cigarettes now?	1 = Yes 2 = No	HAR3
14	Smoking	1 = if HAR1 = 2 2 = if HAR1=1 & HAR3=2 3 = if HAR1=1 & HAR3=1	SMOKE
15	Serum Cholesterol	mg/100ml	TCP
16	High Blood Pressure	0 if PEPMNK1R $\leq$ 140 1 if PEPMNK1R > 140	HBP

subjects having complete data and assume that the relationships would not change had all subjects been used. Because it is our intention in this book to demonstrate the use of logistic regression analysis with complex survey data rather than to obtain precise population parameter estimates, we will follow this simple approach.

A logistic regression model was fit containing HSAGEIR, HSSEX, two dummy variables for DMARACER, BMPWTLBS BMPHTIN and two dummy variables for SMOKE. The model was fit using STATA's svylogit command with dependent variable HBP, pweight=WTPFHX6, strata=SDPSTRA6 and psu=SDPPSU6. Note that in this first analysis complete data are available on only 16,963 of the original 17,030

Table 6.2 Estimated Coefficients, Standard Errors, *z*-Scores, Two-Tailed *p*-Values and 95% Confidence Intervals for a Logistic Regression Model for the NHANES III Study with Dependent Variable = HBP (n = 16,963)

Variable	Coeff.	Std. Err.	z	P > |z|	95 % CI
HSAGEIR	0.081	0.0025	32.49	< 0.001	0.076, 0.086
HSSEX	0.204	0.0755	2.70	0.009	0.052, 0.356
RACE2	0.558	0.0744	7.51	< 0.001	0.409, 0.708
RACE3	0.044	0.3005	0.15	0.885	−0.560, 0.647
BMPWTLBS	0.012	0.0008	13.90	< 0.001	0.010, 0.013
BMPHTIN	−0.059	0.0126	−4.70	< 0.001	−0.085, −0.034
SMK2	−0.076	0.0950	−0.81	0.425	−0.267, 0.114
SMK3	0.061	0.1051	0.58	0.564	−0.150, 0.272
CONSTANT	−4.257	0.8040	−5.30	< 0.001	−5.873, −2.641

subjects and these represent 176.9 million adults. The results are shown in Table 6.2.

We assess the overall significance of the model via the multivariable Wald test and adjusted Wald test for the significance of the eight regression coefficients in the model. For the model in Table 6.2 the value of the test in (6.11) is

$$W = \hat{\boldsymbol{\beta}}' \left[\widehat{\text{Var}}\left(\hat{\boldsymbol{\beta}}\right)_{8\times 8} \right]^{-1} \hat{\boldsymbol{\beta}} = 1806 ,$$

where $\hat{\boldsymbol{\beta}}$ is the vector of 8 estimated slope coefficients and $\widehat{\text{Var}}\left(\hat{\boldsymbol{\beta}}\right)_{8\times 8}$ is the 8×8 sub-matrix computed using (6.10). The significance level of the test is $\text{P}\left[\chi^2(8) \ge 1806\right] < 0.001$. The value of s for the adjusted Wald test is $98 - 49 = 49$ and the adjusted Wald test from (6.12) is

$$F = \frac{(49 - 8 + 1)}{49 \times 8} \times 1806 = 193.50$$

and $p = \text{P}\left[\text{F}(8, 42) \ge 193.50\right] < 0.001$. Both tests indicate that at least one of the coefficients may be different from zero.

The results in Table 6.2 indicate, on the basis of the individual p-values for the Wald statistics, that smoking may not be significant. As we noted, the function in equation (6.8) is not a true likelihood function. Thus, we cannot use the partial likelihood ratio test to compare a smaller model to a larger model. In this case we must test for the significance of the coefficients of excluded covariates using a multivariable Wald test based on the estimated coefficients and estimated covariance matrix from the larger model.

Application of the Wald test to assess the significance of the coefficient for smoking (SMK2 and SMK3) from the model in Table 6.2 uses the vector of estimated coefficients

$$\hat{\boldsymbol{\beta}}' = (-0.0764019,\ 0.0610105),$$

and the 2×2 sub-matrix of estimated variances and covariances obtained from the full 8×8 matrix, not shown, computed using (6.10)

$$\widehat{\text{Var}}\left(\hat{\boldsymbol{\beta}}\right)_{2\times2} = \begin{bmatrix} 0.009018 & 0.004660 \\ 0.004660 & 0.011036 \end{bmatrix}.$$

The Wald test statistic is

$$W = \hat{\boldsymbol{\beta}}'\left[\widehat{\text{Var}}\left(\hat{\boldsymbol{\beta}}\right)_{2\times2}\right]^{-1}\hat{\boldsymbol{\beta}} = 1.8177,$$

with a p-value obtained as $P\left[\chi^2(2) \geq 1.8177\right] = 0.403$. The adjusted Wald test is

$$F = \frac{(49-2+1)}{49\times2}\times1.8177 = 0.8903$$

and $p = P\left[F(2,48) \geq 0.8903\right] = 0.4172$. We note that the p-value for the adjusted Wald test is slightly larger than that of the Wald test; however, neither is significant. Thus, both tests indicate that we do not have sufficient evidence to conclude that the coefficients for SMOKE are significantly different from zero. We now fit the reduced model.

The results of fitting the model deleting SMK2 and SMK3 are shown in Table 6.3. The first thing we do is to compare the magnitude of the coefficients in Table 6.3 to those in Table 6.2 to check for confounding due to the excluded covariates. As can be seen there is virtually no difference in the two sets of coefficients suggesting that smoking

Table 6.3 Estimated Coefficients, Standard Errors, z-Scores, Two-Tailed p-Values and 95% Confidence Intervals for a Reduced Logistic Regression Model for the NHANES III Study with Dependent Variable = HBP (n = 16,964)

Variable	Coeff.	Std. Err.	z	P > |z|	95 % CI
HSAGEIR	0.080	0.0027	30.04	< 0.001	0.075, 0.085
HSSEX	0.194	0.0791	2.45	0.018	0.035, 0.353
RACE2	0.572	0.0710	8.05	< 0.001	0.429, 0.714
RACE3	0.052	0.3007	0.17	0.863	–0.552, 0.656
BMPWTLBS	0.011	0.0008	13.61	< 0.001	0.010, 0.013
BMPHTIN	–0.059	0.0127	–4.65	< 0.001	–0.084, –0.033
CONSTANT	–4.211	0.7940	–5.30	< 0.001	–5.807, –2.616

is not a confounder of the relationship between any of the remaining covariates and high blood pressure.

We note that the number of subjects with complete covariate data in the smaller model is 16,964. Since the increase is so small we are not going to worry about fitting the models, full and reduced, to different data. If the sample sizes were substantially different then we would first fit the smaller model on the smaller sample size to have more comparable results. Then we would fit the model on the larger data set to see what, if any, changes occur in estimates of the coefficients.

Following the guidelines we have established in previous chapters, at this point in the analysis we would:

- determine whether the continuous covariates in the model are linear in the logit.
- determine whether there are any significant interactions among the independent variables in the model.
- assess model calibration and discrimination through goodness-of-fit tests and area under the ROC curve.
- examine the case-wise diagnostic statistics to identify poorly fit and influential covariate patterns.

Unfortunately, none of these procedures is readily available when modeling data from complex sample surveys. Thomas and Rao (1987) consider chi-square goodness-of-fit tests and Roberts, Rao and Kumar (1987) extend the diagnostics discussed in Chapter 5 to the survey sampling setting. However, the tests and diagnostic statistics have not, as yet, been implemented into any of the commonly available packages. The computations required to obtain the tests, measures of leverage, h, and

Table 6.4 Estimated Coefficients, Standard Errors, z-Scores, Two-Tailed p-Values and 95% Confidence Intervals for a Logistic Regression Model Containing the Variables in Table 6.3 but Assuming the Data Come From a Simple Random Sample. Dependent Variable = HBP (n = 16,964)

Variable	Coeff.	Std. Err.	z	P > |z|	95 % CI
HSAGEIR	0.070	0.0014	49.86	< 0.001	0.067, 0.072
HSSEX	0.090	0.0613	1.48	0.140	−0.030, 0.211
RACE2	0.477	0.0509	9.36	< 0.001	0.377, 0.576
RACE3	0.092	0.1431	0.64	0.522	−0.189, 0.372
BMPWTLBS	0.008	0.0006	13.74	< 0.001	0.007, 0.010
BMPHTIN	−0.045	0.0085	−5.31	< 0.001	−0.062, −0.028
CONSTANT	−3.872	0.5293	−7.32	< 0.001	−4.909, −2.834

the contribution to fit, ΔX^2, are not trivial and require considerable skill in programming matrix calculations. In addition the version of Cook's distance is not an easily computed function of leverage and contribution to fit. Thus, at present, there is little in the way of model checking and fit assessment that can be done within packages like STATA and SUDAAN and we use the model in Table 6.3 as our final model.

Statistical analyses of survey data that take the survey design (stratification and clustering) and statistical weights into consideration are generally called "design-based." When such features are ignored and the data are handled as if they arose from a simple random sample, the resulting statistical analyses are termed "model-based." One approach that analysts have used when dealing with survey data is to estimate parameters using design-based methods but to use model-based methods to perform other functions. For example, in this analysis, determination of linearity of the logit for the continuous covariates in the model, assessment of model calibration and examination of diagnostic statistics could be carried out by treating the data as if they resulted from a simple random sample. Any discoveries made in those analyses would then be implemented in the final design-based analysis. For example, using fractional polynomial analysis (which, as currently implemented in STATA, does not take into account survey features) reveals that HSAGEIR is not linear in the logit and that the appropriate transformation is to include two terms, x and x^3. We also find that BMPWTLBS is not linear in the logit with recommended transformation $\ln(x)$. This knowledge, obtained from the model-based analysis may then be implemented into the more appropriate design-based analysis to obtain the slope coefficients and estimated odds ratios. Since no methods currently exist in standard software packages for assessing fit of logistic models derived from complex sample surveys, a similar strategy may be used to carry out goodness of fit testing. Coefficients from the design-based model are used to obtain probabilities of the response for each subject in the study. STATA facilitates this process by allowing the external coefficients (obtained from the design-based analysis) to be imported and used to compute probabilities of response for which goodness-of-fit may be assessed in a model-based environment assuming the data arose from a simple random sample. Exercises at the end of this chapter will allow readers to practice some of these methods.

To illustrate the fact that design-based and model-based procedures may result in different parameter estimates, we present in Table 6.4 the model corresponding to the one presented in Table 6.3, but using a model-based analysis that ignores the survey features of stratification, clustering and unique statistical weights. Although in this example both modeling approaches produce similar coefficients and associated p-values, this would not necessarily be true in general – especially if the sample sizes were somewhat smaller. It should also be noted that for *linear estimates* such as means, totals and proportions, design-based standard errors are typically much larger than model-based standard errors. In fact, for linear estimates, the design effect (defined as the ratio of the variance under design-based analysis to the variance under simple random sampling) is typically much larger than 1. This measure reflects the inflation in variance that occurs due to homogeneity within clusters and can be expressed as $1+(n-1)\rho_y$, where ρ_y is the intracluster correlation coefficient and n is the average number of units in the sampled cluster. These intracluster correlation coefficients can range from small negative values (when the data within clusters are highly heterogeneous) to unity (when the data in clusters are highly correlated). Only when the data are highly heterogeneous within clusters will the design effect be less than 1. However, as described by Neuhaus and Segal (1993), design effects for *regression coefficients* can be expressed as $1+(n-1)\rho_x\rho_y$. Note that in this expression the intracluster correlation coefficient for the independent variable is multiplied by the intracluster correlation coefficient for the dependent variable, both of which are by definition smaller than 1. As a result, the design effect will be smaller than that seen for means, totals, or proportions. We also note that since

Table 6.5 Estimated Odds Ratios and 95% Confidence Intervals for Variables in Table 6.3 Using "Design-Based" versus "Model Based" Analysis (n = 16,964)

	"Design-Based" Analysis"		"Model-Based" Analysis	
Variable	Odds Ratio	95% CI	Odds Ratio	95% CI
HSAGEIR	1.083	1.077, 1.089	1.072	1.069, 1.075
HSSEX	1.214	1.036, 1.423	1.095	0.971, 1.235
RACE2	1.771	1.536, 2.043	1.611	1.458, 1.780
RACE3	1.053	0.576, 1.928	1.096	0.828, 1.451
BMPWTLBS	1.012	1.010, 1.013	1.008	1.007, 1.010
BMPHTIN	0.943	0.919, 0.967	0.956	0.940, 0.972

ρ_x and ρ_y are not necessarily in the same direction, the product of the intracluster correlation coefficients could be negative and the resulting design effect could be smaller than 1.

The estimated odds ratios for the model covariates and their 95 percent confidence intervals under both design-based and model-based scenarios are presented in Table 6.5. In this table we simply included all continuous variables as linear terms. The interpretation of the estimated odds ratios and confidence intervals in Table 6.5 is the same as in earlier chapters. For example, the odds of high blood pressure for males is estimated to be 1.21 times higher (95% confidence interval: 1.04, 1.44) than the odds of high blood pressure for females, controlling for age, race, weight, and height. Although not dramatic in this example, the effect of ignoring the survey features is clear since this odds ratio would only have been 1.10 (95% confidence interval: 0.97, 1.24) if the data were incorrectly treated as a simple random sample. One interprets the remaining odds ratios and confidence intervals in Table 6.5 in a similar manner.

In summary, we fit logistic regression models to data obtained from complex sample surveys via an approximate likelihood that incorporates the known sampling weights. We assess the overall model significance as well as tests of subsets of coefficients using multivariable Wald or adjusted Wald tests. However, the interpretation of odds ratios from a fitted model is the same as for models fit to less-complicated sampling plans. We note that work needs to be done to make available the methods of assessing fit and case-wise diagnostics obtained from complex sample surveys to the typical user of logistic regression software.

EXERCISES

1. Use the data presented in Breslow and Zhao (1988) to perform the analyses reported in that paper and the analysis reported in Fears and Brown (1986).

2. Fit the model in Table 6.3 using the suggested fractional polynomials for HSAGEIR and BMPWTLBS. Use this new, non-linear model to provide appropriate odds ratio estimates. Use STATA to import the coefficients from the design-based analysis to compute probabilities of response and assess goodness-of-fit.

3. Use the data from the NHANES III survey described in Table 6.1 to find the best model for assessing factors associated with high cholesterol (defined as TCP > 230 mg/100mL). Prepare a table of estimated odds ratios and 95 percent confidence intervals for all covariates in the final model. Compare results for the design-based versus the model-based analysis. Assuming the data resulted from a simple random sample, determine whether the continuous covariates in the model are linear in the logit, determine whether there are any significant interactions among the independent variables in the model, assess model calibration and discrimination, and identify poorly fit and influential covariate patterns. Develop your final design-based model taking into consideration all of these aspects of model development.

CHAPTER 7

Logistic Regression for Matched Case-Control Studies

7.1 INTRODUCTION

An important special case of the stratified case-control study discussed in Chapter 6 is the matched study. A discussion of the rationale for matched studies may be found in epidemiology texts such as Breslow and Day (1980), Kleinbaum, Kupper, and Morgenstern (1982), Schlesselman (1982), Kelsey, Thompson, and Evans (1986) and Rothman and Greenland (1998). In this study design, subjects are stratified on the basis of variables believed to be associated with the outcome. Age and sex are examples of commonly used stratification variables. Within each stratum, samples of cases ($y=1$) and controls ($y=0$) are chosen. The number of cases and controls need not be constant across strata, but the most common matched designs include one case and from one to five controls per stratum and are thus referred to as $1-M$ matched studies.

In this chapter we develop the methods for analyzing general matched studies. Greater detail is provided for the 1–1 design as it is the most common type of matched study. We also illustrate the methods for the general 1–M matched study with data from a 1–3 design.

We begin by providing some motivation and rationale for the need for special methods for the matched study. In Chapter 6 it was noted that we could handle the stratified sample by including the design variables created from the stratification variable in the model. This approach works well when the number of subjects in each stratum is large. However, in a typical matched study we are likely to have few subjects per stratum. For example, in the 1–1 matched design with n case-control pairs we have only two subjects per stratum. Thus, in a fully stratified analysis with p covariates, we would be required to estimate $n+p$ parameters consisting of the constant term, the p slope coefficients for the covariates and the $n-1$ coefficients for the stratum-specific design variables using a sample of size $2n$. The optimality properties of the method of maximum likelihood, derived by letting the sample size become large, hold only when the number of parameters remains fixed. This is clearly not the case in any 1–M matched study. With the fully stratified analysis, the number of parameters increases at the same rate as the sample size. For example, with a model containing one dichotomous covariate it can be shown (see Breslow and Day (1980)) that the bias in the estimate of the coefficient is 100% when analyzing a matched 1–1 design via a fully stratified likelihood. If we regard the stratum-specific parameters as nuisance parameters, and if we are willing to forgo their estimation, then we can use methods for conditional inference to create a likelihood function that yields maximum likelihood estimators of the slope coefficients in the logistic regression model which are consistent and asymptotically normally distributed. The mathematical details of conditional likelihood analysis may be found in Cox and Hinkley (1974).

Suppose that there are K strata with n_{1k} cases and n_{0k} controls in stratum $k, k=1,2,\ldots,K$. We begin with the stratum-specific logistic regression model

$$\pi_k(\mathbf{x})=\frac{e^{\alpha_k+\boldsymbol{\beta}'\mathbf{x}}}{1+e^{\alpha_k+\boldsymbol{\beta}'\mathbf{x}}}, \tag{7.1}$$

where α_k denotes the contribution to the logit of all terms constant within the kth stratum (i.e., the matching or stratification variable(s)). In this chapter, the vector of coefficients, $\boldsymbol{\beta}$, contains only the p slope coefficients, $\boldsymbol{\beta}'=(\beta_1,\beta_2,\ldots,\beta_p)$. It follows from the results in Chapter 3 that each slope coefficient gives the change in the log-odds for a one unit increase in the covariate holding all other covariates constant in every stratum. This is important to keep in mind as the steps, to be described, in developing a conditional likelihood result in a model that does not look like a logistic regression model yet it contains the coefficient vector, $\boldsymbol{\beta}$. The fact that the model does not look like a logistic regression model leads new users to think that estimated coefficients must be modified in some way before they can be used to estimate odds ratios. This is not the case and we pay particular attention in this chapter to estimation and interpretation of odds ratios.

The conditional likelihood for the kth stratum is obtained as the probability of the observed data conditional on the stratum total and the total number of cases observed, the sufficient statistic for the nuisance parameter. In this setting it is the probability of the observed data relative to the probability of the data for all possible assignments of n_{1k} cases and n_{0k} controls to $n_k = n_{1k} + n_{0k}$ subjects. The number of possible assignments of case status to n_{1k} subjects among the n_k subjects, denoted here as c_k, is given by the mathematical expression

$$c_k = \binom{n_k}{n_{1k}} = \frac{n_k!}{n_{1k}!(n_k - n_{1k})!}.$$

Let the subscript j denote any one of these c_k assignments. For any assignment we let subjects 1 to n_{1k} correspond to the cases and subjects $n_{1k}+1$ to n_k to the controls. This is indexed by i for the observed data and by i_j for the jth possible assignment. The conditional likelihood is

$$l_k(\boldsymbol{\beta}) = \frac{\prod_{i=1}^{n_{1k}} P(\mathbf{x}_i | y_i = 1) \prod_{i=n_{1k}+1}^{n_k} P(\mathbf{x}_i | y_i = 0)}{\sum_{j=1}^{c_k} \left\{ \prod_{i_j=1}^{n_{1k}} P(\mathbf{x}_{ji_j} | y_{i_j} = 1) \prod_{i_j=n_{1k}+1}^{n_k} P(\mathbf{x}_{ji_j} | y_{i_j} = 0) \right\}}. \tag{7.2}$$

The full conditional likelihood is the product of the $l_k(\boldsymbol{\beta})$ in (7.2) over the K strata, namely,

$$l(\boldsymbol{\beta}) = \prod_{k=1}^{K} l_k(\boldsymbol{\beta}). \tag{7.3}$$

If we assume that the stratum-specific logistic regression model in (7.1) is correct then application of Bayes theorem to each $P(\mathbf{x}|y)$ term in (7.2) yields

$$l_k(\boldsymbol{\beta}) = \frac{\prod_{i=1}^{n_{1k}} e^{\boldsymbol{\beta}'\mathbf{x}_i}}{\sum_{j=1}^{c_k}\prod_{i_j=1}^{n_{1k}} e^{\boldsymbol{\beta}'\mathbf{x}_{ji_j}}} \tag{7.4}$$

Note that when we apply Bayes theorem all terms of the form $\exp(\alpha_k)/(1+\exp(\alpha_k+\boldsymbol{\beta}'\mathbf{x}))$ appear equally in both the numerator and denominator of equation (7.2) and thus cancel out. Algebraic simplification yields the function shown in equation (7.4) where $\boldsymbol{\beta}$ is the only unknown parameter. The conditional maximum likelihood estimator for $\boldsymbol{\beta}$ is that value that maximizes equation (7.3) when $l_k(\boldsymbol{\beta})$ is as shown in equation (7.4). Except in one special case it is not possible to express the likelihood in (7.4) in a form similar to the unconditional likelihood in equation (1.4). However, as we noted earlier the coefficients have not been modified and thus have the same interpretation as those in equation (7.1).

Software to perform the necessary calculations is available in many packages. For example STATA has a special conditional logistic regression command. In SAS, one must use a modification of the proportional hazards regression command, PHREG. The calculations for this chapter were performed using STATA's clogit command. Since not all packages have special commands for matched studies we show in the next section how one may use a standard logistic regression software package to perform the calculations for the $1-1$ matched design.

7.2 LOGISTIC REGRESSION ANALYSIS FOR THE 1–1 MATCHED STUDY

The most frequently used matched design is one in which each case is matched to a single control, thus there are two subjects in each stratum. To simplify the notation, let $\mathbf{x}_{1k}$ denote the data vector for the case and $\mathbf{x}_{0k}$ the data vector for the control in the kth stratum or pair. Using this notation, the conditional likelihood for the kth stratum from equation (7.4) is

$$l_k(\boldsymbol{\beta}) = \frac{e^{\boldsymbol{\beta}'\mathbf{x}_{1k}}}{e^{\boldsymbol{\beta}'\mathbf{x}_{1k}} + e^{\boldsymbol{\beta}'\mathbf{x}_{0k}}} \tag{7.5}$$

As we described in the previous section when we made the decision not to estimate stratum-specific covariate effects (i.e., the intercepts) we changed the likelihood from one modeling the probability of the outcome to one modeling the probability of the covariate values. Given specific values for $\boldsymbol{\beta}$, $\mathbf{x}_{1k}$ and $\mathbf{x}_{0k}$, equation (7.5) is the probability that the subject identified as the case is in fact the case under the assumptions that: (1) we have two subjects one of whom is the case and (2) the logistic regression model in equation (7.1) is the correct model. For example suppose we have a model with a single dichotomous covariate and $\beta = 0.8$. If the observed data are $x_{1k} = 1$ and $x_{0k} = 0$ then the value of equation (7.5) is

$$l_k(\beta = 0.8) = \frac{e^{0.8\times 1}}{e^{0.8\times 1} + e^{0.8\times 0}} = 0.690 .$$

Thus, the probability is 0.69 that a subject with $x = 1$ is the case compared to a subject with $x = 0$. On the other hand, if $x_{1k} = 0$ and $x_{0k} = 1$ then

$$l_k(\beta = 0.8) = \frac{e^{0.8\times 0}}{e^{0.8\times 0} + e^{0.8\times 1}} = 0.310$$

and the probability is 0.31 that a subject with $x = 0$ is the case compared to a subject with $x = 1$.

It also follows from equation (7.5) that if the data for the case and the control are identical, $\mathbf{x}_{1k} = \mathbf{x}_{0k}$, then $l_k(\boldsymbol{\beta}) = 0.5$ for any value of $\boldsymbol{\beta}$ (i.e., the data for the case and control are equally likely under the model). Thus, case-control pairs with the same value for any covariate are *uninformative* for estimation of that covariate's coefficient. We use the term *uninformative* to describe the fact that the value of the covariate does not help distinguish which subject is more likely to be the case. This tends to occur most frequently with dichotomous covariates where common values, often called concordant pairs, are most likely to occur. A fact not discussed in this chapter, which can be found in Breslow and Day (1980), is that the maximum likelihood estimator of the coefficient for a dichotomous covariate in a univariable conditional logistic regression model fit to 1–1 matched data is the log of the ratio of discordant pairs. The practical significance of this is that the estimator may be based on a small fraction of the total number of possible pairs. We feel it is good practice to form the 2×2 table cross-classifying case versus control for all dichotomous covariates in order to determine the number of discordant pairs. This is essentially a univariable logistic regression and univariable analyses of all covariates should be among the first steps in any model building process. The reader should be aware that if there are not both types of pairs, $(x_{1k}=1, x_{0k}=0)$ and $(x_{1k}=0, x_{0k}=1)$, present in the data then the estimator is undefined. In this case software packages will either remove the covariate from the model or give an impracticably large coefficient and standard error. This is the same zero cell problem discussed in Chapter 4, Section 5. The same type of problem can occur for polychotomous covariates but it involves more complex relationships than simply a zero frequency cell in the cross-classification of case versus control (Breslow and Day (1980)).

As we noted, not all software packages have specific commands for maximizing the conditional log-likelihood. It is possible, with some data manipulation, to use a standard logistic regression package to maximize the full conditional log-likelihood for the 1−1 design. We begin by re-expressing equation (7.5) by dividing its numerator and denominator by $e^{\boldsymbol{\beta}'\mathbf{x}_{0k}}$ yielding

$$
\begin{aligned}
l_k(\boldsymbol{\beta}) &= \frac{e^{\boldsymbol{\beta}'(\mathbf{x}_{1k}-\mathbf{x}_{0k})}}{1+e^{\boldsymbol{\beta}'(\mathbf{x}_{1k}-\mathbf{x}_{0k})}} \\
&= \frac{e^{\boldsymbol{\beta}'\mathbf{x}_k^*}}{1+e^{\boldsymbol{\beta}'\mathbf{x}_k^*}} .
\end{aligned}
\tag{7.6}
$$

The expression on the right side of equation (7.6) is the usual logistic regression model with the constant term set equal to zero, $\beta_0 = 0$, and data vector equal to the data value of the case minus the data value of the control, $\mathbf{x}_k^* = (\mathbf{x}_{1k} - \mathbf{x}_{0k})$. It follows that the full conditional likelihood may be expressed as

$$
\begin{aligned}
l(\boldsymbol{\beta}) &= \prod_{k=1}^{K} \frac{e^{\boldsymbol{\beta}'\mathbf{x}_k^*}}{1+e^{\boldsymbol{\beta}'\mathbf{x}_k^*}} \\
&= \prod_{k=1}^{K} \left[\frac{e^{\boldsymbol{\beta}'\mathbf{x}_k^*}}{1+e^{\boldsymbol{\beta}'\mathbf{x}_k^*}}\right]^{y_k} \left[\frac{1}{1+e^{\boldsymbol{\beta}'\mathbf{x}_k^*}}\right]^{1-y_k} ,
\end{aligned}
$$

where $y_k = 1$.

This observation allows us to use standard logistic regression software to compute the conditional maximum likelihood estimates and obtain estimated standard errors of the estimated coefficients. To do this we define the sample size as the number of case-control pairs, use as covariates the differences $\mathbf{x}_k^*$, set the values of the response variable equal to 1, $y_k = 1$, and exclude the constant term from the model. Thus, from a computational point of view, the 1–1 matched design may be fit using any logistic regression program. However, the logistic regression package must allow the user to exclude the constant term and it must also allow a setting where the outcome is constant. For example STATA's logit command does allow the user to exclude the constant term but does not permit the outcome to be constant. SAS's procedure LOGISTIC allows both.

We have found that the process of creating the differences and setting the outcome equal to 1 can be confusing to new users. As we noted earlier in this chapter, it is important to distinguish between the model being fit to the data and the computational manipulations used to apply standard logistic regression software. The process becomes less confusing when considering modeling strategies if we focus on terms in the logistic regression model first and then perform the computations necessary to obtain the parameter estimates. A few examples should serve to illustrate this point.

First consider a dichotomous independent variable coded zero or one. It generates a single coefficient in the logit, regardless of whether we enter the variable as a design variable or treat it as if it were continuous. It follows that the difference variable, x^*, computed as the difference between the two dichotomous variables in the pair, may take on one of three possible values, (–1, 0, or 1). If we mistakenly thought of x^* as being the actual data we would have created two design variables. This should not be done. Instead, the correct method is to treat x^* as if it were continuous in the model.

As a second example, consider a variable such as race, coded at three levels. To correctly model this variable in the 1–1 matched design we would create, for each case and control in a pair, the values of the two design variables representing race. Then we would compute the difference between the case and control for each of these two design variables and treat each of these differences as if they were continuous.

The same process is followed for any categorical covariate. For example, suppose we wished to examine the scale in the logit of a continuous variable. One approach illustrated in Chapter 4 is to create design variables corresponding to the quartiles of the distribution and then plot their estimated coefficients. (Note: the quartiles come from the distribution in the combined sample of $2K$ observations.) In the matched study we would do the same thing, with the one intermediate step of calculating the difference between the three design variables for case-control pairs. The software package may not have the option to indicate that these differences in design variables are from the same variable. Thus, we have to be sure that all three are included in any model we fit.

One other point to keep in mind is that since the differences between variables used to form strata are zero for all strata, they do not enter any model in main effects form. However, we may include interaction terms between stratification variables and other covariates, as differences in these are likely not to be zero.

In summary, the conceptual process for modeling matched data is identical to that already illustrated for unmatched data. If we develop our modeling strategies in the matched 1–1 design as if we had an unmatched design and then use the conditional likelihood, we will always be proceeding correctly.

7.3 AN EXAMPLE OF THE USE OF THE LOGISTIC REGRESSION MODEL IN A 1–1 MATCHED STUDY

For illustrative purposes a 1–1 matched data set was created from the low birth weight data by randomly selecting for each woman who gave birth to a low birth weight baby, a mother of the same age who did not give birth to a low birth weight baby. For three of the young mothers (age less than 17) it was not possible to identify a match since there were no remaining mothers of normal weight babies of that age. The data set consists of 56 age matched case-control pairs. With the exception of the number of first trimester visits, which has been excluded by us due to its lack of importance in the earlier analysis, the variables are the same as those in the low birth weight data set described in Table 1.6. In this example the number of prior pre-term deliveries has been coded as a yes (1)–no (0) variable. Thus, at the initial stage of model building we have available the following variables: race (RACE), smoking status (SMOKE), presence of hypertension (HT), presence of uterine irritability (UI), presence of previous pre-term delivery (PTD), and the weight of the mother at the last menstrual period (LWT). The variable AGE is available when we evaluate interactions. The data are available on the two web sites noted in the Preface in a file named LOWBWT11.DAT.

As we noted earlier, the model building in this chapter is done using STATA's clogit command. Thus, at this point we do not need to create the difference variables that are required if another package is used.

The results of fitting univariable models are displayed in Table 7.1. Only the coefficients for SMOKE and PTD are significant at the five percent level. The frequencies of the discordant pairs in the last column indicate that "thin data" may be a problem for each of the nominal scale covariates but HT more so than the other covariates. We will need to pay close attention to the estimated standard errors and confidence interval widths in our multivariable models.

Before fitting multivariable models we note that the "intercept only" model (or base model) for assessing overall significance in the 1–1 design is a model with likelihood

$$l(\boldsymbol{\beta}=\mathbf{0})=\prod_{k=1}^{K}0.5=(0.5)^K,$$

a value usually not presented in computer output.

Since there are only six variables eligible for inclusion in the model, we begin model development with all variables in the model. Table 7.2 presents the results of fitting this model. We see in Table 7.2 that neither design variable for RACE is significant. In addition, the value of the partial likelihood ratio test for the exclusion of RACE is $G = 0.885$ which, with 2 degrees-of-freedom, yields a p-value of 0.642. However, RACE may be a confounder of the effects of the other variables in the model. To assess this, we display the results of fitting the model without RACE in Table 7.3. Comparing the estimated coefficients in Table 7.2 and Table 7.3, we see that RACE seems to only modestly confound the association for LWT, whose coefficient changes by 16.7%. Because the change is not too substantial, and because we have a small sample size (56 pairs), we choose to exclude RACE from the model. We proceed to the next step in which we identify the correct scale for LWT.

In general the same methods discussed in Chapter 4 can be used with matched studies. We demonstrate the use of fractional polynomials and design variables. The smoothed scatterplots cannot be used unless we ignore the case-control matches. We comment on breaking the matches at the end of the section.

Table 7.1 Univariable Logistic Regression Models for the 1-1 Matched Low Birth Weight Data, $n = 56$ Pairs

Variable	Coeff.	Std. Err.	$\widehat{\text{OR}}$	95 %CI	Discordant Pairs $(n_{10}, n_{01})^{+}$
LWT	−0.009	0.0062	0.91*	(0.81, 1.03)	#
SMOKE	1.012	0.4129	2.75	(1.22, 6.18)	(22, 8)
RACE_2	0.087	0.5233	1.09	(0.39, 3.04)	#
RACE_3	−0.029	0.3968	0.97	(0.45, 2.11)	#
PTD	1.322	0.5627	3.75	(1.24, 11.30)	(15, 4)
HT	0.847	0.6901	2.33	(0.60, 9.02)	(7, 3)
UI	1.099	0.5774	3.00	(0.97, 9.30)	(12, 4)

+: Discordant Pairs: n_{10} = frequency of $(x_1 = 1, x_0 = 0)$,

*: Odds ratio for a 10 pound increase in weight

#: Not relevant

n_{01} = frequency of $(x_1 = 0, x_0 = 1)$ and $\widehat{\text{OR}} = n_{10}/n_{01}$

The results of using fractional polynomials to examine the scale of LWT are presented in Table 7.4. The first p-value, 0.050, in the 5th column indicates that treating LWT as linear in the logit offers a significant improvement over the model not including LWT. The best single term model uses LWT^3 but the second p-value, 0.334, indicates that this transformation is not significantly better than the linear model. The best two-term model uses LWT^3 and $LWT^3 \times \ln(LWT)$. The p-value

Table 7.2 Estimated Coefficients, Estimated Standard Errors, Wald Statistics and Two-Tailed p-Values for the Model Containing All Covariates

Variable	Coeff.	Std. Err.	z	P>|z|
LWT	−0.018	0.0101	−1.82	0.068
RACE_2	0.571	0.6896	0.83	0.407
RACE_3	−0.025	0.6992	−0.04	0.971
SMOKE	1.401	0.6278	2.23	0.026
PTD	1.808	0.7887	2.29	0.022
HT	2.361	1.0861	2.17	0.030
UI	1.402	0.6962	2.01	0.044

Log likelihood = −25.7943

Table 7.3 Estimated Coefficients, Estimated Standard Errors, Wald Statistics and Two-Tailed p-Values for the Model Excluding RACE

Variable	Coeff.	Std. Err.	z	P>|z|
LWT	−0.015	0.0081	−1.85	0.064
SMOKE	1.480	0.5620	2.63	0.008
PTD	1.671	0.7468	2.24	0.025
HT	2.329	1.0025	2.32	0.020
UI	1.345	0.6938	1.94	0.053

Log likelihood = −26.2369

comparing the best two-term fractional polynomial model to the best one-term model is 0.860. The value of the likelihood ratio chi-square test comparing the linear model to the best two-term model is $G = 52.474 - 51.239 = 1.235$, which with 3 degrees-of-freedom yields a p-value of 0.745. Thus, we conclude that the fractional polynomial analysis supports treating LWT as linear in the logit.

The second method we illustrate to assess the scale of LWT is based on design variables. This method is described in detail in Chapter 4. In the setting of a matched study the three design variables are created using the quartiles of the combined distribution of LWT $(n = 112)$. The quartiles as computed by STATA are $Q_1 = 106.5$, $Q_2 = 120$ and $Q_3 = 136.5$. We use the first quartile as the reference group.

The model was fit using all the variables shown in Table 7.3 except LWT, which is replaced by the three design variables for quartiles. The estimated coefficients for the three design variables are given in Table

Table 7.4 Summary of the Use of the Method of Fractional Polynomials for LWT

	df	Deviance	G for Model vs. Linear	Approx. p-Value	Powers
Not in model	0	56.299			
Linear	1	52.474	0.000	0.050*	1
$J = 1$	2	51.541	0.933	0.334+	3
$J = 2$	4	51.239	1.235	0.860#	3, 3

* Compares linear model to model without LWT

+ Compares the $J = 1$ model to the linear model

Compares the $J = 2$ model to the $J = 1$ model

Table 7.5 Results of the Quartile Analyses of LWT from the Multivariable Model Containing the Variables Shown in the Table 7.3

Quartile	1	2	3	4
Midpoint	93.25	113.25	128.25	188.75
Coeff.	0.0	−0.399	−0.443	−0.889
95 % CI		(−1.69, 0.90)	(−1.76, 0.87)	(−2.11, 0.34)

7.5.

Even though none of the three estimated coefficients in Table 7.5 is significant their decreasing trend does lend support for linearity in the logit. This is more easily seen by plotting the coefficients versus the midpoints of the quartiles as shown in Figure 7.1

Thus on the basis of the fractional polynomial analysis and the confirming evidence from the plot in Figure 7.1 we decide to model LWT as linear in the logit.

The next step in model development is to assess the possibility of interactions among the variables. A number of clinically plausible in-

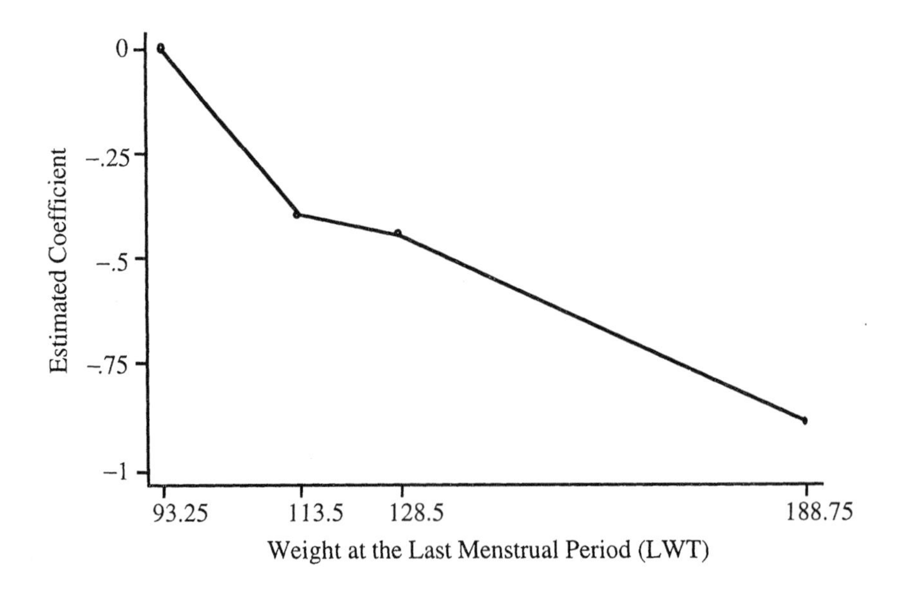

Figure 7.1 Plot of the estimated coefficients for the quartiles of LWT versus the midpoint of the quartile.

Table 7.6 Likelihood Ratio Test Statistic (G) and p-Value for Interactions of Interest When Added to the Main Effects Model in Table 7.3

Interaction	G	p
AGE×LWT	0.50	0.477
AGE×SMOKE	0.01	0.910
AGE×PTD	0.05	0.818
AGE×HT	0.35	0.557
AGE×UI	1.12	0.290
LWT×SMOKE	0.18	0.671
LWT×PTD	0.06	0.800
LWT×HT	0.03	0.868
LWT×UI	0.03	0.868
SMOKE×PTD	< 0.01	> 0.900
SMOKE×HT	0.39	0.532
SMOKE×UI	0.15	0.699
PTD×HT	*	*
PTD×UI	2.56	0.110
HT×UI	*	*

*: Model could not be fit due to zero cells

teractions may be created from the variables in the model. In addition we examine whether the matching variable, age, interacts with any of the variables in the model. The potential interaction variables are shown in the first column of Table 7.6. The remaining columns present the likelihood ratio test and its p-value comparing the model containing the interaction to the main effects model in Table 7.3. Since each interaction generates a single covariate we do not include the degrees-of-freedom in Table 7.6.

The results in Table 7.6 indicate that none of the interactions is significant at the 5 percent level. Thus we conclude that we do not need to include any interactions in our model. We note that two interactions, PTD×HT and HT×UI, generated at least one cell with a zero frequency among the discordant pairs and thus the model could not be fit. See Chapter 4, Section 5 for a more detailed discussion of this type of numerical problem.

We now move to model assessment using the main effects model in Table 7.3.

7.4 ASSESSMENT OF FIT IN A MATCHED STUDY

The approach to assessing the fit of a logistic regression model in the 1-1 matched design is identical to that described in Chapter 5 for unmatched designs. We begin by forming a measure of residual variation, and then use it to explore the sensitivity of the fit to individual case-control pairs. In the 1-1 matched study the likelihood function is defined in terms of the conditional probability of allocation of observed covariates to the case and control within each stratum. As discussed in Section 7.2 the value of the outcome variable when we use the differences method is $y=1$ for all strata (case-control pairs). This corresponds to assigning a conditional probability of 1 to the observed allocation of covariate values to the components of the pairs. The fitted value is the estimate of this conditional probability under the assumption that the logistic regression model is correct. The number of covariate patterns is always the number of pairs or strata. This implies that $m=1$ for all patterns and measures that were based on m-asymptotics in the unmatched case cannot be used in 1-1 matched designs. For example, it is not possible to extend the Hosmer-Lemeshow chi-square goodness-of-fit statistic to the 1-1 matched study design. Zhang (1999) proposes an overall goodness-of-fit test. However, it is not available in any package and it is not easy to compute. Thus, we do not discuss it in this section.

Moolgavkar, Lustbader and Venzon (1985), and Pregibon (1984) have extended the ideas of Pregibon (1981) to matched studies. These authors show that, for 1-1 matched studies, the logistic regression diagnostics may be computed in the same manner as shown in Chapter 5 for unmatched studies. Bedrick and Hill (1996) present methods for assessing model fit within individual matched sets. Their methods require exact methods not available in most software packages. Moolgavkar, Lustbader and Venzon and Pregibon show that we may calculate leverage, h, standardized residuals, r_s, and the measures $\Delta\hat{\beta}$, ΔX^2 and ΔD using the formulae shown in equations (5.12) - (5.16), where the difference variable, $\mathbf{x}^* = \mathbf{x}_1 - \mathbf{x}_0$, replaces $\mathbf{x}$ and we use the logistic model shown in equation (7.6).

The Pearson residual is

$$r = \frac{(y-\hat{\pi})}{\left[\hat{\pi}(1-\hat{\pi})\right]^{1/2}}$$

and, since in the difference formulation $y=1$, this simplifies to

$$r=\sqrt{\frac{(1-\hat{\pi})}{\hat{\pi}}},$$

where $\hat{\pi}$ is the value of equation (7.6) using the estimated parameters. In this situation large residuals are only possible when the fitted value, $\hat{\pi}$, is small. This was the same situation as in the unmatched studies except that poor fit was also possible when $y=0$ and the fitted value was large. Other than this, the expected behavior of the diagnostics as a function of the fitted values is the same for the 1–1 matched study.

At this time we are not aware of a package that has an option to calculate the diagnostic statistics for any matched design. Thus, the only option available to us is to create a data set containing the difference variables and calculate the diagnostic statistics using a standard logistic regression routine. We used SAS's logistic regression procedure to calculate the diagnostics for the example in this section.

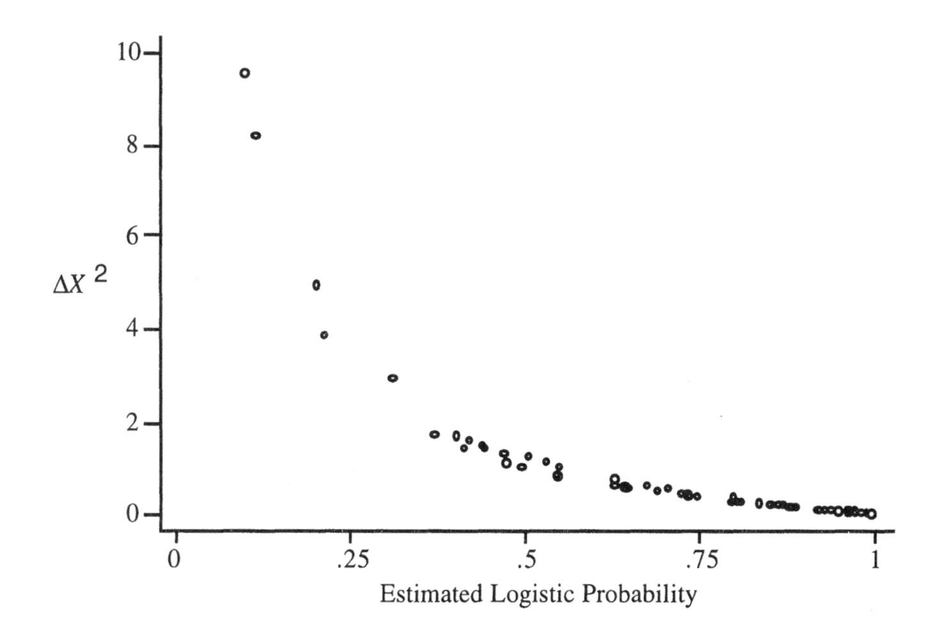

Figure 7.2 Plot of ΔX^2 versus the estimated probability from the fitted model in Table 7.3.

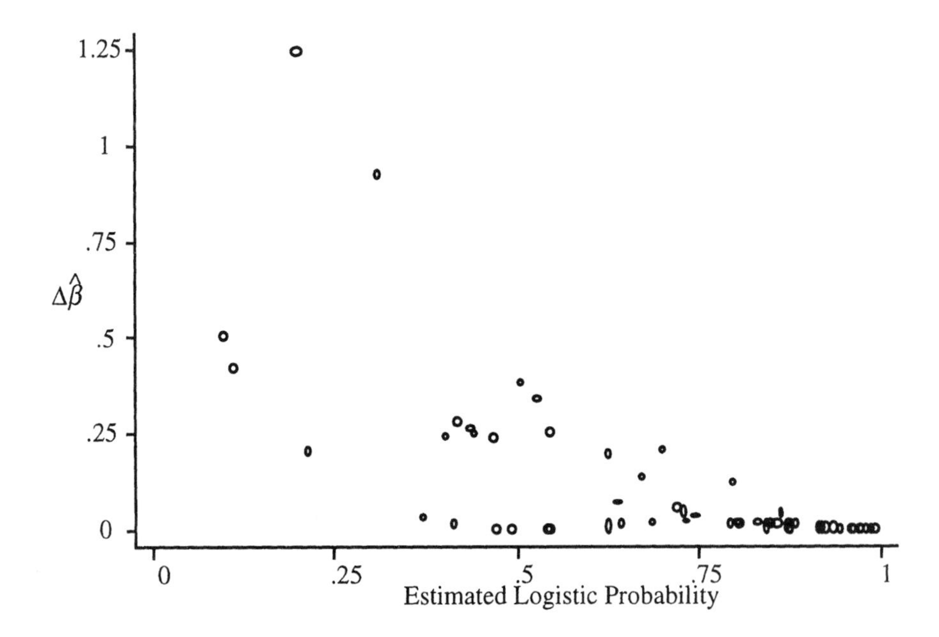

Figure 7.3 Plot of $\Delta\hat{\boldsymbol{\beta}}$ versus the estimated logistic probability from the fitted model in Table 7.3.

To illustrate the use of the diagnostics, we apply them to assess the fit of the model in Table 7.3. The change in the Pearson chi-square, ΔX^2, and the change in the deviance, ΔD, due to deleting a particular pair, show essentially the same thing so we only present the plot for ΔX^2. Plots of ΔX^2 and $\Delta\hat{\boldsymbol{\beta}}$ versus the fitted values, $\hat{\pi}$, are shown in Figure 7.2 and Figure 7.3.

In Figure 7.2 we see, as expected, that ΔX^2 increases as $\hat{\pi}$ decreases. Two points have much larger values, greater than 8.0, than the other points and three others have values between about 3.0 and 5.0.

The plot of $\Delta\hat{\boldsymbol{\beta}}$ in Figure 7.3 shows two values larger than about 0.7 and two more with values between about 0.4 and 0.5. Each of these four values corresponds to one of the five values identified in Figure 7.2.

In Figure 7.4 we plot ΔX^2 versus $\hat{\pi}$ with the size of the plotting symbol proportional to $\Delta\hat{\boldsymbol{\beta}}$. The purpose of this plot is to better show the relationship between $\Delta\hat{\boldsymbol{\beta}}$, ΔX^2 and the leverage, h. We can see in

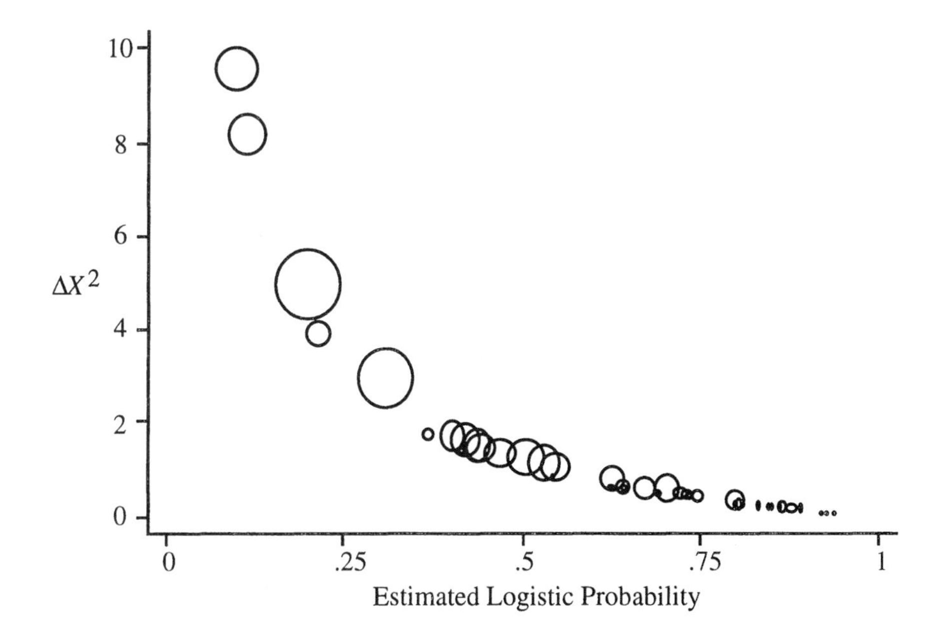

Figure 7.4 Plot of ΔX^2 versus the estimated logistic probability from the fitted model in Table 7.3 with the size of the plotting symbol proportional to $\Delta\hat{\beta}$.

Figure 7.4 that the two pairs with the largest value of $\Delta\hat{\boldsymbol{\beta}}$ occur in that region of the estimated probability scale where leverage is expected to be the largest, namely, $0.2-0.4$. Thus for these two points large $\Delta\hat{\boldsymbol{\beta}}$ is in fact due to moderately large ΔX^2 and leverage. The next two largest circles in Figure 7.4 correspond to the two pairs with the largest values of ΔX^2.

The next step in the analysis is to identify the four pairs with large values of $\Delta\hat{\boldsymbol{\beta}}$ and list their respective data along with the values of the diagnostic statistics. This step is shown in Table 7.7. The results of fitting the model with each pair deleted are shown in Table 7.8.

Specifically, the data from pair 9 show that the control is 48 pounds lighter than the case and the control smoked during the pregnancy and the case did not smoke. The negative sign of the coefficient for LWT indicates that heavier women are less likely to have a low weight birth. Thus, deleting a pair where the control is much lighter should and does increase the protective effect of weight. On the other hand, the coefficient for SMOKE is positive indicating that women who

Table 7.7 Pair, Data, Estimated Probability, and the Three Diagnostic Statistics $\Delta\hat{\beta}$, ΔX^2, and Leverage (h) for Four Extreme Pairs

Pair	Cont/ Case	LWT	SMK	PTD	HT	UI	$\hat{\pi}$	$\Delta\hat{\beta}$	ΔX^2	h
9	0	100	1	0	0	0	0.10	0.50	9.53	0.05
	1	148	0	0	0	0				
16	0	169	0	1	0	1	0.31	0.92	2.92	0.24
	1	120	1	0	0	0				
27	0	95	0	0	1	0	0.20	1.25	4.97	0.20
	1	130	1	0	0	0				
34	0	90	1	1	0	0	0.11	0.42	8.19	0.05
	1	128	0	1	0	0				

smoke are more likely to have a low weight birth. Deleting a pair where the control smoked and the case did not should, and in fact does, increase the coefficient for SMOKE. The actual decrease in the Pearson chi-square is less than expected from the value of ΔX^2 in Table 7.7.

When we delete pair 16 we see that the coefficients for PTD and UI change by over 25 percent. Examining the data in Table 7.7 we see that in this pair the control had both a prior pre-term delivery and uterine irritability. Since the coefficients for these covariates are positive these conditions are more consistent with being a case. Thus deleting this pair increases the two coefficients. The actual change in the Pearson chi-square is 4.39, a bit larger than expected from the value in Table 7.7.

The effects of deleting pair 27 are large changes, at least 40 percent, in the coefficients for LWT and HT. The pattern in LWT is similar to pair 9, a control much lighter than the case. Thus deletion of pair 27 also increases the protective effect of LWT. The pattern in HT is similar to PTD and UI seen in pair 16. Here the control is hypertensive and the case is not, thus deletion increases the effect of HT. The fairly large change in the coefficient for HT, 52.8 percent, is due to the fact that only 10 of the 56 pairs are discordant in this covariate.

The results when we delete pair 34 are nearly identical to those for pair 9. The large change in the coefficient for LWT is due to the fact that the case is 38 lbs. heavier than the control. Also, the control smoked while the case did not smoke during the pregnancy. The

Table 7.8 Estimated Coefficients from Table 7.3, Estimated Coefficients Obtained When Deleting Selected Pairs, Percent Change from the All Data Model and Values of Pearson Chi-Square Statistic

Data	LWT	SMOKE	PTD	HT	UI	X^2
All	−0.015	1.480	1.671	2.329	1.345	50.76
Delete 9	−0.019	1.878	1.883	2.719	1.498	48.52
% Change	30.0	26.9	12.7	16.7	11.4	
Delete 16	−0.013	1.391	2.11	2.407	1.762	46.37
% Change	−16.6	−6.0	26.4	3.3	31.0	
Delete 27	−0.021	1.389	1.807	3.559	1.511	49.00
% Change	39.6	−6.1	8.2	52.8	12.3	
Delete 34	−0.018	1.855	1.863	2.669	1.487	50.31
% Change	24.6	25.4	11.5	14.6	10.5	

change in the Pearson chi-square is small and is much less than that expected from the value of ΔX^2 in Table 7.7.

The results from deleting each of the four pairs provide good examples of why it is important to not rely completely on the values of the diagnostic statistics to predict model changes. We feel it is quite important to go through the process of refitting the model, deleting potentially influential and or poorly fit subjects. Our experience has shown that $\Delta\hat{\beta}$ tends to underestimate the actual changes one sees in coefficients. The behavior of the diagnostic for fit, ΔX^2, is not consistent, sometimes overestimating the actual change in X^2, and in other instances underestimating the change.

As in any model building process the final decision on whether to include or exclude any data depends on an assessment of the clinical plausibility of the data. This decision, as always, should be made in consultation with subject matter experts. We proceed to model interpretation using the full data estimates from Table 7.3.

We present the estimated odds ratios and 95 percent confidence intervals in Table 7.9. These point and interval estimates are computed using the estimated coefficients and standard errors in Table 7.3 in exactly the same manner as described in Chapter 3, namely by exponentiating the coefficient and the endpoints of its confidence interval.

Table 7.9 Estimated Odds Ratios and 95% Confidence Intervals for Model in Table 7.3

Variable	Odds Ratio	95 % CI
Weight at Last Menstrual Period*	0.9	0.73, 1.01
Smoking During Pregnancy	4.4	1.46, 13.21
History of Pre-Term Delivery	5.3	1.23, 22.97
History of Hypertension	10.3	1.44, 73.28
Presence of Uterine Irritability	3.8	0.99, 14.95

*: Odds ratio for a 10 pound increase in weight

The interpretation of estimates and confidence intervals for odds ratios is also the same as described in Chapter 3. Specifically, the odds ratio for weight at last menstrual period estimates a 10 percent reduction in risk of a low weight baby per 10 pound increase in weight. The confidence interval suggests that there could be as much as a 27 percent decrease in risk or there could be no reduction in risk. The odds ratio for a history of smoking during pregnancy suggests that women who do smoke are 4.4 times more likely to have a low weight baby than women who do not smoke and the increase in risk could be as little as 1.5 times or as high as 13.2 times with 95 percent confidence. Having a history of pre-term delivery increases the risk of a low weight baby by 5.3 times and it could be as little as 1.2 times or as much as 23 times with 95 percent confidence. Women with a history of hypertension have a 10-fold increase in risk of a low weight baby and the increase in risk could be as little as 1.4 times or as much as 73.3 times with 95 percent confidence. Presence of uterine irritability carries a 3.8 fold increase in risk of a low weight baby. The confidence interval suggests that there could be no increase to as much as a 15-fold increase in risk.

We note that the confidence intervals are quite wide, especially for history of hypertension. This is due to the fact, described in the previous section and shown in Table 7.1, that there are relatively few discordant pairs in this data set.

In summary, the actual process of model building, assessment of fit and interpretation of odds ratios estimated from the final model is the same for the 1–1 matched case-control study as in any unmatched study. The only difference is that we use a conditional likelihood that eliminates stratum-specific effects from the basic logistic regression model.

In closing this section we note that many investigators break the matched pairs and proceed with the standard analysis as described in Chapters 4 and 5. Lynn and McCulloch (1992) provide some theoretical and simulation-based evidence for breaking the matches when the sample size is large. However, we believe that if data have been collected using a specific matched sampling design then the analysis must have as its foundation the stratum-specific likelihood shown in equation (7.2) and the full likelihood in equation (7.3).

We believe that investigators have used what is really an incorrect analysis for two basic reasons. First, the investigator is not comfortable with the conditional likelihood approach. He/she thinks that somehow the model has been changed and one cannot use estimated coefficients to estimate odds ratios in the usual manner. Second, until recently the analysis had to be performed using difference variables, a cumbersome and tedious data management task. We hope that the presentation of the example in this section convinces investigators that a matched analysis is no more difficult than an unmatched analysis. While software is available, developers need to bring current routines for matched analyses to the same level as their programs for unmatched analyses, especially in the area of diagnostic statistics for assessment of model adequacy and fit.

7.5 AN EXAMPLE OF THE USE OF THE LOGISTIC REGRESSION MODEL IN A 1–*M* MATCHED STUDY

The general approach to the analysis of the 1–*M* matched design and, for that matter, general matched or highly stratified designs is similar to that of the 1–1 matched design. As we demonstrated in the previous two sections the 1–1 matched design may be fit using software for unconditional logistic regression in some, but not all, packages. However, for the 1–*M* design we need software, such as STATA's clogit command, that maximizes a more general conditional likelihood. The need for special software can be seen if we examine the contribution to the likelihood for an individual stratum. In this section, to keep notation simple and to a minimum, we consider a design where $M = 3$. The extension of the methods to other matched designs is not difficult. We let the value of the covariates for the case in stratum k be denoted by $\mathbf{x}_{k1}$ and the values for the three controls be denoted $\mathbf{x}_{k2}, \mathbf{x}_{k3}$, and $\mathbf{x}_{k4}$. The contribution to the likelihood for this stratum is obtained by evaluating the expression shown in equation (7.4) and is

$$l_k(\boldsymbol{\beta}) = \frac{e^{\boldsymbol{\beta}'\mathbf{x}_{k1}}}{e^{\boldsymbol{\beta}'\mathbf{x}_{k1}} + e^{\boldsymbol{\beta}'\mathbf{x}_{k2}} + e^{\boldsymbol{\beta}'\mathbf{x}_{k3}} + e^{\boldsymbol{\beta}'\mathbf{x}_{k4}}} \ . \tag{7.7}$$

The interpretation of equation (7.7) is same as we described in Section 7.2 for equation (7.5). Given the value of the coefficients it gives the probability that the subject with data $\mathbf{x}_{k1}$ is the case relative to three controls with data $\mathbf{x}_{k2}, \mathbf{x}_{k3}$, and $\mathbf{x}_{k4}$. We note that if the covariates are identical for all four subjects then the stratum is uninformative for estimation of the coefficients as $l_k(\boldsymbol{\beta}) = 0.25$ for any value of $\boldsymbol{\beta}$. For an individual covariate there must be at least one control that has a value different from the case or the stratum is uninformative for that specific coefficient. Unfortunately there are no simple expressions involving discordant pairs for the estimator of the coefficient for a dichotomous covariate in a univariable model. One descriptive statistic that is useful for visually assessing the potential for "thin data" for a dichotomous covariate is the 2 by $M+1$ table cross-classifying the case versus the sum of the covariate for the controls. The strata that would not contribute to the analysis would correspond to the counts in the (0,0) and (1,M) cells. As always, we feel it is good practice to fit univariable models and use the estimated standard errors and confidence intervals as indirect evaluation for "thin data".

It is not possible to express the right side of equation (7.7) in the form of an unconditional logistic regression model. Hence, to perform an analysis of a 1–M matched design we must use software that obtains maximum likelihood estimators from a likelihood function whose component terms are like those in equation (7.7). We use STATA's clogit command to fit the models in this section.

To provide a data set for an example and exercises we present a subset of data from a large study on benign breast disease whose results have been published. The original data are from a hospital-based case-control study designed to examine the epidemiology of fibrocystic breast disease. Cases included women with a biopsy-confirmed diagnosis of fibrocystic breast disease identified through two hospitals in New Haven, Connecticut. Controls were selected from among patients admitted to the general surgery, orthopedic, or otolaryngologic services at the same two hospitals. Trained interviewers administered a standardized structured questionnaire to collect information from each subject [see Pastides, et al (1983) and Pastides, et al (1985)].

A code sheet for the data is given in Table 7.10. Data are provided on 50 women who were diagnosed as having benign breast disease and 150 age-matched controls, with three controls per case. Matching was based on the age of the subject at the time of interview. The data are available on the two web sites described in the Preface in a file named BBDM13.DAT.

We consider covariates measuring regular medical check-ups

Table 7.10 Description of Variables in the Benign Breast Disease 1–3 Matched Case-Control Study

Variable	Description	Codes/Values	Name
1	Stratum	1 – 50	STR
2	Observation within Stratum	1 = Case	OBS
		2 – 4 = Control	
3	Age at Interview	Years	AGMT
4	Final Diagnosis	1 = Case	FNDX
		0 = Control	
5	Highest Grade in School	5 – 20	HIGD
6	Degree	0 = None	DEG
		1 = High School	
		2 = Jr. College	
		3 = College	
		4 = Masters	
		5 = Doctoral	
7	Regular Medical Check-ups	1 = Yes	CHK
		2 = No	
8	Age at First Pregnancy	Years	AGP1
9	Age at Menarche	Years	AGMN
10	No. of Stillbirths, Miscarriages etc.	0 – 7	NLV
11	Number of Live Births	0 – 11	LIV
12	Weight of the Subject At Interview	Pounds	WT
13	Age at Last Menstrual Period	Years	AGLP
14	Marital Status	1 = Married	MST
		2 = Divorced	
		3 = Separated	
		4 = Widowed	
		5 = Never Married	

Table 7.11 Univariable Logistic Regression Models for the 1–3 Matched Benign Breast Disease Study, n = 50 Strata

Variable	Coeff.	Std. Err.	$\widehat{OR}$	95 %CI
CHK	−1.245	0.3815	0.29	(0.14, 0.61)
AGMN	0.472	0.1110	2.57*	(1.66, 3.97)
WT	−0.035	0.0086	0.70+	(0.59, 0.83)
MST_2	−0.358	0.5605	0.70	(0.23, 2.10)
MST_4	−0.751	0.7904	0.47	(0.10, 2.22)
MST_5	1.248	0.6059	3.48	(1.06, 11.43)

*: Odds ratio for a 2 year increase

+: Odds ratio for a 10 pound increase

(CHK), age at menarche (AGMN), weight at the interview (WT) and marital status (MST) as an example of model building in a 1–3 matched study. We leave model development using all study variables as an exercise. Tabulation of the frequency distribution of marital status showed that only 6 subjects reported being separated. These subjects are combined with the 20 subjects who reported their status as divorced. The results of fitting the univariable models are shown in Table 7.11.

The results in Table 7.11 show that having regular check-ups and increasing weight significantly reduce the odds of having benign breast disease. Increased age at menarche significantly increases the odds of benign breast disease. The results for marital status suggest that subjects who were ever married have the same odds as those currently married. Women who were never married have significantly increased odds for benign breast disease compared to women who are currently married. These results suggest that a more parsimonious model might use a dichotomous covariate ever-never married. However, we begin by fitting the multivariable model containing all the covariates as coded in Table 7.11. These results are shown in Table 7.12.

The results in Table 7.12 agree, at least in terms of direction of effect and significance, with the univariable models in Table 7.11. The marital status covariate presents some problems in that only 12 of the 200 subjects report never being married. However there is a significant increase in risk and, as such, we proceed using the dichotomous covariate never married, NVMR (0 = ever married, 1 = never married). The results of fitting this model are show in Table 7.13.

Table 7.12 Estimated Coefficients, Estimated Standard Errors, Wald Statistics and Two-Tailed *p*-Values for the Multivariable Model

Variable	Coeff.	Std. Err.	z	P>|z|
CHK	−1.122	0.4474	−2.51	0.012
AGMN	0.356	0.1292	2.76	0.006
WT	−0.028	0.0100	−2.84	0.004
MST_2	−0.203	0.6473	−0.31	0.754
MST_4	−0.493	0.8173	−0.60	0.548
MST_5	1.472	0.7582	1.94	0.052

Log likelihood = −45.2148

The results in Table 7.13 indicate that each of the covariates is significant. In addition, when we compare the values of the coefficients in Table 7.13 to those in Table 7.12 we see that recoding marital status did not introduce any confounding. Also, the z-score for the dichotomous covariate NVMR is about the same order of magnitude as the z-scores for the other covariates. This provides some evidence that there is adequate data. Thus we proceed to examine the scale of the continuous covariates AGMN and WT.

The methods to check for the scale of a continuous variable in the logit for a 1–*M* design are the same as those illustrated in Section 7.3 for the 1–1 matched study, (i.e., fractional polynomials and quartile design variables). When we use these methods we find that there is substantial evidence for keeping both AGMN and WT continuous and lin-

Table 7.13 Estimated Coefficients, Estimated Standard Errors, Wald Statistics and Two-Tailed *p*-Values for the Multivariable Model Using NVMR

Variable	Coeff.	Std. Err.	z	P>|z|
CHK	−1.161	0.4470	−2.59	0.009
AGMN	0.359	0.1279	2.81	0.005
WT	−0.028	0.0100	−2.83	0.005
NVMR	1.593	0.7361	2.17	0.030

Log likelihood = − 45.4390

ear in the logit. Since the methods are the same and their results do not demonstrate anything new, we feel that little is to be gained by presenting the analysis in detail. We move to considering the need to add interactions to the model in Table 7.13.

We examine the need for interactions using the same method used for the 1–1 design in the Section 7.3. We feel that there is some clinical plausibility for interactions involving each of the covariates in the model as well as ones with the matching variable. We fit models adding each of these to the main effects model. Two interactions, AGMT×AGMN and AGMT×NVMR, were significant at approximately the 0.09 level. Since they were not significant at the 5 percent level and since we have only 50 strata, we chose not to include them in the model. Hence we move to model assessment and fit using the main effects model in Table 7.13.

7.6 METHODS FOR ASSESSMENT OF FIT IN A 1–*M* MATCHED STUDY

The approach to assessment of fit in the 1–*M* matched study is similar to that used in the 1–1 matched study in that it is based on extensions of regression diagnostics for the unconditional logistic regression model. The mathematics required to develop these statistics is at a higher level than other sections of the book. Hence, less sophisticated mathematical readers may wish to skip this section and proceed to Section 7.7 where the use of the diagnostic statistics is explained and illustrated. These diagnostic statistics are derived for a general matched design by Moolgavkar, Lustbader and Venzon (1985) and Pregibon (1984). These authors illustrate the use of the diagnostics only for the 1–1 matched design. We showed in Section 7.4 that the diagnostics for the 1–1 matched design could be computed using logistic regression software for the conditional model. Unfortunately, currently available software for logistic regression in the 1–*M* matched design does not compute these same diagnostic statistics. The methods needed to obtain them are, in principle, easy to apply; in practice, the computations necessary to calculate leverage values are tedious. Once the leverage values are obtained, the values of the other diagnostic statistics are calculated via simple transformations of available or easily computed quantities. To simplify the notation somewhat we present the methods for the case when $M = 3$; that is, $M + 1 = 4$.

The first step is to transform the observed values of the covariate vector by centering them about a weighted stratum-specific mean. That is, we compute for each stratum, k, and each subject within each stratum, j,

$$\tilde{\mathbf{x}}_{kj} = \mathbf{x}_{kj} - \sum_{l=1}^{4} \mathbf{x}_{kl}\hat{\theta}_{kl},$$

where

$$\hat{\theta}_{kj} = \frac{e^{\mathbf{x}'_{kj}\hat{\boldsymbol{\beta}}}}{\sum_{l=1}^{4} e^{\mathbf{x}'_{kj}\hat{\boldsymbol{\beta}}}}$$

and note that $\sum_{j=1}^{4}\hat{\theta}_{kj} = 1$. Let $\tilde{\mathbf{X}}$ be the $n = 4K$ by p matrix whose rows are the values of $\tilde{\mathbf{x}}_{kj}$, $k = 1,2,\ldots,K$ and $j = 1,2,3,4$. Let $\mathbf{U}$ be an n by n diagonal matrix with general diagonal element $\hat{\theta}_{kj}$. It may be shown that the maximum likelihood estimate, $\hat{\boldsymbol{\beta}}$, once obtained can be re-computed via the equation

$$\hat{\boldsymbol{\beta}} = \left(\tilde{\mathbf{X}}'\mathbf{U}\tilde{\mathbf{X}}\right)^{-1}\tilde{\mathbf{X}}'\mathbf{U}\mathbf{z},$$

where $\mathbf{z}$ is the vector $\mathbf{z} = \tilde{\mathbf{X}}'\hat{\boldsymbol{\beta}} + \mathbf{U}^{-1}\left(\mathbf{y} - \hat{\boldsymbol{\theta}}\right)$, $\mathbf{y}$ is the vector of values of the outcome variable ($y = 1$ for the case and $y = 0$ for the controls), and $\hat{\boldsymbol{\theta}}$ is the vector whose components are $\hat{\theta}_{kj}$. Recall that $\hat{\theta}_{kj}$ is, under the assumption of a logistic regression model, the estimated conditional probability that subject j within stratum k is a case.

Thus we may recompute the maximum likelihood estimate for the conditional logistic regression model using a linear regression program allowing case weights. We use the vector $\tilde{\mathbf{x}}_{kj}$ as values of the independent variables,

$$z_{kj} = \tilde{\mathbf{x}}'_{kj}\hat{\boldsymbol{\beta}} + \frac{y_{kj} - \hat{\theta}_{kj}}{\hat{\theta}_{kj}}$$

as the values of the dependent variable, and case weight $\hat{\theta}_{kj}$, for $k = 1, 2, \ldots, K$, $j = 1, 2, 3, 4$. It follows that the diagonal elements of the hat matrix computed by the linear regression are the leverage values we need, namely

$$h_{kj} = \hat{\theta}_{kj} \tilde{\mathbf{x}}'_{kj} \left(\tilde{\mathbf{X}}' \mathbf{U} \tilde{\mathbf{X}} \right)^{-1} \tilde{\mathbf{x}}_{kj} \; . \tag{7.8}$$

We note that one must pay close attention to how weights are handled in the statistical package used for the weighted linear regression. For example, SAS's regression procedure outputs the values as defined in equation (7.8). STATA users need to multiply the leverage values created following the weighted regression by $\hat{\theta}_{kj} / \bar{\theta}$ to obtain the leverage values defined in equation (7.8), where $\bar{\theta} = \sum_{k=1}^{K} \sum_{j=1}^{M+1} \hat{\theta}_{kj} / [K(M+1)]$ is the mean of the estimated logistic probabilities.

The Pearson residual is

$$r_{kj} = \frac{\left(y_{kj} - \hat{\theta}_{kj} \right)}{\sqrt{\hat{\theta}_{kj}}} ,$$

and the Pearson chi-square is

$$X^2 = \sum_{k=1}^{K} \sum_{j=1}^{M+1} \frac{\left(y_{kj} - \hat{\theta}_{kj} \right)^2}{\hat{\theta}_{kj}} \; .$$

The standardized Pearson residual is

$$r_{skj} = \frac{\left(y_{kj} - \hat{\theta}_{kj} \right)}{\left[\hat{\theta}_{kj} \left(1 - h_{kj} \right) \right]^{1/2}} \; .$$

In keeping with the diagnostics for the unmatched design we define the square of the standardized residual as the lack of fit diagnostic

$$\Delta X^2_{kj} = r^2_{skj} \; , \tag{7.9}$$

and the influence diagnostic as

$$\Delta\hat{\boldsymbol{\beta}}_{kj} = \Delta X_{kj}^2 \frac{h_{kj}}{1-h_{kj}} \quad . \tag{7.10}$$

We feel that the most informative way to view the diagnostic statistics is via a plot of their values versus the fitted values, $\hat{\theta}_{kj}$. These plots are similar to those used in Chapter 5 to assess graphically the fit of the unconditional logistic regression model and those used in Section 7.4 for the conditional logistical regression model in the 1–1 matched design. Examples of these plots are presented in the next section where we assess the fit of the model in Table 7.13.

Moolgavkar, Lustbader and Venzon (1985) and Pregibon (1984) suggest that one use the stratum-specific totals of the two diagnostics, ΔX^2 and $\Delta\hat{\boldsymbol{\beta}}$ to assess what effect the data in an entire stratum have on the fit of the model. These statistics are computed as quadratic forms involving not only the leverage values for the subjects in the stratum but also those terms in the hat matrix that account for the correlation among the fitted values. An easily computed approximation to these statistics is obtained by ignoring the off diagonal elements in the hat matrix. We feel that the approximations are likely to be accurate enough for practical purposes. For the kth stratum these are

$$\Delta X_k^2 = r_{sk}^2 = \sum_{j=1}^{4} r_{skj}^2 \tag{7.11}$$

and

$$\Delta\hat{\boldsymbol{\beta}}_k = \sum_{j=1}^{4} \Delta\hat{\boldsymbol{\beta}}_{kj} \quad . \tag{7.12}$$

Strata with large values of these statistics would be judged to be poorly fit and/or have large influence respectively. One can use a box plot or a plot of their values versus stratum number to identify those strata with exceptionally large values. For these strata the individual contributions to these quantities should be examined carefully to determine whether cases and/or controls are the cause of the large values.

We note that the diagnostic statistics described in this section can be used instead of the diagnostics described in Section 7.4 for the 1-1 matched study. However, we feel that the diagnostic statistics described in Section 7.4 are easier to compute since one only needs the difference variables and the diagnostics can be obtained from available logistic regression software. In addition, the diagnostics in Section 7.4 yield one value per stratum. The mathematical relationships between the diagnostic statistics in Section 7.4 and the ones described in this section are quite complex. For example, the stratum totals described in equations 7.10 and 7.11 are not arithmetically equal to the values of ΔX^2 and $\Delta\hat{\beta}$ used in Section 7.4. While it may appear that we have two sets of different diagnostic statistics they do identify the same strata as being poorly fit or influential. Thus from a practical point of view one may use either set to assess model adequacy.

In identifying poorly fit or influential subjects deletion of the case in a stratum is tantamount to deletion of all subjects in the stratum. Without a case a stratum contributes no information to the likelihood function. If some but not all controls are deleted in a specific stratum then the stratum may still have enough information to contribute to the likelihood function. A final decision on exclusion or inclusion of cases (entire strata) or controls should be based on the clinical plausibility of the data.

7.7 AN EXAMPLE OF ASSESSMENT OF FIT IN A 1–*M* MATCHED STUDY

As an example we assess the fit of the model fit to the 1–3 matched data from the Benign Breast Disease Study in Table 7.13. The steps in the process are the same as those demonstrated for 1–1 matched studies.

We begin by examining descriptive statistics for the three diagnostic statistics defined in equations (7.8) to (7.10). These analyses show (output not presented) that there is one subject, the case in stratum 12, with an extraordinarily large value for the fit diagnostic statistic, $\Delta X^2_{12,1} = 84.546$. This value is so large that it completely obscures the ability to assess the magnitude of the diagnostic statistic for the other subjects. Thus, we decide to exclude this subject from the usual plots of the diagnostic statistics versus the fitted values. The subject is included when we assess the effect of deletion of subjects on the estimated coefficients in Table 7.14 and Table 7.15 below.

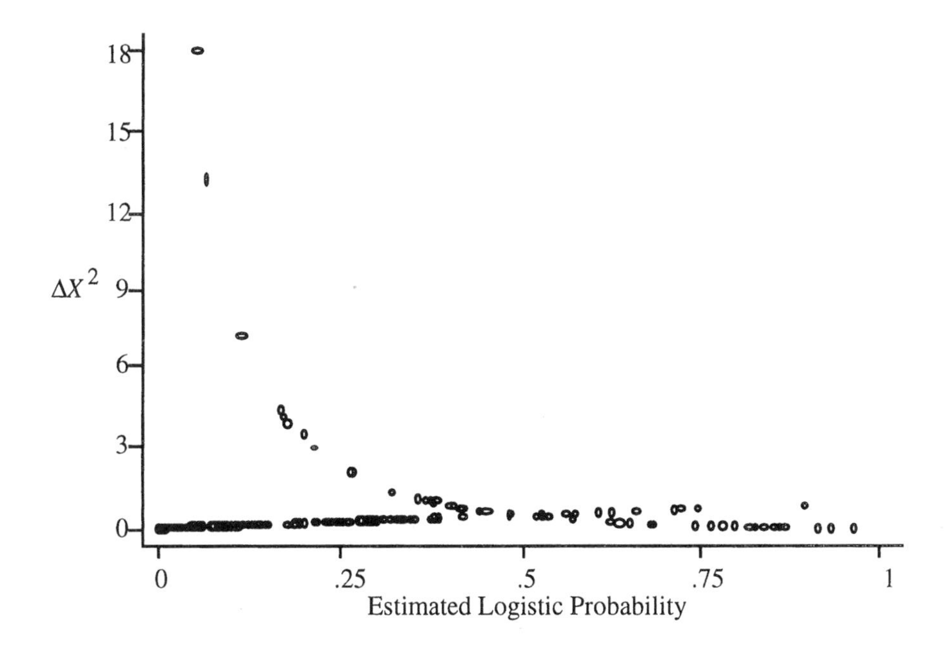

Figure 7.5 Plot of ΔX^2 versus the estimated logistic probability from the fitted model in Table 7.13.

In Figure 7.5 we plot the fit diagnostic, ΔX^2, versus the fitted values, $\hat{\theta}$. We see that there are three cases that are poorly fit with $\Delta X^2 > 6$. These large values occur in the region where $\hat{\theta} < 0.2$ and reflect cases whose estimated probability of being the case is much smaller than the fitted values for the controls in that stratum. Note that the values of ΔX^2 for the controls correspond to the points beginning at the origin $(0,0)$ rising ever so slightly and end at about 0.9.

Next, we examine the plot of the influence diagnostic statistics versus the estimated logistic probabilities shown in Figure 7.6. We note one point with $\Delta\hat{\beta} \approx 0.74$, lies well above all the other points. We note three more points, with values between about 0.2 and 0.25, warrant further examination and evaluation since they lie away from the other points.

In order to explore better the relationship between fit, influence and leverage we plot ΔX^2 versus the estimated logistic probabilities with the size of the plotting symbol proportional to $\Delta\hat{\beta}$ in Figure 7.7. Here we see that the poorly fit case with the largest influence dominates the

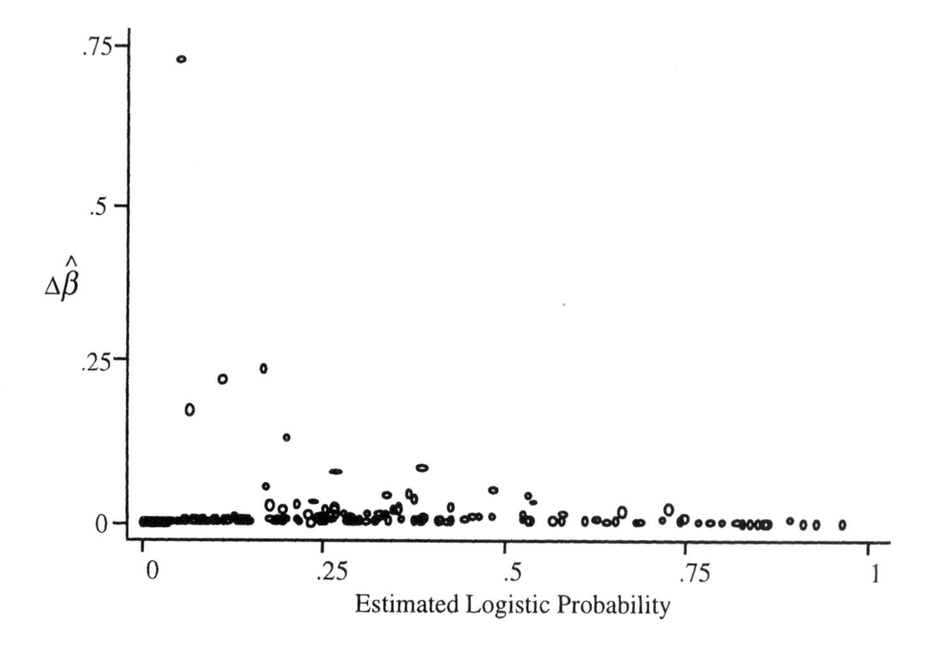

Figure 7.6 Plot of $\Delta\hat{\beta}$ versus the estimated logistic probability from the fitted model in Table 7.13.

plot. Also, two of the three additional points identified in Figure 7.6 correspond to the other two poorly fit cases seen in Figure 7.5. The fourth influential point is not one identified as being especially poorly fit. In all we have four cases and their controls from these plots to examine in more detail (there are 5 all together). As we noted in Chapter 5, for fitted logistic regression models leverage typically tends not to be too large when the fitted values are less than 0.2. This is the case with the fitted model in Table 7.13. Thus we conclude that the large values of $\Delta\hat{\boldsymbol{\beta}}$ are due primarily to large values of ΔX^2, with a small to modest contribution of leverage.

We examined the diagnostic statistics defined in equations (7.11) and (7.12) in two ways. First, we plotted them versus the corresponding stratum number. Second, we examined them using box and whisker plots. Both plots identified the same five strata corresponding to the strata for the four points identified in Figures 7.5 to 7.7 and the dominant point excluded from these plots. The data for the case and three controls in these strata are presented in Table 7.14 along with the values of the diagnostic statistics.

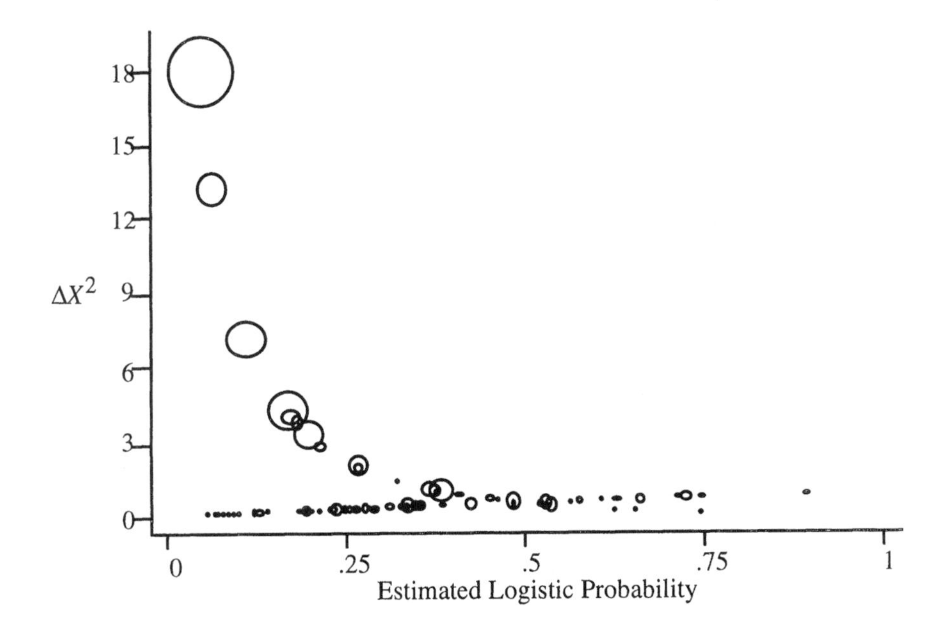

Figure 7.7 Plot of ΔX^2 versus the estimated logistic probability from the fitted model in Table 7.13 with the size of the plotting symbol proportional to $\Delta\hat{\boldsymbol{\beta}}$.

The next step is to exclude the subjects from the analysis and assess the effect of the deletions on the fitted model. Each of the five subjects identified as being poorly fit or influential is a case. Thus, when we delete the subject only the controls are left. Hence, the outcome does not vary and the stratum is deleted from the analysis. The results of fitting the model deleting the strata in Table 7.14 are presented in Table 7.15.

Interpreting the results of deleting an individual stratum in a 1–*M* matched study is more difficult than for the 1–1 design. The reason is that each subject in the stratum has his/her own diagnostic statistics yet it is likely that only one subject is influential and/or poorly fit. The reason(s) for a particular subject's influence and/or lack of fit depend(s) on the relationship between his/her covariates and those of the other *M* subjects in the stratum.

We begin with stratum 12. The case in stratum 12 is the poorest fit with $\Delta X^2_{12,1} = 84.5$. The results in Table 7.15 show that when stratum 12 is deleted $\Delta\hat{\boldsymbol{\beta}}\%$ is between about 24 and 26 percent for AGMN and WT. The estimated logistic probabilities indicate that the third control has the

Table 7.14 Stratum, Data, Estimated Probability, and the Three Diagnostic Statistics $\Delta\hat{\boldsymbol{\beta}}$, ΔX^2 and Leverage (h) for Five Influential and Poorly Fit Strata

STR	Case / Cont	CHK	AGMN	WT	NVMR	$\hat{\theta}$	$\Delta\hat{\beta}$	ΔX^2	h
10	1	2	12	105	0	0.11	0.22	7.21	0.03
	0	1	13	115	0	0.39	0.01	0.40	0.02
	0	2	12	120	0	0.07	<.01	0.07	0.02
	0	1	16	150	0	0.42	0.02	0.45	0.05
12	1	2	10	170	0	0.01	0.71	84.5	0.01
	0	1	13	140	0	0.26	0.02	0.27	0.06
	0	1	11	240	0	0.01	<.01	0.01	0.01
	0	2	16	100	0	0.72	0.02	0.75	0.03
18	1	2	14	135	0	0.05	0.73	17.9	0.04
	0	1	14	132	1	0.89	0.01	0.90	0.01
	0	1	11	205	0	0.01	<.01	0.01	0.01
	0	1	10	127	0	0.05	<.01	0.05	0.03
24	1	2	15	145	0	0.07	0.17	13.2	0.01
	0	1	13	140	0	0.12	<.01	0.12	0.01
	0	1	17	155	0	0.33	0.01	0.34	0.03
	0	1	15	116	0	0.48	0.01	0.49	0.02
31	1	2	16	156	0	0.17	0.24	4.35	0.05
	0	2	12	161	0	0.03	<.01	0.03	0.01
	0	1	13	150	0	0.22	<.01	0.22	0.01
	0	1	13	115	0	0.58	0.01	0.59	0.02

highest probability of being the case and the case has the lowest (within round off). The lack of fit diagnostic is large due to the fact $y_{12,1} = 1$ and $\hat{\theta}_{12,1} \approx 0.01$. The change in the coefficients for AGMN and WT are large and positive because we have eliminated a case that has data favoring control status (young age at menarche and heavier weight) and a control that has data favoring case status (older age at menarche and lighter weight). There is nothing particularly abnormal about the data for any of the subjects in stratum 12. Thus, we conclude that we should not delete the stratum solely on the basis of large ΔX^2.

We see that the results of deleting stratum 18 are similar to those for stratum 12 in Table 7.15. The case in this stratum has the largest

Table 7.15 Estimated Coefficients from Table 7.13, Estimated Coefficients Obtained When Deleting a Selected Stratum, Percent Change from the All Data Model and the Pearson Chi-Square

Data	CHK	AGMN	WT	NVMR	X^2
All	−1.161	0.359	−0.028	1.593	186.34
Delete 10	−1.342	0.404	−0.025	1.685	178.45
Pct. Change	15.6	12.6	−9.7	5.76	
Delete 12	−1.241	0.452	−0.035	1.679	101.51
Pct. Change	6.8	25.7	23.9	5.4	
Delete 18	−1.479	0.368	−0.029	2.247	168.13
Pct. Change	27.3	2.4	3.7	41.0	
Delete 24	−1.366	0.366	−0.030	1.687	172.36
Pct. Change	17.6	2.1	5.3	5.9	
Delete 31	−1.349	0.312	−0.031	1.664	181.38
Pct. Change	16.2	−13.1	12.5	4.5	

value of $\Delta\hat{\beta}\%$ and ΔX^2 is also quite large. In particular, we see that the coefficient for NVMR changes by over 40 percent when the stratum is deleted. The reason is one of the controls was never married while the case was married. Additional analysis showed that 38 of the 50 strata had all four subjects with $NVMR = 0$ and in the remaining 12 only one subject was never married. In essence we have removed $1/12^{\text{th}}$ of the data available for estimating the coefficient for NVMR. The coefficient for CHK changes by about 27 percent. In stratum 18 the case did not have regular check-ups, $CHK = 2$, while each of the three controls did have them. The fact that the estimated coefficient is negative implies that not having regular check-ups is consistent with control status. Thus, deleting a stratum that, in a sense, goes against the model induces change in the direction of the effect, negative. As is the case with stratum 12 all data appear to be reasonable and there is no reason, other than statistical, to exclude stratum 18.

The diagnostic statistics for stratum 24 indicate that case is poorly fit. The estimated logistic probability for the case is small and each of the controls has larger estimated probabilities (two of them much larger). The coefficient for CHK has $\Delta\hat{\beta}\%$ around 18 percent. The coefficient decreases since the case in this stratum did not have regular check-ups. Changes in the other coefficients are less than 6 percent.

Table 7.16 Estimated Odds Ratios and 95% Confidence Intervals for Model in Table 7.13

Variable	Odds Ratio	95 % CI
Regular Medical Check-ups	0.31	0.130, 0.752
Age at Menarche+	2.05	1.243, 3.386
Weight*	0.75	0.620, 0.917
Never Married	4.9	1.162, 20.821

+: Odds ratio for a 2 year increase in age

*: Odds ratio for a 10 pound increase in weight

Again, there is nothing unusual about the data, we just have a case that looks a bit more like a control and vice-versa.

The diagnostic statistics for stratum 10 and stratum 31 are similar. The case is modestly influential and somewhat poorly fit. The estimated logistic probability for the case is not too large and two of the controls have much larger estimated probabilities. The coefficient for CHK has $\Delta\hat{\beta}\%$ around 16 percent. The coefficient decreases since the case in both strata did not have regular check-ups. Changes in the other coefficients are less than 15 percent. Again, there is nothing unusual about the data.

In summary, the diagnostic statistics have proven quite useful for identifying subjects whose data may be influential in estimating coefficients and/or may lead to a fitted value more consistent with the opposite of the observed outcome. In order to interpret the results in the $1-M$ design we must simultaneously consider the estimated logistic probability of the case and all controls as well as the differences in their data.

We conclude the section by presenting, in Table 7.16, the odds ratios obtained from the fitted model in Table 7.13. These results suggest that women who do not have regular medical check-ups are at 69 percent lower risk of having benign breast disease than women who do and this could be as much as a 87 percent decrease or as little as a 25 percent decrease with 95 percent confidence. This may seem like an odd result at first; but it may be a function of the fact that detection is increased with increasing medical check-ups. Women who experience older age at menarche are at greater risk of benign breast disease and the risk doubles for every two year increase. Increasing weight decreases the risk of benign breast disease at the rate of about 25 percent per 10 pound increase. This could be between 8 and 38 percent with 95 percent confidence. Never being married significantly increases the risk of disease by a factor of 5 but this factor could be as low as 1.2 based on the 95 percent confidence interval. *Note*: In a case such as this where the right endpoint is quite large we tend to focus more on the point estimate and left endpoint. The right endpoint is the one most susceptible to large estimated variance due to limited data.

EXERCISES

1. Data from the 1–3 matched Benign Breast Disease Study are used in this chapter to illustrate methods for a 1–*M* matched study. The data are described in Table 7.10. Find the best logistic regression model for a 1–1 matched design using the first of the three controls. (*Note*: It would have been possible to use any one of the three controls. Designation of the first control was arbitrary.)

2. The example in Sections 7.5 through 7.7 used only a few of the variables available in the Benign Breast Disease Study. Repeat the modeling using all the covariates.

In each Exercise the steps in fitting the model should include: (1) a complete univariable analysis, (2) an appropriate selection of variables for a multivariable model (this should include scale identification for continuous covariates and assessment of the need for interactions), (3) an assessment of fit of the multivariable model, (4) preparation and presentation of a table containing the results of the final model (this table should contain point and interval estimates for all relevant odds ratios), and (5) conclusions from the analysis.

CHAPTER 8

Special Topics

8.1 THE MULTINOMIAL LOGISTIC REGRESSION MODEL

8.1.1 Introduction to the Model and Estimation of the Parameters

In the previous seven chapters we focused on the use of the logistic regression model when the outcome variable is dichotomous or binary. The model can be easily modified to handle the case where the outcome variable is nominal with more than two levels. For example, consider a study of choice of a health plan from among three plans offered to the employees of a large corporation. The outcome variable has three levels indicating which plan, A, B or C, is chosen. Possible covariates might include gender, age, income, family size and others. The goal is to model the odds of plan choice as a function of the covariates and to express the results in terms of odds ratios for choice of different plans. McFadden (1974) proposed a modification of the logistic regression model and called it a *discrete choice model*. As a result the model is frequently referred to as the discrete choice model in business and econometric literature while it is called the *multinomial*, *polychotomous* or *polytomous* logistic regression model in the health and life sciences. We use the term *multinomial* in this text.

We could use an outcome variable with any number of levels to illustrate the extension of the model and methods. However, the details are most easily illustrated with three categories. Further generalization to more than three categories is a problem more of notation than of concept. Hence, in the remainder of this section, we consider only the situation where the outcome variable has three categories.

When one considers a regression model for a discrete outcome variable with more than two responses, one must pay attention to the measurement scale. In this section, we discuss the logistic regression model for the case in which the outcome is nominal scale. We discuss logistic regression models for ordinal scale outcomes in the next section.

We assume that the categories of the outcome variable, Y, are coded 0, 1, or 2. In practice one should check that the software package that is going to be used allows a zero code since we have used packages that require that the codes begin with 1. Recall that the logistic regression model we use for a binary outcome variable is parameterized in terms of the logit of $Y=1$ versus $Y=0$. In the three outcome category model we need two logit functions. We have to decide which outcome categories to compare. The obvious extension is to use $Y=0$ as the referent or baseline outcome and to form logits comparing $Y=1$ and $Y=2$ to it. We show later in this section that the logit function for $Y=2$ versus $Y=1$ is the difference between these two logits.

To develop the model, assume we have p covariates and a constant term, denoted by the vector, **x**, of length $p+1$ where $x_0=1$. We denote the two logit functions as

$$\begin{aligned} g_1(\mathbf{x}) &= \ln\left[\frac{P(Y=1|\mathbf{x})}{P(Y=0|\mathbf{x})}\right] \\ &= \beta_{10}+\beta_{11}x_1+\beta_{12}x_2+\cdots+\beta_{1p}x_p \\ &= \mathbf{x}'\boldsymbol{\beta}_1 \end{aligned} \tag{8.1}$$

and

$$\begin{aligned} g_2(\mathbf{x}) &= \ln\left[\frac{P(Y=2|\mathbf{x})}{P(Y=0|\mathbf{x})}\right] \\ &= \beta_{20}+\beta_{21}x_1+\beta_{22}x_2+\cdots+\beta_{2p}x_p \\ &= \mathbf{x}'\boldsymbol{\beta}_2. \end{aligned} \tag{8.2}$$

It follows that the conditional probabilities of each outcome category given the covariate vector are

$$\mathrm{P}(Y=0|\mathbf{x})=\frac{1}{1+e^{g_1(\mathbf{x})}+e^{g_2(\mathbf{x})}}, \tag{8.3}$$

$$\mathrm{P}(Y=1|\mathbf{x})=\frac{e^{g_1(\mathbf{x})}}{1+e^{g_1(\mathbf{x})}+e^{g_2(\mathbf{x})}}, \tag{8.4}$$

and

$$\mathrm{P}(Y=2|\mathbf{x})=\frac{e^{g_2(\mathbf{x})}}{1+e^{g_1(\mathbf{x})}+e^{g_2(\mathbf{x})}}. \tag{8.5}$$

Following the convention for the binary model, we let $\pi_j(\mathbf{x})=\mathrm{P}(Y=j|\mathbf{x})$ for $j=0,1,2$. Each probability is a function of the vector of $2(p+1)$ parameters $\boldsymbol{\beta}'=(\boldsymbol{\beta}_1',\boldsymbol{\beta}_2')$.

A general expression for the conditional probability in the three category model is

$$\mathrm{P}(Y=j|\mathbf{x})=\frac{e^{g_j(\mathbf{x})}}{\sum_{k=0}^{2}e^{g_k(\mathbf{x})}},$$

where the vector $\boldsymbol{\beta}_0=\mathbf{0}$ and $g_0(\mathbf{x})=0$.

To construct the likelihood function we create three binary variables coded 0 or 1 to indicate the group membership of an observation. We note that these variables are introduced only to clarify the likelihood function and are not used in the actual multinomial logistic regression analysis. The variables are coded as follows: if $Y=0$ then $Y_0=1$, $Y_1=0$, and $Y_2=0$; if $Y=1$ then $Y_0=0$, $Y_1=1$, and $Y_2=0$; and if $Y=2$ then $Y_0=0$, $Y_1=0$, and $Y_2=1$. We note that no matter what value Y takes on, the sum of these variables is $\sum_{j=0}^{2}Y_j=1$. Using this notation it follows that the conditional likelihood function for a sample of n independent observations is

$$l(\boldsymbol{\beta})=\prod_{i=1}^{n}\left[\pi_0(\mathbf{x}_i)^{y_{0i}}\,\pi_1(\mathbf{x}_i)^{y_{1i}}\,\pi_2(\mathbf{x}_i)^{y_{2i}}\right].$$

Taking the log and using the fact that $\sum y_{ji} = 1$ for each i, the log-likelihood function is

$$L(\boldsymbol{\beta}) = \sum_{i=1}^{n} y_{1i} g_1(\mathbf{x}_i) + y_{2i} g_2(\mathbf{x}_i) - \ln\left(1 + e^{g_1(\mathbf{x}_i)} + e^{g_2(\mathbf{x}_i)}\right). \quad (8.6)$$

The likelihood equations are found by taking the first partial derivatives of $L(\boldsymbol{\beta})$ with respect to each of the $2(p+1)$ unknown parameters. To simplify the notation somewhat, we let $\pi_{ji} = \pi_j(\mathbf{x}_i)$. The general form of these equations is:

$$\frac{\partial L(\boldsymbol{\beta})}{\partial \beta_{jk}} = \sum_{i=1}^{n} x_{ki}\left(y_{ji} - \pi_{ji}\right) \quad (8.7)$$

for $j = 1,2$ and $k = 0,1,2,\ldots,p$, with $x_{0i} = 1$ for each subject.

The maximum likelihood estimator, $\hat{\boldsymbol{\beta}}$, is obtained by setting these equations equal to zero and solving for $\boldsymbol{\beta}$. The solution requires the same type of iterative computation that is used to obtain the estimate in the binary outcome case.

The matrix of second partial derivatives is required to obtain the information matrix and the estimator of the covariance matrix of the maximum likelihood estimator. The general form of the elements in the matrix of second partial derivatives is as follows:

$$\frac{\partial^2 L(\beta)}{\partial \beta_{jk} \partial \beta_{jk'}} = -\sum_{i=1}^{n} x_{k'i} x_{ki} \pi_{ji}\left(1 - \pi_{ji}\right) \quad (8.8)$$

and

$$\frac{\partial^2 L(\beta)}{\partial \beta_{jk} \partial \beta_{j'k'}} = \sum_{i=1}^{n} x_{k'i} x_{ki} \pi_{ji} \pi_{j'i} \quad (8.9)$$

for j and $j' = 1,2$ and k and $k' = 0,1,2,\ldots,p$. The observed information matrix, $\mathbf{I}(\hat{\boldsymbol{\beta}})$, is the $2(p+1)$ by $2(p+1)$ matrix whose elements are the negatives of the values in equations (8.8) and (8.9) evaluated at $\hat{\boldsymbol{\beta}}$. The estimator of the covariance matrix of the maximum likelihood estimator is the inverse of the observed information matrix,

$$\hat{\text{Var}}(\hat{\boldsymbol{\beta}}) = \mathbf{I}(\hat{\boldsymbol{\beta}})^{-1}.$$

A more concise representation for the estimator of the information matrix may be obtained by using a form similar to the binary outcome case. Let the matrix $\mathbf{X}$ be the n by $p+1$ matrix containing the values of the covariates for each subject, let the matrix $\mathbf{V}_j$ be the n by n diagonal matrix with general element $\hat{\pi}_{ji}(1-\hat{\pi}_{ji})$ for $j=1,2$ and $i=1,2,3,\ldots,n$, and let $\mathbf{V}_3$ be the n by n diagonal matrix with general element $\hat{\pi}_{1i}\hat{\pi}_{2i}$. The estimator of the information matrix may be expressed as

$$\hat{\mathbf{I}}(\hat{\boldsymbol{\beta}}) = \begin{bmatrix} \hat{\mathbf{I}}(\hat{\boldsymbol{\beta}})_{11} & \hat{\mathbf{I}}(\hat{\boldsymbol{\beta}})_{12} \\ \hat{\mathbf{I}}(\hat{\boldsymbol{\beta}})_{21} & \hat{\mathbf{I}}(\hat{\boldsymbol{\beta}})_{22} \end{bmatrix}, \tag{8.10}$$

where

$$\hat{\mathbf{I}}(\hat{\boldsymbol{\beta}})_{11} = (\mathbf{X}'\mathbf{V}_1\mathbf{X}),$$

$$\hat{\mathbf{I}}(\hat{\boldsymbol{\beta}})_{22} = (\mathbf{X}'\mathbf{V}_2\mathbf{X}),$$

and

$$\hat{\mathbf{I}}(\hat{\boldsymbol{\beta}})_{12} = \hat{\mathbf{I}}(\hat{\boldsymbol{\beta}})_{21} = -(\mathbf{X}'\mathbf{V}_3\mathbf{X}).$$

8.1.2 Interpreting and Assessing the Significance of the Estimated Coefficients

Data from a study undertaken to assess factors associated with women's knowledge, attitude, and behavior toward mammography have been made available to us by Dr. J. Zapka of the Division of Preventive and Behavioral Medicine, University of Massachusetts Medical School. Results from the full study may be found in Zapka, Stoddard, Maul, and Costanza, (1991), Costanza, Stoddard, Gaw, and Zapka, (1992) and Zapka, Hosmer, Costanza, Harris and Stoddard, (1992). The data used in this text are a subset of the data from the main study and have been

Table 8.1 Code Sheet for the Variables in the Mammography Experience Study

Variable	Description	Codes/Values	Name
1	Identification Code	1-412	OBS
2	Mammography Experience	0 = Never 1 = Within One Year 2 = Over One Year Ago	ME
3	"You do not need a mammogram unless you develop symptoms"	1 = Strongly Agree 2 = Agree 3 = Disagree 4 = Strongly Disagree	SYMPT
4	Perceived benefit of mammography*	5 - 20	PB
5	Mother or Sister with a history of breast cancer	0 = No 1 = Yes	HIST
6	"Has anyone taught you how to examine your own breasts: that is BSE?"	0 = No 1 = Yes	BSE
7	"How likely is it that a mammogram could find a new case of breast cancer?"	1 = Not likely 2 = Somewhat likely 3 = Very likely	DETC

*The variable PB is the sum of five scaled responses, each on a four point scale. (Women with low values perceive the greatest benefit of mammography.

modified to preserve subject confidentiality. The data are described in Table 8.1 and may be obtained from the two web sites cited in the Preface.

To simplify the discussion of the estimation and interpretation of odds ratios in the multinomial outcome setting we need to generalize the notation used in the binary outcome case to include the outcomes being compared as well as the values of the covariate. We assume that the outcome labeled with $Y=0$ is the reference outcome. The subscript on the odds ratio indicates which outcome is being compared to the reference outcome. The odds ratio of outcome $Y=j$ versus outcome $Y=0$ for covariate values of $x=a$ versus $x=b$ is

$$\mathrm{OR}_j(a,b)=\frac{\mathrm{P}(Y=j|x=a)/\mathrm{P}(Y=0|x=a)}{\mathrm{P}(Y=j|x=b)/\mathrm{P}(Y=0|x=b)}.$$

In the special case when the covariate is binary, coded 0 or 1, we simplify the notation further and let $\mathrm{OR}_j = \mathrm{OR}_j(1,0)$.

We begin by considering a model containing a single dichotomous covariate coded 0 or 1. In the binary outcome model the estimated slope coefficient is identical to the log-odds ratio obtained from the 2 by 2 table cross-classifying the outcome and the covariate. As we noted in the previous section, when the outcome has three levels there are two logit functions. We define these functions in such a way that the two estimated coefficients, one from each logit function, are equal to the log-odds ratios from the pair of 2 by 2 tables obtained by cross-classifying the $y = j$ and $y = 0$ outcomes by the covariate with $y = 0$ as the reference outcome value.

As a specific example, consider the cross-classification of mammography experience (ME) by HIST displayed in Table 8.2. When we use $ME = 0$ as the reference outcome the two odds ratios calculated from Table 8.2 are

$$\widehat{\mathrm{OR}}_1 = \frac{19 \times 220}{85 \times 14} = 3.51$$

and

$$\widehat{\mathrm{OR}}_2 = \frac{11 \times 220}{63 \times 14} = 2.74.$$

The results of fitting a three-category logistic regression model, using STATA's mlogit command, to these data are presented in Table 8.3. We obtain the values in Table 8.3, labeled $\widehat{\mathrm{OR}}$, by exponentiating the estimated slope coefficients. We note that they are identical to the

Table 8.2 Cross-Classification of Mammography Experience (ME) by Family History of Breast Cancer (HIST) and Estimated Odds Ratios Using Never as the Reference Outcome Value

	HIST			
ME	No (0)	Yes (1)	Total	$\widehat{\mathrm{OR}}$
Never (0)	220	14	234	1.0
Within 1 Year (1)	85	19	104	3.51
Over 1 Year (2)	63	11	74	2.74
Total	368	44	412	

Table 8.3 Results of Fitting the Logistic Regression Model to the Data in Table 8.2

Logit	Variable	Coeff.	Std. Err.	$\hat{OR}$	95 % CI
1	HIST	1.256	0.3747	3.51	1.685, 7.321
	Constant	−0.951	0.1277		
2	HIST	1.009	0.4275	2.74	1.187, 6.342
	Constant	−1.250	0.1429		

Log-likelihood = -- 396.1700

values obtained from Table 8.2. As is the case in the binary outcome setting with a dichotomous covariate, the estimated standard error of the coefficient is the square root of the sum of the inverse of the cell frequencies. For example, the estimated standard error of the coefficient for HIST in the first logit is

$$\hat{SE}\left(\hat{\beta}_{11}\right)=\left[\frac{1}{19}+\frac{1}{220}+\frac{1}{85}+\frac{1}{14}\right]^{0.5}=0.3747,$$

which is identical to the value in Table 8.3.

The endpoints of the confidence interval are obtained in exactly the same manner as for the binary outcome case. First we obtain the confidence interval for the coefficient, the endpoints of which are then exponentiated to obtain the confidence interval for the odds ratio. For example, the 95% CI for the odds ratio of $ME=1$ versus $ME=0$ shown in Table 8.3 is calculated as follows:

$$\exp(1.256 \pm 1.96 \times 0.3747)=(1.685, 7.321).$$

The endpoints for the confidence interval for $ME=2$ versus $ME=0$ in Table 8.3 are obtained in a similar manner.

We interpret each estimated odds ratios and its corresponding confidence interval as if it came from a binary outcome setting. In some cases it may further support the analysis to compare the magnitude of the two estimated odds ratios. This can be done with or without the support of tests of equality.

The interpretation of the effect of family history on frequency of screening is as follows: (1) The odds among women with a family history of breast cancer having a mammogram within the last year is 3.5 times greater than the odds among women without a family history. In other words, women with a family history of breast cancer are 3.5 times more likely to be frequent users of mammography screening than are women without a family history of breast cancer. The confidence interval indicates that the odds could be a little as 1.7 times or as much as 7.3 times larger with 95 percent confidence. (2) The odds among women with a family history of breast cancer of having a mammogram more than one year ago is 2.7 times greater than women without a family history. Put another way, women with a history of breast cancer are 2.7 times as likely to have had a mammogram over one year ago than are women without a family history of breast cancer. The odds could be a little as 1.2 times or as much as 6.3 times larger with 95 percent confidence. Thus we see that having a family history of breast cancer is a significant factor in use of mammography screening.

We note that the test of the equality of the two odds ratios, $\text{OR}_1 = \text{OR}_2$, is equivalent to a test that the log-odds for $ME = 2$ versus $ME = 1$ is equal to zero. The simplest way to obtain the point and interval estimate is from the difference between the two estimated slope coefficients in the logistic regression model. For example, using the frequencies in Table 8.2 and the estimated coefficients from Table 8.3 we have

$$\begin{aligned}\hat{\beta}_{21} - \hat{\beta}_{11} &= 1.009 - 1.256 \\ &= -0.247 \\ &= \ln\left(\frac{11 \times 85}{19 \times 63}\right).\end{aligned}$$

The estimator of the variance of the difference between the two coefficients, $\hat{\beta}_{21} - \hat{\beta}_{11}$, is

$$\widehat{\text{Var}}\left(\hat{\beta}_{21} - \hat{\beta}_{11}\right) = \widehat{\text{Var}}\left(\hat{\beta}_{21}\right) + \widehat{\text{Var}}\left(\hat{\beta}_{11}\right) - 2\widehat{\text{Cov}}\left(\hat{\beta}_{21}, \hat{\beta}_{11}\right).$$

We obtain values for the estimates of the variances and covariances from a listing of the estimated covariance matrix, which is an option in most, if not all, packages. As described in Section 8.1 the form of this matrix is a little different from the covariance matrix in the binary setting. There are two matrices containing the estimates of the variances and covariances of the estimated coefficients in each logit and a third containing the estimated covariances of the estimated coefficients from the different logits. The matrix for the model in Table 8.3 is shown in Table

Table 8.4 Estimated Covariance Matrix for the Fitted Model in Table 8.3

		Logit 1		Logit 2	
		HIST	Constant	HIST	Constant
Logit 1	HIST	0.1404			
	Constant	−0.0163	0.0163		
Logit 2	HIST	0.0760		0.1828	
	Constant	−0.0045	0.0045	−0.0204	0.0204

Log-likelihood = −396.1700

8.4, where Logit 1 is the logit function for $ME=1$ versus $ME=0$ and Logit 2 is the logit function for $ME=2$ versus $ME=0$.

Using the results in Table 8.4 we obtain the estimate of the variance of the difference in the two estimated coefficients as

$$\widehat{\text{Var}}\left(\hat{\beta}_{21}-\hat{\beta}_{11}\right)=0.1404+0.1828-2\times 0.0760=0.1712.$$

The endpoints of a 95 percent confidence interval for this difference are

$$-0.247\pm 1.96\times\sqrt{0.1712}=(-1.058,0.564).$$

Since the confidence interval includes zero we cannot conclude that the log odds for $ME=1$ is different from the log odds for $ME=2$. Equivalently, we can express these results in terms of odds ratios by exponentiating the point and interval estimates. This yields the odds ratio for $ME=2$ versus $ME=1$ as $\widehat{\text{OR}}=0.781$ and a confidence interval of $(0.347,1.758)$. The interpretation of this odds ratio is that the odds of less recent use is 22 percent lower than the odds of recent use among women with a family history of breast cancer, i.e., $\widehat{\text{OR}}_2 \approx 0.78\times\widehat{\text{OR}}_1$.

In practice, if there was no difference in the separate odds ratios over all model covariates then we might consider pooling outcome categories 1 and 2 into a binary ("ever" versus "never") outcome. We return to this question following model development in the next section.

We note that in a model with many covariates the extra computations required for these auxiliary comparisons could become a burden. In this setting, procedures like STATA's test or lincom are quite helpful.

A preliminary indication of the importance of the variable may be obtained from the two Wald statistics; but as is the case with any multi-degree of freedom variable, we should use the likelihood ratio test to assess significance. For example, to test for the significance of the coefficients for HIST we compare the log-likelihood from the model containing HIST to the log-likelihood for the model containing only the two constant terms, one for each logit function. Under the null hypothesis that the coefficients are zero, minus twice the change in the log-likelihood follows a chi-square distribution with 2 degrees of freedom. In the example, the log-likelihood for the constant only model is $L_0 = -402.5990$. The value of the statistic is

$$G = -2 \times [-402.5990 - (-396.1700)] = 12.86,$$

which yields a p-value of 0.002. Thus, from a statistical point of view, the variable HIST is significantly associated with a woman's decision to have a mammogram.

In general, the likelihood ratio test for the significance of the coefficients for a variable has degrees of freedom equal to the number of outcome categories minus one times the degrees of freedom for the variable in each logit. For example, if we have a four category outcome variable and a covariate that is modeled as continuous then the degrees of freedom is $(4-1)\times 1 = 3$. If we have a categorical covariate coded at five levels then the covariate has four design variables within each logit and the degrees of freedom for the test are $(4-1)\times(5-1)=12$. This is easy to keep track of if we remember that we are modeling one logit for comparing the reference outcome category to each other outcome category.

For a polychotomous covariate we expand the number of odds ratios to include comparisons of each level of the covariate to a reference level for each possible logit function. To illustrate this we consider the variable DETC modeled via two design variables using the value of 1 (not likely) as the reference covariate value. The cross-classification of ME by DETC is given in Table 8.5. Using the value of $ME = 0$ as the reference outcome category and $DETC = 1$ as the reference covariate value, the four odds ratios are as follows:

Table 8.5 Cross-Classification of Mammography Experience (ME) by DETC

	DETC			
ME	1	2	3	Total
Never (0)	13	77	144	234
Within 1 Year (1)	1	12	91	104
Over 1 Year (2)	4	16	54	74
Total	18	105	289	412

$$\widehat{\text{OR}}_1(2,1) = \frac{12 \times 13}{77 \times 1} = 2.03,$$

$$\widehat{\text{OR}}_1(3,1) = \frac{91 \times 13}{144 \times 1} = 8.22,$$

$$\widehat{\text{OR}}_2(2,1) = \frac{16 \times 13}{77 \times 4} = 0.68,$$

and

$$\widehat{\text{OR}}_2(3,1) = \frac{54 \times 13}{144 \times 4} = 1.22.$$

The results of fitting the logistic regression model to these data are presented in Table 8.6.

We see that exponentiation of the estimated logistic regression coefficients yields the odds ratios formed from 2 by 2 tables obtained

Table 8.6 Results of Fitting the Logistic Regression Model to the Data in Table 8.5

Logit	Variable	Coeff.	Std. Err.	$\widehat{\text{OR}}$	95 % CI
1	DETC_2	0.706	1.0831	2.03	0.242, 16.928
	DETC_3	2.106	1.0463	8.22	1.057, 63.864
	Constant	−2.565	1.0377		
2	DETC_2	−0.393	0.6344	0.68	0.195, 2.341
	DETC_3	0.198	0.5936	1.22	0.381, 3.901
	Constant	−1.179	0.5718		

Log-likelihood = −389.2005

from the main 3 by 3 contingency table. The odds ratios for logit 1 are obtained from the 2 by 3 table containing the rows corresponding to $ME = 0$ and $ME = 1$ and the 3 columns. The odds ratios for logit 2 are obtained from the 2 by 3 table containing the rows corresponding to $ME = 0$ and $ME = 2$ and the 3 columns.

To assess the significance of the variable DETC, we calculate minus twice the change in the log-likelihood relative to the constant only model. The value of the test statistic is

$$G = -2 \times [-402.5990 - (-389.2005)] = 26.80$$

which, with 4 degrees of freedom, yields a p-value of less than 0.001.

Thus, we would conclude that a woman's opinion on the ability of a mammogram to detect a new case of breast cancer is significantly associated with her decision to have a mammogram. Examining the estimated odds ratios and their confidence intervals we see that the association is strongest when comparing the women who have had a mammogram within the last year, $ME = 1$, to those who have never had one, and comparing the not likely to very likely response. The interpretation is that the odds of having a mammogram within the last year among women who feel that a mammogram is very likely to detect a new case of breast cancer is 8.22 times larger than the odds among women who feel that it is not likely that a mammogram can detect a new case of breast cancer. All other estimated odds ratios have confidence intervals that include 1.0. The fact that the confidence interval estimates for the logit of $ME = 1$ versus $ME = 0$ are quite wide is a function of the cell with one subject in Table 8.5. This follows from the fact that the estimated standard errors are equal to the square root of the sum of the inverse of the cell counts. For example, the estimated standard error of the coefficient for the log odds of $DETC = 3$ versus $DETC = 1$ in first logit is

$$\widehat{\text{SE}}\left(\hat{\beta}_{12}\right) = \left[\frac{1}{91} + \frac{1}{13} + \frac{1}{144} + \frac{1}{1}\right]^{0.5} = 1.0463.$$

We could compare the two sets of odds ratios over the responses of DETC in the manner similar to that illustrated for HIST to determine whether the two logit functions are different. In results not presented, this test with two degrees-of-freedom has $p = 0.045$. Thus we conclude that for DETC we should not combine the $ME = 1$ and $ME = 2$ outcome categories. The fact that this result is different from the result for HIST further indicates that a decision to collapse response categories to obtain a simpler outcome variable should not be made until we do a thorough modeling of the data.

Continuous covariates that are modeled as linear in the logit have a single estimated coefficient in each logit function. This coefficient, when exponentiated, gives the estimated odds ratio for a change of one unit in the variable. Thus, remarks in Chapter 3 about knowing what a single unit is, and estimation of odds ratios for a clinically meaningful change apply directly to each logit function in the multinomial logistic regression model.

8.1.3 Model-Building Strategies for Multinomial Logistic Regression

In principle, the strategies and methods for multivariable modeling with a multinomial outcome variable are identical to those for the binary outcome variable discussed in Chapter 4. The theory for stepwise selection of variables has been worked out and is available in some packages. However, the method is not currently available in many of the other widely distributed statistical software packages, such as STATA. To illustrate modeling and interpretation of the results, we proceed with an analysis of the data from the mammography study.

Model building in the mammography study is simplified by the fact that there are only five independent variables and 412 subjects. We do have a few decisions to make regarding how some of the variables are going to be entered into the model. In particular, the variable SYMPT is coded at four levels on an ordinal scale. Traditionally, variables of this type have either been analyzed as if they were continuous or categorical. We begin the model-building process with SYMPT coded into three design variables, using the "strongly agree" response as the reference value. The variable DETC is coded at three levels and is ordinal scale. We decided to treat it as poylchotomous with two design variables using the "not likely" response as the reference value. The rationale for coding these ordinal scale variables into design variables rather than treating them as if they were continuous is that the coefficients for the design variables may be plotted to assess the functional form of the two logits over the categories. Initially, we treat the variable PB as if it were continuous and linear in the logits. The results of fitting the full multivariable model are given in Table 8.7

Examination of the Wald statistics in Table 8.7 suggests that, with the possible exception of the variable DETC, each of the variables may

Table 8.7 Estimated Coefficients, Estimated Standard Errors, Wald Statistics and Two-Tailed p-Values for the Full Multivariable Model Fit to the Mammography Experience Data

Logit	Variable	Coeff.	Std. Err.	z	P>|z|
1	SYMPT_2	0.110	0.9228	0.12	0.905
	SYMPT_3	1.925	0.7776	2.48	0.013
	SYMPT_4	2.457	0.7753	3.17	0.002
	PB	−0.219	0.0755	−2.91	0.004
	HIST	1.366	0.4375	3.12	0.002
	BSE	1.292	0.5299	2.44	0.015
	DETC_2	0.017	1.2619	0.02	0.988
	DETC_3	0.904	1.1268	0.80	0.422
	Constant	−2.999	1.5392	−1.95	0.051
2	SYMPT_2	−0.290	0.6441	−0.45	0.652
	SYMPT_3	0.817	0.5398	1.51	0.130
	SYMPT_4	1.132	0.5477	2.07	0.039
	PB	−0.148	0.0764	−1.94	0.052
	HIST	1.065	0.4594	2.32	0.020
	BSE	1.052	0.5150	2.04	0.041
	DETC_2	−0.924	0.7137	−1.30	0.195
	DETC_3	−0.691	0.6871	−1.01	0.315
	Constant	−0.986	1.1118	−0.89	0.375

Log-likelihood = −346.9510

contribute to the model. For the moment we keep all variables in the model while we examine the coding of the variable SYMPT.

The two estimated coefficients for the design variable SYMPT_2, which estimate the log odds for agree versus the reference value of strongly agree, suggest that these two categories are similar since neither Wald statistic is significant. The sign and magnitude of the estimated coefficients for the design variables SYMPT_3 and SYMPT_4 suggest that the log odds of disagree and strongly disagree differ from strongly agree and are of similar magnitude within each of the two logit functions. To examine this further we performed a Wald test for the equality of the coefficients for SYMPT_3 and SYMPT_4 within each of the logits. The p-values for the two tests are $p = 0.070$ in the first logit and $p = 0.335$ in the second logit. These results suggest that we could use a simpler model that dichotomizes SYMPT into two levels, coded 0 =

Table 8.8 Estimated Coefficients, Estimated Standard Errors, Wald Statistics and Two-Tailed p-Values for the Model Fit Using SYMPTD to the Mammography Experience Data

Logit	Variable	Coeff.	Std. Err.	z	P>|z|
1	SYMPTD	2.095	0.4574	4.58	<0.001
	PB	−0.251	0.0729	−3.44	0.001
	HIST	1.293	0.4335	2.98	0.003
	BSE	1.293	0.5263	2.36	0.018
	DETC_2	0.090	1.1610	0.08	0.938
	DETC_3	0.973	1.1263	0.86	0.388
	Constant	−2.704	1.4344	−1.89	0.059
2	SYMPTD	1.121	0.3572	3.14	0.002
	PB	−0.168	0.0742	−2.27	0.023
	HIST	1.014	0.4538	2.24	0.025
	BSE	1.029	0.5140	2.00	0.045
	DETC_2	−0.902	0.7146	−1.26	0.207
	DETC_3	−0.670	0.6876	−0.97	0.330
	Constant	−0.999	1.0720	−0.93	0.351

Log-likelihood = −348.7480

strongly agree or agree and 1 = disagree or strongly disagree. The results of fitting the simpler model are shown in Table 8.8. The new dichotomous variable is labeled SYMPTD in the output. Our decision to use SYMPTD involved four separate Wald tests: Two tested that the coefficients for SYMPT_2 are zero and two more tested the equality of the coefficients for SYMPT_3 and SYMPT_4. The overall four degree-of-freedom combined Wald test obtained using STATA's test command with the accumulate option yields $W = 3.58$ with $p = 0.466$. This Wald test is equivalent to the likelihood ratio test comparing the model in Table 8.8 to the model in Table 8.7. The value of the test is

$$G = -2[-348.7480 - (-346.9510)] = 3.5940,$$

which, with four degrees-of-freedom, yields a $p = 0.464$. Thus we conclude that the more complicated model is no better than the simpler model using SYMPTD.

Table 8.9 Estimated Coefficients, Estimated Standard Errors, Wald Statistics and Two-Tailed *p*-Values for the Model Fit Excluding DETC to the Mammography Experience Data

Logit	Variable	Coeff.	Std. Err.	z	P>|z|
1	SYMPTD	2.230	0.4520	4.94	<0.001
	PB	−0.283	0.0713	−3.96	<0.001
	HIST	1.297	0.4293	3.02	0.003
	BSE	1.221	0.5210	2.34	0.019
	Constant	−1.789	0.8471	−2.11	0.035
2	SYMPTD	1.153	0.3514	3.28	0.001
	PB	−0.158	0.0712	−2.22	0.027
	HIST	1.061	0.4527	2.35	0.019
	BSE	0.960	0.5072	1.89	0.058
	Constant	−1.742	0.8087	−2.15	0.031

Log-likelihood = −353.0190

The next step is to evaluate the role of DETC in the model. The results in Table 8.8 show that none of the four Wald statistics is significant. We fit a model excluding DETC and the results are shown in Table 8.9. The greatest change in a coefficient is 12.7 percent for PB in logit 1. The next largest change is six percent. The value of the likelihood ratio test of the model in Table 8.9 versus the model in Table 8.8 is $G = 8.5421$ which, with four degrees-of-freedom, gives $p = 0.074$. The fact that DETC is at most a marginal confounder of only one coefficient and that the likelihood ratio test is not significant at the 0.05 level suggests that we should exclude DETC. However, one other possibility is to explore collapsing DETC into two categories. We note that the reference group, $DETC = 1$, has a frequency of only 18. Thus we form a dichotomous variable DETCD, by combining the two categories "not likely" and "somewhat likely". Note that the new dichotomous variable is equal to the design variable DETC_3. The results of fitting this model are shown in Table 8.10. We see that coefficient for DETCD has $p = 0.019$ in logit 1 and $p = 0.720$ in logit 2. The likelihood ratio test of the model in Table 8.10 versus the model in Table 8.9 is $G = 6.9055$ which, with two degrees-of-freedom, yields $p = 0.032$. Thus DETCD contributes significantly to the model. In addition, the likelihood ratio test of the model in Table 8.10 versus the model in Table 8.8 has $G = 1.6467$ which, with two degrees-of-freedom, yields $p = 0.441$. This

Table 8.10 Estimated Coefficients, Estimated Standard Errors, Wald Statistics and Two-Tailed *p*-Values for the Model Fit Using DETCD to the Mammography Experience Data

Logit	Variable	Coeff.	Std. Err.	z	P>|z|
1	SYMPTD	2.095	0.4574	4.58	<0.001
	PB	−0.249	0.0725	−3.44	0.001
	HIST	1.310	0.4336	3.02	0.003
	BSE	1.237	0.5254	2.35	0.019
	DETCD	0.885	0.3562	2.35	0.019
	Constant	−2.624	0.9264	−2.83	0.005
2	SYMPTD	1.127	0.3564	3.16	0.002
	PB	−0.154	0.0726	−2.12	0.034
	HIST	1.063	0.4528	2.35	0.019
	BSE	0.956	0.5073	1.88	0.056
	DETCD	0.114	0.3182	0.36	0.720
	Constant	−1.824	0.8551	−2.13	0.033

Log-likelihood = −349.5663

indicates that the model that uses all three categories of DETC is not better than the model using the dichotomous grouped covariate DETCD. The largest change in a coefficient when comparing these two models is eight percent for PB in logit 2. This indicates that use of the dichotomous covariate gives as good adjustment of the effects of the other covariates as the full three-category covariate. Based on these results and tests we decide to use the dichotomous variable DETCD. The next step is to assess the scale of PB.

In theory we have the same methods available to check the scale of a continuous covariate in a multinomial logistic regression model as in a binary model. However, not all methods have been fully implemented in software packages. In particular, the only method we can use with current multinomial logistic regression software is the design variable approach. An alternative is to approximate the fit of a multinomial logistic model by fitting separate binary models. Begg and Gray (1984) proposed this approach. For example, in a three group problem we would fit a model for $Y=1$ versus $Y=0$, ignoring the $Y=2$ data, using a standard logistic regression package for a binary outcome variable and then fit separately a model for $Y=2$ versus $Y=0$, ignoring the $Y=1$ data. Begg and Gray show that the estimates of the logistic regres-

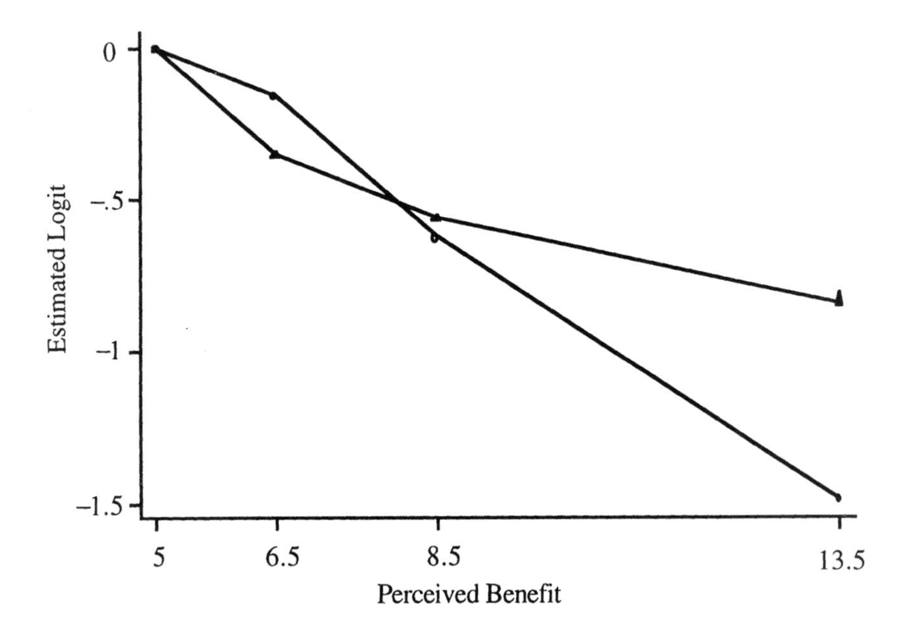

Figure 8.1 Plot of the estimated logistic regression coefficients for the quartile design variables created from PB for Logit 1 (o) and Logit 2 (Δ).

sion coefficients obtained in this manner are consistent, and under many circumstances the loss in efficiency is not too great. It has been our experience that the coefficients obtained from separately fit logistic models are close to those from the multinomial fit. This suggests that the individualized fitting approach can be useful for scale selection for continuous covariates. In particular we can use the software for fractional polynomials and scatterplot smoothing discussed and illustrated in Chapter 3.

We note that since we can approximate a full multinomial logistic model by separate binary logistic models, this opens up the possibility of performing variable selection using the stepwise or best subsets approaches discussed in Chapter 3. If at all possible, final inferences should be based on estimated coefficients and estimated standard errors from fitting the multinomial logistic regression model.

To assess the scale of the variable PB we begin by using the method of creating design variables. The values of PB are integers and range from 5 to 17, with relatively few values exceeding 10. PB was broken into four approximate quartiles corresponding to the values of 5, 6–7,

8–9 and ≥ 10. The three design variables are formed using $PB = 5$ as the reference value. A plot of the estimated logistic regression coefficients for the three design variables from the two logit functions is shown in Figure 8.1. We note that the fitted model contained all the other covariates.

The polygons for the two logit functions in Figure 8.1 show strong evidence that the logits are linear in PB. In addition we checked the scale of PB using fractional polynomials by fitting separate binary logistic regression models. Although not shown, the results indicated that neither the best $m = 1$ nor the best $m = 2$ term fractional polynomial model improved on the linear model. Hence, we choose to include the variable PB in the model as continuous and linear in each of the two logit functions.

As is the case when the outcome variable is binary, the next step in model development is to assess the need to include interaction terms in the model. In the mammography experience study each pair of variables creates a clinically plausible interaction. None of the 10 interactions contributed significantly to the model. Thus, we take as our preliminary final model the one displayed in Table 8.10.

To illustrate the Begg and Gray method of fitting individual logistic regression models we refit the model shown in Table 8.10. The results of this fit along with the maximum likelihood fit are given in Table 8.11. In Table 8.11 the columns labeled ILR give the estimated coefficients and estimated standard errors from the individualized logistic regressions and the maximum likelihood estimates are given in the columns labeled MLE.

Comparing the pairs of columns in Table 8.11, one set for estimated coefficients and the other for estimated standard errors, we see that the method of individual logistic regressions proposed by Begg and Gray provides a good approximation to both the estimates of the coefficients and estimates of the standard errors. Thus, in the absence of software capable of fitting a multinomial logistic regression model, we could use the results of individual logistic regressions, realizing of course that the resulting estimates are approximations to the maximum likelihood estimates.

One problem that we were not faced with in the binary outcome case but which can be an issue in a multinomial logistic regression model occurs when a covariate is significant for some but not all logit functions. If we model using the principle that we would like to minimize the number of parameters, then we should force the coefficients to be zero in some logit functions and estimate their values for the other

Table 8.11 Comparison of the Maximum Likelihood Estimates, MLE, and the Estimates from Individual Logistic Regression Fits, ILR

		MLE	ILR	MLE	ILR
Logit	Variable	Coeff.	Coeff.	Std. Err.	Std. Err.
1	SYMPTD	2.095	2.091	0.4574	0.4651
	PB	−0.249	−0.243	0.0725	0.0738
	HIST	1.310	1.385	0.4336	0.4683
	BSE	1.237	1.363	0.5254	0.5339
	DETCD	0.885	0.853	0.3562	0.3655
	Constant	−2.624	−2.765	0.9264	09422
2	SYMPTD	1.127	1.153	0.3564	0.3566
	PB	−0.154	−0.154	0.0726	0.0726
	HIST	1.063	1.098	0.4528	0.4593
	BSE	0.956	0.953	0.5073	0.5097
	DETCD	0.114	0.099	0.3182	0.3191
	Constant	−1.824	−1.838	0.8551	0.8600

logit functions. This strategy is not possible with currently available multinomial logistic regression software, but can be accommodated using the individualized logistic regression approach. As in all modeling situations clinical considerations should play an important role in variable selection.

Finally, if the analysis is performed via individual logistic regressions, we may employ currently available software and use the variable selection strategies described in Chapter 4 for each logit function.

8.1.4 Assessment of Fit and Diagnostics for the Multinomial Logistic Regression Model

As with any fitted model, before we use it to make inferences, we should assess its overall fit and examine the contribution of each subject to the fit. In multinomial logistic regression, the multiple outcome categories make this a more difficult problem than was the case with a model for a binary outcome variable. When we model a binary outcome variable we have a single fitted value, the estimated logistic probability of the outcome being present, $\mathrm{P}(Y=1|\mathbf{x})$. When the outcome variable has three categories we have two estimated logistic probabilities, the estimated probabilities of categories 1 and 2, $\mathrm{P}(Y=1|\mathbf{x})$ and $\mathrm{P}(Y=2|\mathbf{x})$. Lesaffre (1986) and Lesaffre and Albert (1989) have proposed extensions of tests for goodness-of-fit and logistic regression diagnostics to the multinomial logistic regression model. However, these methods are not that easy to calculate using available software. Thus, until software developers add these methods to their packages we recommend assessing fit and calculating logistic regression diagnostics using the individual logistic regressions approach of Begg and Gray.

For an outcome variable with three categories, we suggest assessing the fit of the two logistic regression models and then integrating the results, usually descriptively, to make a statement about the fit of the multinomial logistic regression model. The procedure for assessing the fit of each individual logistic regression model is described in Chapter 5. Integration of the results requires thoughtful consideration of the effects of influential and poorly fit covariate patterns on each logit function. In particular, covariate patterns that are influential for only one logit should be examined closely with due consideration to clinical issues before they are excluded from analyses. While this process requires more computation than for a single logistic regression model for a binary outcome variable, there is nothing new conceptually.

We illustrate the methods by considering assessment of fit of the multinomial logistic regression model shown in Table 8.10 for the mammography experience study. Summary goodness-of-fit statistics are presented in Table 8.12 for each of the individual logistic regression models. Recall that logit model 1 refers to the logistic regression comparing the women who had a mammogram within a year of the interview ($ME=1$) to the women who never had a mammogram ($ME=0$) and logit model 2 compares the women who had a mammogram over 1 year prior to the interview ($ME=2$) to the women who never had a mammogram ($ME=0$). These statistics are calculated using the observed covariate patterns generated by the variables in the model. For logit model 1 there were $J=74$ patterns and for logit model 2 there

Table 8.12 Summary Goodness-of-Fit Statistics (p-value) for the Individual Logistic Regressions

Logit	HL ($\hat{C}$)	Pearson (X^2)	Stukel (S)
1	12.20 (0.142)	67.84 (0.996)	1.02 (0.601)
2	9.62 (0.293)	68.83 (0.733)	1.86 (0.393)

were $J = 75$ patterns.

The Hosmer-Lemeshow statistics in the first column of Table 8.12 have values of 12.20 (df = 8, p = 0.142) and 9.62 (df = 8, p = 0.293). The second column of Table 8.12 contains the values of the Pearson chi-square statistic computed by covariate pattern. The reported p-values are calculated using the normal approximation discussed in Chapter 5. The third column of Table 8.12 contains the values of Stukel's test (see Chapter 5 for details on how it is calculated). The p-values for all three tests in both logits are not significant, indicating good overall fit of the model.

The leverage, h, and diagnostic statistics $\Delta\hat{\boldsymbol{\beta}}$, ΔX^2, and ΔD defined in equations (5.12) and (5.14)–(5.16) were calculated for each covariate pattern for each of the two individually fit logistic regression models. Plots similar to those shown in Chapter 5 identified several patterns with large values for one or more statistics. Information for these patterns is summarized in Table 8.13. The quantity $P\#$ is an arbitrary designation for covariate pattern within each individually fit model. Its value depends on the order in which the covariate patterns are formed. Pattern numbers are provided to facilitate discussion of the values of the diagnostic statistics.

Examining the diagnostic statistics for logit model 1 we see why ΔX^2 is quite large for covariate pattern 4. The estimated logistic probability is small yet the observed probability, y_j/m_j, is 0.5. This difference generates an extremely large Pearson residual. The deviance residual, while not quite as large, is also considered significant. The extremely small value for the leverage is the primary reason that $\Delta\hat{\boldsymbol{\beta}}$ is not especially large. Based on the clinical plausibility of the observed covariates we have little reason to exclude the two subjects in covariate pattern 4.

Covariate pattern 63 for logit model 1 presents a new challenge in assessing the fit of a model. The responses to the variables SYMPTD, HIST, BSE, and DETCD in this pattern are what we might call a "modal" response. The observed pattern in these variables represents a woman who disagrees ($SYMPTD = 1$) with the statement, "You don't need a mammogram unless you develop symptoms," has no family history of disease ($HIST = 0$), has been taught breast self-examination ($BSE = 1$), and believes that it is very likely for mammography to detect a new case of breast cancer ($DETCD = 1$). In fact, 149 of the 338 subjects used in fitting logit model 1 had this particular response to these variables. The single remaining variable to differentiate outcome among these subjects is the scaled variable PB and covariate pattern 63 corresponds to $PB = 9$. For this covariate pattern the value of $\Delta\hat{\beta} = 1.733$, which is quite large. This agrees with the expectations set out in Table 5.3 of a pattern with moderate leverage and change in Pearson chi-square. We would not want to discard data on 18 subjects representing a fairly common response pattern without first trying to improve the model. On the other hand, we have little other additional information in the covariates. Addition of all interaction terms of PB with the other main effects did not significantly improve the model for logit 1. There are seven other covariate patterns with the same "modal" response as pattern 63. The logistic regression model fits each of these other patterns adequately. For these patterns the value of PB ranged from 5 to 12 so $PB = 9$ for pattern 63 is not an extreme response.

Before considering the diagnostic statistics for logit model 2, we point out the fact that covariate pattern 63 in logit model 1 provides an excellent example of why diagnostics should be calculated by covariate patterns formed from the main effects in a model rather than for individual cases. Had we considered the 18 subjects with covariate pattern

Table 8.13 Covariate Pattern, Data, Observed Outcome (y_j), Number (m_j), Estimated Logistic Probability $\hat{\pi}$, and the Value of the Three Diagnostic Statistics $\Delta\hat{\beta}$, ΔX^2, ΔD and Leverage (h), for Influential or Poorly Fit Covariate Patterns from Each Individual Logistic Regression Model

	Logit 1		Logit 2		
Data / P #	4	63	62	63	66
SYMPTD	0	1	1	1	1
PB	6	9	9	10	10
HIST	0	0	1	0	0
BSE	0	1	1	0	1
DETCD	0	1	0	0	1
y_j	1	11	3	1	2
m_j	2	18	3	1	19
$\hat{\pi}$	0.015	0.345	0.496	0.098	0.237
$\Delta\hat{\beta}$	0.543	1.733	0.956	0.264	0.999
ΔX^2	33.585	7.037	3.818	9.490	2.534
ΔD	5.820	6.586	5.268	4.781	3.014
h	0.016	0.198	0.200	0.027	0.282

63 individually an entirely different picture would emerge. First, the leverage for each of the 18 subjects would be $0.011 = 0.198/18$. For the 11 subjects with the response present the diagnostic statistics would have had values of $\Delta X^2 = 1.92$ and $\Delta\hat{\beta} = 0.021$, which would indicate some lack-of-fit but little influence on the estimated coefficients. The 9 subjects with the response absent would have had $\Delta X^2 = 0.533$ and $\Delta\hat{\beta} = 0.006$, which would support an adequate model. Thus, had we considered the data on an individual basis, we would have missed an important source of lack of fit and influence on the estimated coefficients.

Examining the diagnostic statistics in Table 8.13 for logit model 2 we see that pattern 63 is poorly fit. We remind the reader that the computations for the two individual logistic regressions were performed on separate data sets, thus the pattern numbers for the two models do not refer to the same covariate patterns. In the case of pattern 63 the estimated logistic probability is quite small yet the observed probability is 1.0. This yields a large residual. The fact that the leverage is small moderates the effect of the large value of ΔX^2 on the influence measure. The problem is, as was the case with pattern 4 for logit model 1, that the observed outcome is just contrary to the model.

Covariate pattern 66 has the largest values of $\Delta\hat{\beta}$. This pattern also represents the "modal" response described above in our discussion of the fit of logit model 1. In the case of logit model 2 a total of 115 of the 308 subjects used in the analysis had the "modal" response pattern. We note that for pattern 66 $PB = 10$. Covariate pattern 62 has the second largest value of $\Delta\hat{\beta}$. In this case there are only 3 subjects with these values of the four dichotomous covariates and all three had $PB = 9$. Both covariate patterns are influential due to the fact ΔX^2 and leverage are moderately large.

In Table 8.13 we see that the patterns with the largest values of $\Delta\hat{\beta}$ have PB equal to 9 or 10. It appears that the response of subjects with PB in this range is more variable than the logistic model is able to accommodate. The addition of interaction terms involving PB to logit 1 or 2 does not improve either model.

At this point in the analysis we have few options with the available data. No alternative model was able to improve on the model shown in Table 8.10. To explore the effect the three influential covariate patterns have on the model we eliminate the 40 subjects with data corresponding to covariate pattern 63 in logit 1 and patterns 62 and 66 in logit 2. The results are presented in Table 8.14.

Table 8.14 Estimated Coefficients, Estimated Standard Errors, Wald Statistics and Two-Tailed *p*-Values for the Model Fit After Deleting 40 Subjects Corresponding to Covariate Patterns 62, 63 and 66 in Table 8.13

Logit	Variable	Coeff.	Std. Err.	*z*	P>|z|
1	SYMPTD	2.125	0.4633	4.59	<0.001
	PB	−0.216	0.0854	−2.53	0.011
	HIST	1.243	0.4418	2.82	0.005
	BSE	1.271	0.5310	2.39	0.017
	DETCD	0.883	0.3691	2.39	0.017
	Constant	−2.892	1.0416	−2.78	0.005
2	SYMPTD	1.191	0.3610	3.30	0.001
	PB	−0.080	0.0786	−1.02	0.307
	HIST	0.606	0.4952	1.22	0.221
	BSE	1.081	0.5123	2.11	0.035
	DETCD	0.477	0.3404	1.40	0.161
	Constant	−2.664	0.9556	−2.79	0.005

Log-likelihood = −313.7473

Comparing the estimated coefficients in Tables 8.10 and 8.14, we see that in logit 1 the magnitude of the coefficients has not changed. The maximum percent change is 13 percent for PB. However, in logit 2, three of the five coefficients have substantial changes. The coefficient for PB went from −0.154 to −0.080, a decrease of 48 percent. The coefficient for HIST went from 1.063 to 0.606, a decrease of 43 percent. The coefficient for DETCD went from 0.114 to 0.477, an increase of 318 percent. Thus we see that deleting these 40 subjects has a substantial effect on the estimated odds ratios for $ME = 2$ versus $ME = 0$.

The problems encountered in fitting the model to patterns with PB in the middle of its range point out one of the dangers in using summary indices. The variable perceived benefit, PB, was created from five other variables. Considerable variability, and hence, information, about the responses of individual subjects may be lost. In the current example a value of $PB = 9$ or 10 could have been obtained from many possible combinations of responses to its five component variables. In situations when the original data are available a prudent strategy would be to remove the summary variable and consider the individual components as covariates.

Table 8.15 Estimated Odds Ratios and 95 Percent Confidence Intervals for Factors Associated with Use of Mammography Screening

	Mammogram Within One Year versus Never		Mammogram Over One Year Ago versus Never	
Variable	Odds Ratio	95 % CI	Odds Ratio	95 % CI
SYMPTD	8.12	3.314, 19.911	3.09	1.536, 6.208
PB	1.65*	1.239, 2.189	1.36*	1.024, 1,810
HIST	3.71	1.584, 8.669	2.90	1.191, 7.033
BSE	3.44	1.230, 9.649	2.60	0.962, 7.031
DETCD	2.42	1.206, 4.871	1.12	0.601, 2.091

*: Odds ratio for a 2 point decrease in Perceived Benefit

The data for the 40 subjects excluded are not unusual, thus there is no clinical basis for excluding them from the analysis. Thus, based on our assessment of model fit and the diagnostic statistics we conclude that the final model is the one in presented in Table 8.10. Estimated odds ratios and 95 percent confidence intervals based on this model are shown in Table 8.15.

The estimated odds ratios in Table 8.15 show that there is an increase in the odds for mammography screening for both frequency of use categories versus the never category. The point estimates of the odds ratios are numerically larger for recent versus never reflecting the greater disparity in the perceived value of mammography screening between these two groups. In general all five model covariates are associated with increased use of screening.

Specifically for the covariate SYMPTD, women who disagree with the statement "You don't need a mammogram unless you develop symptoms" are 8.1 times more likely to have had a recent mammogram and 3.1 times more likely to have had a less recent mammogram when compared to women who do not disagree with the statement. Both odds ratios are significant since the confidence intervals do not contain 1.0.

Women with a family history of breast cancer are estimated to be 3.7 times more likely to be recent users and 2.9 times more likely to be less recent users of mammography screening when compared to women without a family history. The confidence intervals do not contain 1.0 indicating that the increase is significant.

Having been taught breast self examination is a significant factor in obtaining mammography in the past year with an odds ratio of 3.4. The estimated odds ratio for obtaining a mammogram less recently is

2.6. However the confidence interval indicates that odds ratios between 0.96 and 7 are consistent with the observed data. The results are similar in direction but smaller for belief that it is very likely that a mammogram can detect a new case of breast cancer.

The covariate PB (Perceived Benefit) has a negative coefficient in Table 8.10, suggesting that larger values indicate less belief in the benefit of mammography screening. In a case such as this we could estimate an odds ratio for an increase of 2 points in the score. This odds ratio would be less than 1 and reflect that "less belief" is significantly associated with less frequent use. All of the other covariates in the model have estimated odds ratios greater than 1.0. In order for the odds ratio for PB to be in a similar direction we estimate the effect for a two point decrease, a change of –2.0. Thus the estimate of 1.65 for frequent use is interpreted to mean that for every 2 point decrease in the value of PB the odds for frequent use of mammography screening is estimated to increase 1.65 times. Similarly, the estimated odds ratio of 1.36 for less frequent use indicates that for every 2 point decrease in the score there is a 1.36-fold increase in the odds ratio.

As indicated in the discussion of the results the real challenge when fitting a multinomial logistic regression model is the fact that there are multiple odds ratios for each model covariate. This certainly complicates the discussion. On the other hand, using a multinomial outcome can provide more complete description of the process being studied. For example, if we had combined the two frequency of use categories into an "ever" versus "never" binary outcome then we would have completely missed the gradation in odds ratios seen in Table 8.15. From a statistical point of view, one should not pool the outcome categories unless the estimated coefficients in the logits are not significantly different from each other. In the case of the model in Table 8.10 the multivariable Wald test of the equality of the two logits is $W = 10.87$ which, with 5 degrees-of-freedom, yields $p = 0.054$. Thus we feel that there is little statistical justification for a pooled outcome category analysis.

In summary, fitting and interpreting the results from a multinomial logistic regression model follows the same basic paradigm as was the case for a binary model. The difference is that the user should be aware of the possibility that informative comparative statements may be required for the multiple odds ratios for each covariate.

8.2 ORDINAL LOGISTIC REGRESSION MODELS

8.2.1 Introduction to the Models, Methods for Fitting and Interpretation of Model Parameters

There are occasions when the scale of a multiple category outcome is not nominal but ordinal. Common examples of ordinal outcomes include variables such as extent of disease (none, some, severe), job performance (inadequate, satisfactory, outstanding) and opinion on a political candidate's position on some issue (strongly disagree, disagree, agree, strongly agree). In such a setting one could use the multinomial logistic model described in Section 8.1. This analysis, however, would not take into account the ordinal nature of the outcome and hence the estimated odds ratios may not address the questions asked of the analysis. In this section we consider a number of different logistic regression models that do take the rank ordering of the outcomes into account. Each model we discuss can be fit either directly or with some slight modification of existing statistical software.

It has been our experience that one problem users have with ordinal logistic regression models is that there is more than one logistic regression model to choose from. In the next section we describe and then compare through an example three of the most commonly used models: the adjacent-category, the continuation-ratio and the proportional odds models. There is a fairly large literature considering various aspects of ordinal logistic regression models. A few of the more general references include the text by Agresti (1990), which discusses the three models we consider as well as other more specialized models, and the text by McCullagh and Nelder (1989). Ananth and Kleinbaum (1997), in a review paper, consider the continuation-ratio and the proportional odds models as well as three other less frequently used models: the unconstrained partial-proportional odds model, the constrained partial-proportional odds model and the stereotype logistic model. Greenland (1994) also considers the continuation-ratio, the proportional odds models and the stereotype logistic model.

Assume that the ordinal outcome variable, Y, can take on $K+1$ values coded $0,1,2,\ldots,K$. We denote a general expression for the probability that the outcome is equal to k conditional on a vector, $\mathbf{x}$, of p covariates as $\Pr[Y=k \mid \mathbf{x}]=\phi_k(\mathbf{x})$. If we assume that the model is the multinomial logistic model in Section 8.1 then $\phi_k(\mathbf{x})=\pi_k(\mathbf{x})$ where, for $K=2$ the model is given in equations (8.3) – (8.5). In the context of ordinal logistic regression models the multinomial model is frequently called the *baseline logit model*. This term arises from the fact that the model is usually parametrized so that the coefficients are log-odds comparing category $Y=k$ to a "baseline" category, $Y=0$. As shown in Section 8.1 the fully parametrized baseline logistic regression model has $K\times(p+1)$ coefficients. Under this model the logits, as shown in Section 8.1, are

$$g_k(\mathbf{x})=\ln\left[\frac{\pi_k(\mathbf{x})}{\pi_0(\mathbf{x})}\right]=\beta_{k0}+\mathbf{x}'\boldsymbol{\beta}_k \tag{8.11}$$

for $k=1,2,\ldots,K$.

When we move to an ordinal model we have to decide what outcomes to compare and what the most reasonable model is for the logit. For example, suppose that we wish to compare each response to the next larger response. This model is called the *adjacent-category logistic model*. If we assume that the log odds does not depend on the response and the log odds is linear in the coefficients then the adjacent category logits are as follows:

$$a_k(\mathbf{x})=\ln\left[\frac{\phi_k(\mathbf{x})}{\phi_{k-1}(\mathbf{x})}\right]=\alpha_k+\mathbf{x}'\boldsymbol{\beta} \tag{8.12}$$

for $k=1,2,\ldots,K$. The adjacent-category logits are a constrained version of the baseline logits. To see this we express the baseline logits in terms of the adjacent-category logits as follows:

$$\begin{aligned}\ln\left[\frac{\phi_k(\mathbf{x})}{\phi_0(\mathbf{x})}\right]&=\ln\left[\frac{\phi_1(\mathbf{x})}{\phi_0(\mathbf{x})}\right]+\ln\left[\frac{\phi_2(\mathbf{x})}{\phi_1(\mathbf{x})}\right]+\cdots+\ln\left[\frac{\phi_k(\mathbf{x})}{\phi_{k-1}(\mathbf{x})}\right]\\&=a_1(\mathbf{x})+a_2(\mathbf{x})+\cdots+a_k(\mathbf{x})\\&=(\alpha_1+\mathbf{x}'\boldsymbol{\beta})+(\alpha_2+\mathbf{x}'\boldsymbol{\beta})+\cdots+(\alpha_k+\mathbf{x}'\boldsymbol{\beta})\\&=(\alpha_1+\alpha_2+\cdots+\alpha_k)+k\mathbf{x}'\boldsymbol{\beta}.\end{aligned} \tag{8.13}$$

Thus we see that the model in equation (8.13) is a version of the baseline model in equation (8.11) with intercept $\beta_{k0}=(\alpha_1+\alpha_2+\cdots+\alpha_k)$ and slope coefficients $\boldsymbol{\beta}_k=k\boldsymbol{\beta}$. As we show shortly in an example, an easy way to fit the adjacent-category model is via a constrained baseline logistic model.

Suppose instead of comparing each response to the next larger response we compare each response to all lower responses that is $Y=k$ versus $Y<k$ for $k=1,2,\ldots,K$. This model is called the *continuation-ratio logistic model.* We define the logit for this model as follows:

$$
\begin{aligned}
r_k(\mathbf{x}) &= \ln\left[\frac{P(Y=k|\mathbf{x})}{P(Y<k|\mathbf{x})}\right] \\
&= \ln\left[\frac{\phi_k(\mathbf{x})}{\phi_0(\mathbf{x})+\phi_1(\mathbf{x})+\cdots+\phi_{k-1}(\mathbf{x})}\right] \\
&= \theta_k + \mathbf{x}'\boldsymbol{\beta}_k
\end{aligned} \tag{8.14}
$$

for $k=1,2,\ldots,K$. Under the parametrization in equation (8.14) the continuation-ratio logits have different constant terms and slopes for each logit. The advantage of this unconstrained parametrization is that the model can be fit via K ordinary binary logistic regression models. We demonstrate this fact via an example shortly. We can also constrain the model in equation (8.14) to have a common vector of slope coefficients and different intercepts, namely

$$
r_k(\mathbf{x}) = \theta_k + \mathbf{x}'\boldsymbol{\beta}. \tag{8.15}
$$

Special software is required to fit the model in equation (8.15). For example, Wolfe (1998) has developed a command for use with STATA. We note that it is also possible to define the continuation ratio in terms of $Y=k$ versus $Y>k$ for $k=0,1,\ldots,K-1$. Unfortunately the results one obtains from the two parametrizations are not equivalent. We prefer the formulation given in equations (8.14) and (8.15) since, if $K=1$, each of the models in equations (8.11) to (8.15) simplifies to the usual logistic regression model where the odds ratios compare response $Y=1$ to response $Y=0$.

The third ordinal logistic regression model we consider is the proportional odds model. With this model we compare the probability of an equal or smaller response, $Y \le k$, to the probability of a larger response, $Y>k$,

$$
\begin{aligned}
c_k(\mathbf{x}) &= \ln\left[\frac{P(Y \le k|\mathbf{x})}{P(Y > k|\mathbf{x})}\right] \\
&= \ln\left[\frac{\phi_0(\mathbf{x}) + \phi_1(\mathbf{x}) + \cdots + \phi_k(\mathbf{x})}{\phi_{k+1}(\mathbf{x}) + \phi_{k+2}(\mathbf{x}) + \cdots + \phi_K(\mathbf{x})}\right] \\
&= \tau_k - \mathbf{x}'\boldsymbol{\beta} \qquad (8.16)
\end{aligned}
$$

for $k = 0, 1, \ldots, K-1$. We note that in the case when $K = 1$ the model as defined in equation (8.16) simplifies to the complement of the usual logistic regression model in that it yields odds ratios of $Y = 0$ versus $Y = 1$. We negate the coefficient vector in equation (8.16) to be consistent with software packages such as STATA and other references discussing this model.

The method used to fit each of the models, except the unconstrained continuation-ratio model, is based on an adaptation of the multinomial likelihood and its log shown in equation (8.6) for $K = 2$. The basic procedure involves the following steps: (1) the expressions defining the model specific logits are used to create an equation defining $\phi_k(\mathbf{x})$ as a function of the unknown parameters. (2) The values of a $K+1$ dimensional multinomial outcome, $\mathbf{z}' = (z_0, z_1, \ldots, z_K)$, are created from the ordinal outcome as $z_k = 1$ if $y = k$ and $z_k = 0$ otherwise. It follows that only one value of z is equal to one. The general form of the likelihood for a sample of n independent observations, $(y_i, \mathbf{x}_i)$, $i = 1, 2, \ldots, n$, is

$$
l(\boldsymbol{\beta}) = \prod_{i=1}^{n}\left[\phi_0(\mathbf{x}_i)^{z_{0i}}\,\phi_1(\mathbf{x}_i)^{z_{1i}} \times \cdots \times \phi_K(\mathbf{x}_i)^{z_{Ki}}\right]
$$

where we use " $\boldsymbol{\beta}$" somewhat imprecisely to denote both the p slope coefficients and the K model-specific intercept coefficients. It follows that the log-likelihood function is

$$
L(\boldsymbol{\beta}) = \sum_{i=1}^{n} z_{0i}\ln\left[\phi_0(\mathbf{x}_i)\right] + z_{1i}\ln\left[\phi_1(\mathbf{x}_i)\right] + \cdots + z_{Ki}\ln\left[\phi_K(\mathbf{x}_i)\right]. \qquad (8.17)
$$

We obtain the MLEs of the parameters by differentiating equation (8.17) with respect to each of the unknown parameters, setting each of the $K + p$ equations equal to zero and solving for " $\hat{\boldsymbol{\beta}}$". We obtain the estimator of the covariance matrix of the estimated coefficients in the usual manner by evaluating the inverse of the negative of the matrix of second partial derivatives at "$\hat{\boldsymbol{\beta}}$" .

At this point in the discussion it is not especially worthwhile to show the specific form of $\phi_k(\mathbf{x})$ for each model, the details of the likelihood equations or the matrix of second partial derivatives. Instead, we focus on a simple example to illustrate the use of the models and to aid in the interpretation of the odds ratios that result from each of them. As we noted above, an ordinal scale outcome can arise in a number of different ways. For example, we can create an ordinal outcome by categorizing an observed continuous outcome variable. Alternatively, we may observe categories that we hypothesize have come from categorizing a hypothetical and unobserved continuous outcome. This is often a useful way to envision outcome scales in categories ranging from strongly disagree to strongly agree. Another possibility is that the outcome is a composite of a number of other scored variables. Common examples are health status or extent of disease, which arise from many individual clinical indicators such as the Apgar score of a baby at birth. The Apgar score ranges between 0 and 10 and is the sum of 5 variables, each scored as either 0, 1, or 2.

The example we use comes from the Low Birth Weight Study (see Section 1.6.2) where we form a four category outcome from birth weight (BWT) using cutpoints: 2500g, 3000g and 3500g. This example is not typical of many ordinal outcomes that use loosely defined "low," "medium" or "high" categorizations of some measurable quantity. Instead, here we explicitly derived this variable from a measured continuous variable. We make use of this fact when we show how the proportional odds model can be derived from the categorization of a continuous variable. In addition some of the exercises are designed to extend this discussion. First, we need to give some thought to the assignment of codes to the outcome variable, as this has implications on the definition of the odds ratio calculated by the various ordinal models. The obvious choice is to use the naturally increasing sequence of codes: 0 if $BWT \leq 2500$, 1 if $2500 < BWT \leq 3000$, 2 if $3000 < BWT \leq 3500$ and 3 if $BWT > 3500$. This coding is appropriate if we want low or lower weight as the reference outcome. However this is in the opposite direction of how we modeled low birth weight in earlier chapters. Thus a decreasing sequence of codes might make more sense to use for some ordinal models namely: 3 if $BWT \leq 2500$, 2 if $2500 < BWT \leq 3000$, 1 if $3000 < BWT \leq 3500$ and 0 when $BWT > 3500$. With this coding, the heaviest births are the reference outcome. This is the coding we use for

Table 8.16 Cross-Classification of the Four Category Ordinal Scale Birth Weight Outcome versus Smoking Status of the Mother

Birth Weight	Smoking Status		
Category	No (0)	Yes (1)	Total
0: BWT > 3500	35	11	46
1: 3000 < BWT ≤ 3500	29	17	46
2: 2500 < BWT ≤ 3000	22	16	38
3: BWT ≤ 2500	29	30	59
Total	115	74	189

the outcome variable BWT4 in this section. In truth, the actual coding, for the most part, does not make a difference, as long one is able to figure out how to correct the signs of the coefficients obtained by software packages. We illustrate this with examples.

As a starting point consider the cross-classification of BWT4 versus smoking status of the mother during the pregnancy shown in Table 8.16. The odds ratios for the multinomial or baseline logit model defined in equation (8.11) are

$$\hat{\text{OR}}(1,0)=\frac{17\times 35}{29\times 11}=1.87,$$

$$\hat{\text{OR}}(2,0)=\frac{16\times 35}{22\times 11}=2.31,$$

and

$$\hat{\text{OR}}(3,0)=\frac{30\times 35}{29\times 11}=3.29,$$

where we use $\hat{\text{OR}}(k,0)$ to denote the odds ratio of maternal smoking for $\text{BWT4}=k$ versus $\text{BWT4}=0$. The increase in the odds ratio demonstrates an increase in odds of a progressively lower weight baby among women who smoke during pregnancy.

The adjacent-category model postulates that the log odds of each successively higher comparison of the baseline log odds is a constant multiple of the log odds of $Y=1$ versus $Y=0$. Under the adjacent-category model, the relationship we require is $\ln[\text{OR}(k,0)]=$

$k \times \ln[\text{OR}(1,0)]$. The results of fitting the adjacent-category model via a constrained baseline model are shown in Table 8.17.

We obtain the equations for the adjacent-category logits by using the algebraic relationship between the constrained baseline and adjacent-category models shown in equation (8.13). It follows that the first estimated adjacent-category logit is identical to the first estimated baseline logit, namely

$$\hat{a}_1(SMOKE) = -0.110 + 0.370 \times SMOKE \; .$$

The estimated coefficient for SMOKE in the second adjacent-category logit is the same as in the first. The estimated coefficient for logit 2 in Table 8.17 is twice the value in logit 1 and reflects the constraint placed on the fitted baseline logit model. It follows from equation (8.13) that the estimate of the constant term for the second adjacent-category logit is equal to the difference between the two estimated constant terms in Table 8.17,

$$\hat{\alpha}_2 = \hat{\beta}_{20} - \hat{\beta}_{10} = -0.441 - (-0.110) = -0.331 .$$

Hence the equation for the second adjacent-category logit is

$$\hat{a}_2(SMOKE) = -0.331 + 0.370 \times SMOKE .$$

The equation for the third adjacent-category logit is obtained in a similar manner. In particular the estimated coefficient for SMOKE shown

Table 8.17 Estimated Coefficients, Standard Errors, *z*-Scores, Two-tailed *p*-Values for the Fitted Constrained Baseline Model

Logit	Variable	Coef.	Std. Err.	z	P>|z|
1	SMOKE	0.370	0.1332	2.77	0.006
	constant	−0.110	0.2106	−0.52	0.602
2	SMOKE	0.739	0.2664	2.77	0.006
	constant	−0.441	0.2333	−1.89	0.059
3	SMOKE	1.109	0.3996	2.77	0.006
	constant	−0.175	0.2495	−0.70	0.483

Log-likelihood = −255.6528

in the third logit in Table 8.17 is three times the estimated coefficient for the first logit. It follows from equation (8.13) that the estimate of constant term is $\hat{\alpha}_3 = \hat{\beta}_{30} - \hat{\beta}_{20} = -0.175 - (-0.441) = 0.266$. Hence the third estimated adjacent-category logit is

$$\hat{a}_2(SMOKE) = 0.266 + 0.370 \times SMOKE.$$

Under the adjacent-category model the estimate of the odds ratio for smoking status during pregnancy of the mother is

$$\hat{\text{OR}}(k, k-1) = \exp(0.370) = 1.45$$

for $k = 1,2,3$. The interpretation of this estimate is that the odds of a birth in the next lower weight category among women who smoke during pregnancy are 1.45 times the odds among women who do not smoke.

Since the adjacent-category model is a constrained baseline model we can test that the two models are not different from each other via a likelihood ratio test or multivariable Wald test. The log-likelihood for the fitted baseline model (output not shown) based on the data in Table 8.16 is −255.4859. Thus, the likelihood ratio test is

$$G = -2[-255.6528 - (-255.4859)] = 0.334$$

which, with two degrees-of-freedom, gives $p = \text{P}(\chi^2(2) > 0.334) = 0.846$. The two degrees-of-freedom come from the constraints described above for adjacent-category logits two and three. In general the degrees-of-freedom for this test are $((K+1)-2) \times p$ where $K+1$ is the number of categories and p is the number of covariates in each model. In work not shown we obtained the same result with the Wald test. Thus we cannot say that the adjacent-category model is different from the baseline model. Since the adjacent-category model summarizes the effect of smoking into a single odds ratio we might prefer to use this model. However, this discussion considered only one covariate and the final decision in any practical setting should consider all model covariates as well as an evaluation of model fit.

Next we consider the continuation-ratio model. As shown in equation (8.14) the coefficients for this model yield the log odds for a birth in the next lower weight category relative to all heavier weight catego-

Table 8.18 Estimated Coefficients, Standard Errors, z-Scores, Two-tailed p-Values for the Fitted Unconstrained Continuation-Ratio Model

Logit	Variable	Coef.	Std. Err.	z	P>|z|
1	SMOKE	0.623	0.4613	1.35	0.177
	constant	−0.188	0.2511	−0.75	0.454
2	SMOKE	0.508	0.3991	1.27	0.203
	constant	−1.068	0.2471	−4.32	0.000
3	SMOKE	0.704	0.3196	2.20	0.028
	constant	−1.087	0.2147	−5.06	<0.001

Log-likelihood = -62.8400 + (-77.7436) + (-114.9023)
= -255.4859

ries. The unconstrained model described in equation (8.14) can be fit via a set (3 in this case) of binary logistic regressions. Each fit is based on a binary outcome, y_k^*, defined as follows:

$$y_k^* = \begin{cases} 1 & \text{if } y = k \\ 0 & \text{if } y < k \\ \text{Missing} & \text{if } y > k \end{cases}$$

for $k = 1,2,3$. The results of fitting the unconstrained continuation-ratio logit model containing SMOKE are shown in Table 8.18. The results of the three separate fits are summarized into one single table for purposes of emphasizing that we have fit a single multiple category outcome. This model is, in terms of the number of parameters and log-likelihood, fully equivalent to the unconstrained baseline model. Note that, as shown at the bottom of Table 8.18, the sum of the values of the log-likelihoods from the three separate fits is equal to the log-likelihood from the unconstrained baseline model.

The three estimated coefficients in Table 8.18 are quite similar (all are approximately 0.6). The estimates indicate that the odds of a birth in the next lower weight category relative to higher weight categories among women who smoked during pregnancy is about $1.8 = \exp(0.6)$ times that of women who didn't smoke.

To test for the equality of the three smoking coefficients, we make use of the fact that, as a result of the definition of the model, the three sets of parameter estimates are independent. Thus a simple test for equality is the two degree-of-freedom chi square statistic

$$W^2 = \frac{(0.623-0.508)^2}{\left[(0.4613)^2+(0.3991)^2\right]} + \frac{(0.623-0.704)^2}{\left[(0.4613)^2+(0.3196)^2\right]} = 0.056,$$

which yields $p = \Pr\left[\chi^2(2) > 0.056\right] = 0.972$. Hence we cannot say, at the 0.05 level, that the three coefficients are different and we consider fitting the constrained continuation-ratio logit model in equation (8.15).

The results of fitting this model are shown in Table 8.19. The estimate of the odds ratio for smoking during pregnancy is $1.87 = \exp(0.627)$. The wording of the interpretation is the same as that given for the approximate value from the unconstrained model. This odds ratio is a bit larger than the estimate of 1.45 obtained under the adjacent-category model. The reason is that the reference group for the continuation-ratio model includes all heavier weight categories not just the next highest, which is used in the adjacent-category model.

In general, the continuation-ratio model might be preferred over the baseline and adjacent-category model when the conditioning used in defining and fitting the model makes clinical sense. A common example is one where the number of attempts to pass a test or attain some binary outcome is modeled. The first logit models the log odds of passing the test the first time it is taken. The second logit models the log odds of passing the test on the second attempt given that it was not passed on the first attempt. And this process continues until one is modeling the Kth attempt. Since this is not a common setting we do not consider the model in any more detail. Further elaboration and discussion can be found in the references cited in this section.

Probably the most frequently used ordinal logistic regression model in practice is the constrained cumulative logit model called the proportional odds model given in equation (8.16). Each of the previ-

Table 8.19 Estimated Coefficients, Standard Errors, z-Scores, Two-tailed p-Values for the Fitted Constrained Continuation-Ratio Model

Variable	Coef.	Std. Err.	z	P>|z|
SMOKE	0.627	0.2192	2.86	0.004
constant1	−0.189	0.2204		
constant2	−1.114	0.2129		
constant3	−1.052	0.1862		

Log-likelihood = − 255.5594

ously discussed models for ordinal data compares a single outcome response to one or more reference responses (e.g., $Y=k$ versus $Y=k-1$, or $Y=k$ versus $Y<k$). The proportional odds model describes a less than or equal versus more comparison. For example if the outcome is extent of disease the model gives the log odds of no more severe outcome versus a more severe outcome. The constraint placed on the model is that the log odds does not depend on the outcome category. Thus inferences from fitted proportional odds models lend themselves to a general discussion of direction of response and do not have to focus on specific outcome categories. The results are much simpler to describe than those from any of the unconstrained models but are of about the same order of complexity as results from the other constrained models.

This consistency of effect across response categories in the proportional odds model is similar to that described for the constrained adjacent-category and continuation-ratio models and, as such, should always be tested. Modifications of the proportional odds model that allow one or more covariates to have category-specific effects are discussed by Ananth and Kleinbaum (1997). These "partial" proportional odds models have not, as yet, seen wide use in practice and we do not consider them further in this text.

One way of deriving the proportional odds model is via categorization of an underlying continuous response variable. This derivation is intuitively appealing in that it allows us to use some concepts from linear regression modeling. For example, the cutpoints used to obtain the four category variable BWT4 are 2500, 3000 and 3500 grams. Due to the way packages handle the proportional odds model it turns out to be more convenient to code the ordinal outcome so it increases in the same direction as its underlying continuous response. Thus we define the outcome BWT4N as follows: 0 if $BWT \leq 2500$, 1 if $2500 < BWT \leq 3000$, 2 if $3000 < BWT \leq 3500$ and 3 when $BWT > 3500$ and the specific cutpoints as $cp_1 = 2500$, $cp_2 = 3000$ and $cp_3 = 3500$.

We show in Figure 8.2 a hypothetical line or model, $BWT = \lambda_0 + \beta \times LWT$, that describes mean birth weight as a function of mothers weight at the last menstrual period. The interval-specific coding of the BWT4N outcome variable is shown to the left of the BWT axis as $Y=k$ for $k=0,1,2,3$. The particular values used to obtain the line are $\lambda_0 = 100$ and $\beta = 20$ and are for demonstration purposes only. The actual linear regression of BWT on LWT could have been used; however, the resulting graph would not have had as large a range in the

BWT axis or as steep a slope. It is the idea that is important not the actual numbers.

Suppose that instead of the usual normal errors linear regression model we have a model where the errors follow the logistic distribution. The statistical model for birth weight is $BWT = \lambda_0 + \lambda_1 \times LWT + \sigma \times \varepsilon$, where σ is proportional to the variance and ε follows the standard logistic distribution with cumulative distribution function

$$\mathrm{P}(\varepsilon \leq z) = \frac{e^z}{1+e^z}. \tag{8.18}$$

A concise discussion of this distribution may be found in Evans, Hastings and Peacock (1993).

The regression based on the continuous outcome models the mean of BWT as a function of LWT. In ordinal logistic regression we model the probability that BWT falls in the four intervals defined by the three cutpoints shown in Figure 8.2. For example, we show in Figure 8.3 the underlying logistic distribution for the regression model in Figure 8.2 at

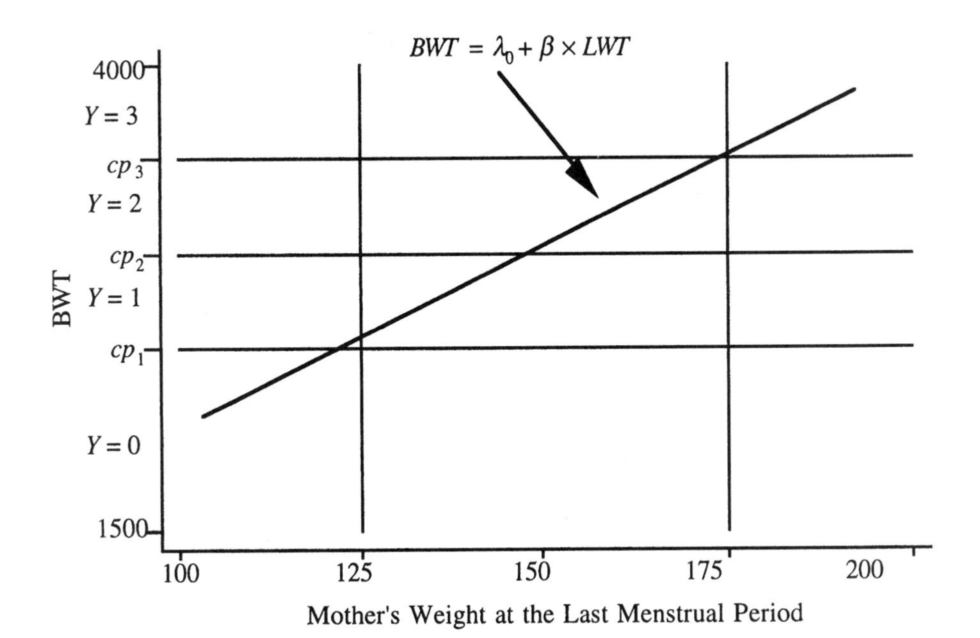

Figure 8.2 Plot of a hypothetical model describing mean birth weight as a function of mother's weight.

$LWT = 125$. The mean is 2600 grams. The probabilities for the four ordinal outcomes are the respective areas under this curve. The area below 2500 is the largest indicating that, among women who weigh 125 pounds, a birth weight less than or equal to 2500 grams ($BWT4N = 0$) is the most likely ordinal outcome. However, at 175 pounds the mean from the regression line is 3600 grams and the probability is largest for the $BWT4N = 3$ ordinal outcome and smallest for the $BWT4N = 0$ ordinal outcome. Under the proportional odds model we model the ratios of cumulative areas defined by the cutpoints.

Consider women who weigh 125 pounds. Under our coding of the four category ordinal variable BWT4N we have

$$\begin{aligned} P(BWT4N = 0 \mid LWT = 125) &= P(BWT \le cp_1 \mid LWT = 125) \\ &= P[(\lambda_0 + 125 \times \lambda_1 + \sigma \times \varepsilon) \le cp_1] \\ &= P\left[\varepsilon \le \frac{cp_1 - (\lambda_0 + 125 \times \lambda_1)}{\sigma}\right] \\ &= P[\varepsilon \le \tau_1 - 125 \times \beta] \end{aligned} \tag{8.19}$$

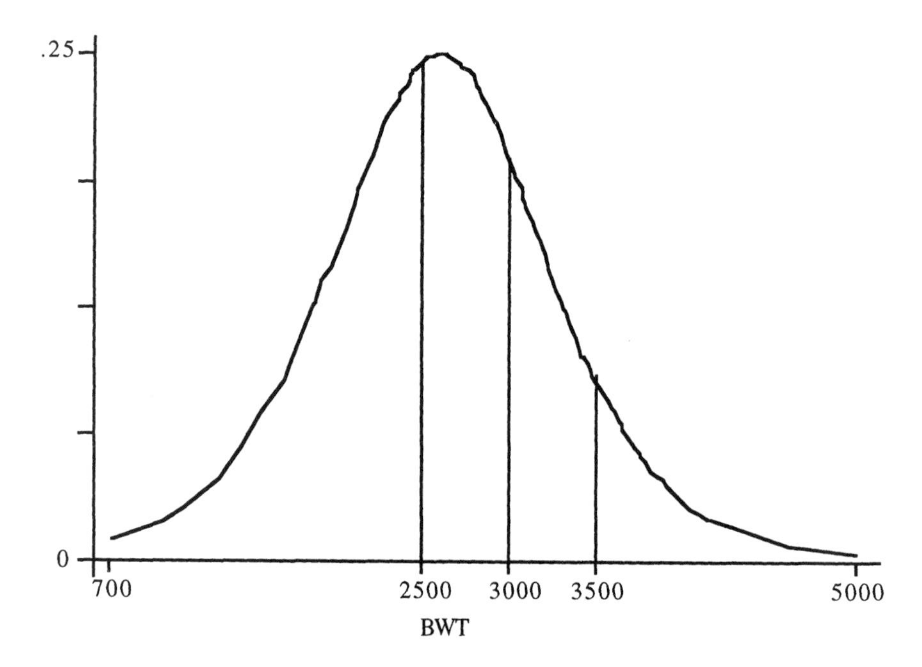

Figure 8.3 Plot of the hypothetical underlying logistic distribution at $LWT = 125$.

where we let $\tau_1 = (cp_1 - \lambda_0)/\sigma$ and $\beta = \lambda_1/\sigma$. Under the assumption of errors with the distribution function in equation (8.18), the probability in equation (8.19) is

$$\mathrm{P}[\varepsilon \le \tau_1 - 125 \times \beta] = \frac{e^{\tau_1 - 125\times\beta}}{1 + e^{\tau_1 - 125\times\beta}}. \tag{8.20}$$

It follows from equation (8.20) that

$$\mathrm{P}(\varepsilon > \tau_1 - 125 \times \beta) = 1 - \mathrm{P}(\varepsilon \le \tau_1 - 125 \times \beta) = \frac{1}{1 + e^{\tau_1 - 125\times\beta}}. \tag{8.21}$$

Hence the log odds of a lighter weight baby at this cutpoint among 125 pound women is

$$\begin{aligned}
\ln\left[\frac{\mathrm{P}(BWT4N \le 0 \mid LWT = 125)}{\mathrm{P}(BWT4N > 0 \mid LWT = 125)}\right] &= \ln\left[\frac{\mathrm{P}(\varepsilon \le \tau_1 - 125 \times \beta)}{\mathrm{P}(\varepsilon > \tau_1 - 125 \times \beta)}\right] \\
&= \ln\left[\frac{\dfrac{e^{\tau_1 - 125\times\beta}}{1 + e^{\tau_1 - 125\times\beta}}}{\dfrac{1}{1 + e^{\tau_1 - 125\times\beta}}}\right] \\
&= \ln\left[e^{\tau_1 - 125\times\beta}\right] \\
&= \tau_1 - 125 \times \beta
\end{aligned} \tag{8.22}$$

which is the proportional odds model in equation (8.16). If we follow the steps in equation (8.19) to equation (8.22) then we obtain identical expressions for the other outcome categories. For example, at the cutpoint cp_3 we have the log odds

$$\begin{aligned}
\ln\left[\frac{\mathrm{P}(BWT4N \le 2 \mid LWT = 125)}{\mathrm{P}(BWT4N > 2 \mid LWT = 125)}\right] &= \ln\left[\frac{\mathrm{P}(\varepsilon \le \tau_3 - 125 \times \beta)}{\mathrm{P}(\varepsilon > \tau_3 - 125 \times \beta)}\right] \\
&= \tau_3 - 125 \times \beta .
\end{aligned}$$

By similar calculations at $BWT4N = 1$ among 175 pound women the log odds is

$$\ln\left[\frac{P(BWT4N \le 1 | LWT = 175)}{P(BWT4N > 1 | LWT = 175)}\right] = \ln\left[\frac{P(\varepsilon \le \tau_2 - 175 \times \beta)}{P(\varepsilon > \tau_2 - 175 \times \beta)}\right]$$
$$= \tau_2 - 175 \times \beta.$$

We can follow the same derivation for any covariate, x, and any number of categories for an ordinal outcome variable, Y, and we obtain as the log odds for as small or smaller outcome the equation

$$\ln\left[\frac{P(Y \le k | x)}{P(Y > k | x)}\right] = \ln\left[\frac{P(\varepsilon \le \tau_{k+1} - x \times \beta)}{P(\varepsilon > \tau_{k+1} - x \times \beta)}\right] = \tau_{k+1} - x \times \beta. \quad (8.23)$$

It follows from equation (8.23) that the log of the odds ratio for $x = x_1$ versus $x = x_0$ is

$$\ln\left[\frac{P(Y \le k | x_1)}{P(Y > k | x_1)}\right] - \ln\left[\frac{P(Y \le k | x_0)}{P(Y > k | x_0)}\right] =$$
$$= (\tau_{k+1} - x_1 \times \beta) - (\tau_{k+1} - x_0 \times \beta)$$
$$= -\beta(x_1 - x_0). \quad (8.24)$$

How we use the results from a package and equation (8.24) to estimate an odds ratio depends on the package used. For example, the results of fitting the proportional odds model in STATA with outcome BWT4N and covariate LWT are shown in Table 8.20. Note that the coefficient for LWT in Table 8.20 is positive reflecting the direction of the association seen in Figure 8.3. Hence increasing values of LWT are associated with increasing values of BWT4N. Thus the output is consistent with the underlying hypothetical continuous outcome model. The negative sign in equation (8.24) reflects the fact that under a positive association the covariate is protective (i.e., negatively associated with smaller values of the ordinal outcome). Hence the estimate of the effect of a 10-pound increase in LWT on the odds ratio for as light or lighter versus a heavier baby is

$$\widehat{OR} = \exp(-0.013 \times 10) = 0.88.$$

Table 8.20 Results of Fitting the Proportional Odds Model to the Four Category Birth Weight Outcome, BWT4N, with Covariate LWT

Variable	Coef.	Std. Err.	z	P>|z|	95% CI
LWT	0.013	0.0043	2.95	0.003	0.004, 0.021
constant1	0.832	0.5686			
constant2	1.707	0.5782			
constant3	2.831	0.6027			

Log-likelihood = −255.1477

This estimate implies a 12 percent reduction in the odds for a lower weight baby per 10-pound increase in weight.

One feature of the proportional odds model that is identical to the binary logistic model is that we can reverse the direction of the model by simply changing the signs of the coefficients. For example, if we are interested in modeling heavier versus lighter weight babies then the estimate of the odds ratio for a 10-pound increase in weight is

$$\widehat{OR} = \exp(0.013 \times 10) = 1.14 .$$

This estimate indicates that there is a 14 percent increase in the odds of a heavier baby per 10-pound increase in weight.

The output from SAS's logistic procedure is identical to Table 8.20 except that the coefficient for LWT is −0.013 since SAS uses a model that does not negate the coefficient, $\boldsymbol{\beta}$, in equation (8.16).

As a second example we fit the model containing smoking status of the mother during pregnancy. Women who smoke during pregnancy tend to have lower weight births thus the association in the conceptual underlying continuous model is negative. The results of fitting this model in STATA are shown in Table 8.21 where the coefficient for SMOKE is negative. Hence the estimate of the odds ratio for a lower versus a heavier weight baby is, from equation (8.24),

$$\widehat{OR} = \exp[-(-0.761)] = 2.14 .$$

The interpretation is that women who smoke during pregnancy have 2.1 times the odds of a lower versus a heavier baby than women who do not smoke. Similar to the discussion for LWT the estimate of the odds ratio for a heavier versus lighter weight baby is

$$\hat{\text{OR}} = \exp(-0.761) = 0.47.$$

The interpretation of this estimate is that the odds of a heavier versus lighter weight baby are 53 percent less for women who smoke during pregnancy.

As with other constrained ordinal models, one should check to see whether the assumption of proportional odds is supported by the data. This is typically done by comparing the fitted proportional odds model to the unconstrained baseline logit model via either a score or likelihood ratio test. The problem with these tests is that the proportional odds model can not be obtained by placing linear constraints on the coefficients in the baseline model. While the tests are not completely statistically correct, they can be used to provide some evidence of model adequacy. The degrees-of-freedom for the comparison is the same as for the test of the adjacent category versus the baseline model, $((K+1)-2)\times p$. For example, the likelihood ratio comparison of the proportional odds model containing smoking status of the mother shown in Table 8.21 to the baseline logit model is

$$G = -2[-255.6725 - (-255.4859)] = 0.373,$$

which, with two degrees-of-freedom, gives $p = \text{P}(\chi^2(2) > 0.373) = 0.830$. The SAS package uses the score test and obtains a p-value based on two degrees-of-freedom of 0.644. Thus, on the basis of either test, we cannot say that the baseline and proportional odds models are different. Thus, at least for the covariate SMOKE, the inferences from the two

Table 8.21 Results of Fitting the Proportional Odds Model to the Four Category Birth Weight Outcome, BWT4N, with Covariate SMOKE

Variable	Coef.	Std. Err.	z	P>|z|	95% CI
SMOKE	−0.761	0.2719	-2.80	0.005	−1.293, −0.228
constant1	−1.116	0.1984			
constant2	−0.248	0.1819			
constant3	0.867	0.1937			

Log-likelihood = −255.6725

models are equivalent.

The choice of what model to ultimately use in any problem should consider which odds ratios are most informative for the problem as well as an assessment of model adequacy. In the next section we consider a more complete analysis of the four category birth weight outcome.

8.2.2 Model Building Strategies for Ordinal Logistic Regression Models

The steps in model building for an ordinal logistic model are the same as described in Chapter 4 for the binary logistic regression model. Unfortunately, however, the full array of modeling tools is not available in software packages.

For ordinal models we think that a sensible approach to model building involves the following steps: Perform the usual purposeful or stepwise selection of main effects. Check for the scale of continuous covariates using design variables in the ordinal model. In addition, one could check for nonlinearity using fractional polynomial analyses with K separate binary regressions of $y=k$ versus $y=0$. Any nonlinear transformation found should, of course, make clinical sense, be reasonably similar across the separate logistic regressions and make a significant improvement over treating the covariate as linear in the ordinal model. Next, check to make sure all omitted covariates are neither significant nor confounders of main effects in the model. Lastly, check the need to include interactions using the usual selection methods. At this point, check any model assumptions of constant coefficients by comparing the constrained model to its unconstrained version. As shown in Section 8.2.1 this can be done via a likelihood ratio comparison of the fitted model versus the baseline model. Diagnostic statistics and goodness-of-fit tests have not been extended for use with ordinal models. Thus one has to use the separate binary regressions approach. The big disadvantage of this approach is that one is really not checking the actual fitted model, only an approximation to it. However this method may help identify influential and poorly fit subjects. In general this approach is a bit ad-hoc and all results should be checked by deleting identified subjects and refitting the ordinal model. Finally, inferential statements based on estimated odds ratios and their confidence intervals should be worded in such a way that it is clear which ordinal model has been used.

Since the basic process is so similar to ones used many times in this text we begin with a model for the four-category birth weight outcome containing the significant main effects. We focus on fitting the proportional odds model. The results of fitting this model are shown in Table 8.22. We include age in the model because of its known clinical importance. In addition we keep the indicator variable for hypertension, HT, in the model because of its clinical importance and the fact that the likelihood ratio test for its significance yields $p = 0.046$ (work not shown).

We used three separate binary regressions to check the scale of AGE and LWT and this analysis supported keeping them linear in the logit. A check for scale using the quartile design variables for LWT yielded three similar coefficients suggesting that we consider recoding LWT into a single dichotomous covariate comparing the first quartile to the other three quartiles. Replacing LWT with this dichotomous covariate yielded a model with a slightly smaller log-likelihood but the coefficient for HT changed by about 18 percent. Thus it seems that keeping LWT continuous in the logit gives us nearly as good a model and provides better adjustment of the effect of HT.

Next we checked for interactions. The only significant interaction is between LWT and HT with $p = 0.044$ for the likelihood ratio test and $p = 0.062$ for the Wald test of the coefficient. However the estimated coefficient for the main effect of HT is $\hat{\beta}_{HT} = -5.648$ with an estimated

Table 8.22 Results of Fitting the Proportional Odds model to the Four Category Birth Weight Outcome, BWT4N

Variable	Coef.	Std. Err.	z	P>|z|	95% CI
AGE	0.001	0.0275	0.02	0.982	−0.053, 0.054
LWT	0.013	0.0049	2.65	0.008	0.003, 0.022
RACE_2	−1.471	0.4347	−3.38	0.001	−2.323, −0.619
RACE_3	−0.869	0.3345	−2.60	0.009	−1.535, −0.213
SMOKE	−0.988	0.3150	−3.14	0.002	−1.605, −0.370
HT	−1.194	0.6122	−1.95	0.051	−2.394, 0.006
UI	−0.913	0.4045	−2.26	0.024	−1.706, −0.120
PTD	−0.822	0.4174	−1.97	0.049	−1.640, −0.004
Constant1	−0.495	0.8798			
Constant2	0.516	0.8817			
Constant3	1.803	0.8914			

Log-likelihood = −235.6504

Table 8.23 Estimates of Odds Ratios for a Lighter versus a Heavier Birth and 95 Percent Confidence Intervals Obtained from the Fitted Proportional Odds Model in Table 8.22

Variable	$\widehat{OR}$	95% CI
AGE	0.99+	0.582, 1.699
LWT	0.88+	0.803, 0.970
RACE_2	4.35	1.857, 10.206
RACE_3	2.38	1.237, 4.641
SMOKE	2.69	1.448, 4.978
HT	3.30	0.994, 10.957
UI	2.49	1.127, 5.507
PTD	2.28	1.004, 5.155

+: Estimate for a 10 year or 10 pound increase

standard error of $\widehat{SE}(\hat{\beta}_{HT}) = 2.5353$ indicating considerable numerical instability largely due to the fact that there are only 12 subjects with $HT = 1$. Thus we decide not to include this interaction in the model.

Next we check the assumption of proportional odds of the fitted model in Table 8.22. The likelihood ratio test comparing the baseline logit model to the proportional odds model is

$$G = -2 \times [(-235.6504) - (-224.7788)] = 21.7432.$$

The degrees-of-freedom for this comparison are $((K+1)-2) \times p = (4-2) \times 8 = 16$ and the approximate p-value is $P[\chi^2(16) \geq 21.7432] = 0.152$. Thus we conclude that we cannot say that the proportional odds assumption does not hold.

To assess the fit and influence of individual subjects we performed three separate full assessments of fit based on the binary logistic regressions of $BWT4N = k$ versus $BWT4N = 0$. The methods are as described and illustrated in detail in Chapter 5. The results of this analysis identified that subjects with identification numbers 98, 132, 133, 138 and 188 are highly influential with $\Delta\hat{\boldsymbol{\beta}} > 1$ in one of the three binary regressions. Examination of the data for these subjects yields no unusual values. However, two of these subjects are among the twelve with a history of hypertension. If we remove all five subjects then we do see changes of greater than 20 percent in three of the eight coefficients. However their removal leads to two cells with zero frequency in the cross-tabulation of BWT4N versus HT. Due to these zero frequency cells we cannot fit the baseline logit model and hence are unable to test the proportional odds assumption with the likelihood ratio comparison. As a result, we decide not to remove any of these subjects and use the model in Table 8.22 as our final model.

We provide estimates of the odds ratios for a lighter versus a heavier birth and their confidence intervals in Table 8.23. We obtain these odds ratios as follows: If we follow the usual method of exponentiating the estimates of the coefficients in Table 8.22 then the odds ratios compare the odds for a heavier versus a lighter birth. A fact we discussed in detail for LWT is that the output from STATA is set up to be consistent with an underlying hypothetical continuous response. Namely, positive coefficients imply a positive association between the covariate and the hypothetical continuous response. In this example the reverse direction is a clinically more relevant comparison, lighter versus heavier birth. Thus the odds ratios we present in Table 8.23 are obtained by exponentiating the negative of the values in Table 8.22.

The results of this analysis show that, after controlling for the age and weight of the mother, race other than white, smoking during pregnancy, history of hypertension, uterine irritability and history of a pre-term delivery increase the odds of a lighter versus heavier birth. The increase in the odds ranges from 4.4 times for black versus white race to 2.3 times for history of pre-term delivery. All the estimates are significant at the five percent level except history of hypertension, which has $p = 0.051$.

We could have selected one of the other ordinal models but we feel that the proportional odds model provides the most useful clinical description of the four category ordinal birth weight variable. We leave fitting the other models as exercises.

8.3 LOGISTIC REGRESSION MODELS FOR THE ANALYSIS OF CORRELATED DATA

Up to this point in the text we have considered the use of the logistic regression model in settings where we observe a single dichotomous response for a sample of statistically independent subjects. However, there are settings where the assumption of independence of responses may not hold for a variety of reasons. For example, consider a study of asthma in children. Suppose the study subjects are interviewed bimonthly for one year. At each interview the mother is asked, during the previous two months, if the child had an asthma attack severe enough to require medical attention, whether the child had a chest cold, and how many smokers lived in the household. The date of the visit is also recorded. The child's age and race are recorded at the first interview. The primary outcome is the occurrence of an asthma attack. However, there is a fundamental lack of independence in the observations due to the fact that we have six measurements on each child. In this example each child represents a cluster of correlated observations of the outcome. The measurements of the presence or absence of a chest cold and the number of smokers residing in the household can change from observation to observation and thus are called *cluster-specific covariates*. The date changes in a systematic way and is recorded to model possible seasonal effects. The child's age and race are constant for the duration of the study and are referred to as *cluster-level covariates*. The terms clusters, subjects, cluster-specific and cluster-level covariates are general enough to describe multiple measurements on a single subject or single measurements on different but related subjects. An example of the latter setting would be a study of all children in a household.

The goals of the analysis in a correlated data setting are, for the most part, identical to those discussed in earlier chapters. Specifically, we are interested in estimating the effect of the covariates on the dichotomous outcome via odds ratios. However, the models and estimation methods are a bit more complicated in the correlated data setting.

There is a large and rapidly expanding literature dealing with statistical research on methods for the analysis of correlated binary data. Most of the research in this area is at a mathematical level that is well beyond this text. However software to fit the more common and established models for correlated binary data is available in major packages such as SAS and STATA. Thus the goal of this section is to introduce the models that can be fit with the major software packages and to discuss the strengths and limitations of these models as well as the interpretation of model parameters. Two accessible review papers that discuss the models we consider are Neuhaus, Kalbfleisch and Hauck (1991) and Neuhaus (1992). Diggle, Liang and Zeger (1994) discuss methods for the analysis of longitudinal data and consider models for binary data. Ashby, Neuhaus, Hauck, Bacchetti, Heilbron, Jewell, Segal and Fusaro (1992) provide a detailed annotated bibliography on methods for analyzing correlated categorical data. Collett's (1991) excellent text on modeling binary data discusses methods for analyzing correlated binary data at a level comparable to this text. Pendergast, Gange, Newton, Lindstrom, Palta and Fisher (1996) also review methods for clustered binary data. Breslow and Clayton (1993) consider mixed models for generalized linear models. Agresti, Booth, Hobert and Caffo (2000) present a summary of different methods for the analysis of correlated binary data via random effects models, one of which we discuss in this section. Their paper considers other models and different data settings where random effects models can be effectively used. Coull and Agresti (2000) consider extensions of the logistic-normal mixed model considered in this section. Rosner (1984) and Glynn and Rosner (1994) consider specialized models for the analysis of paired binary outcomes.

The basic approach with correlated binary data is to try to mimic the usual normal errors linear mixed effects model. Suppose we are in a setting with m subjects (or clusters) and n_i observations per subject. We denote the dichotomous outcome variable as Y_{ij} and the vector of covariates as $\mathbf{x}'_{ij} = (1, x_{1ij}, x_{2ij}, \ldots, x_{pij})$ for the jth observation in the ith cluster. Note that some of the covariates may be constant within subject and some may change from observation to observation. At this point we do not use different notation for each. The most frequently used subject- or cluster-specific logistic model is the logistic-normal model. In general, the model is referred to in the literature as a "cluster-specific" model as this term is a bit more general than "subject-specific". It describes the case of multiple observations on a single subject and single observations on related subjects. For the most part we use the term cluster-specific rather than subject-specific. Under this model, the correlation among individual responses within a cluster is accounted for by adding a cluster-specific random effect term to the logit. The equation for the logit is

$$g(\mathbf{x}_{ij}, \alpha_i, \boldsymbol{\beta}_s) = \alpha_i + \mathbf{x}'_{ij}\boldsymbol{\beta}_s \qquad (8.25)$$

where it is assumed that the random effects follow a normal distribution with mean zero and constant variance, i.e., $\alpha_i \sim \mathrm{N}(0, \sigma_\alpha^2)$. In practice the random effect terms are unobserved and this leads to complications when we consider estimation of the regression coefficients, $\boldsymbol{\beta}_s$. We have added the subscript s to indicate that the coefficients apply to a logistic regression model that is specific to subjects with random effect α_i. Suppose that in our hypothetical asthma study the coefficient for having had a chest cold in the previous two months is $\ln(2)$. The cluster-specific interpretation is that having a cold doubles the odds of a specific child having a severe asthma attack in the next two months. Alternatively, the odds among children with the same value of the unobserved random effect who had a cold is two times that of those with the same value of the unobserved random effect who did not have a cold. The interpretation applies to a specific child or specific unobserved group of asthmatic children, not to broad groups of asthmatic children. Since the covariate "having had a cold" can change from month to month, the within subject interpretation provides a clear estimate of the increase in the odds for a specific subject. On the other hand, suppose that race is a dichotomous covariate coded as either white or non-white and its coefficient is $\ln(2)$. The cluster-specific interpretation is that a non-white child with random effect α_i has odds of a severe asthma attack that is twice the odds of a white child with the same random effect. Since both the race and random effect are constant within subject and cannot change, this odds ratio is not likely to be useful in practice. These two simple examples illustrate that the logistic-normal model is most likely to be useful for inferences about covariates whose values can change at the subject level.

The effect of the term α_i in equation (8.25) is to increase the correlation among responses within a cluster relative to the correlation between clusters. The basic idea is that the underlying logistic probabilities for observations of the outcome in a cluster have a common value of α_i. Thus their outcomes will be more highly correlated than the correlation among outcomes when the α_i's are different. The greater the difference in the values of the α_i's the greater the within relative to the between cluster correlation. The heterogeneity in the α_i's is simply a function of their variance σ_α^2. Thus the within-cluster correlation increases with increasing σ_α^2.

An alternative to the cluster-specific model in equation (8.25) is the population average model. Under this model we average, in a sense, over the statistical distribution of the random effect and assume that this process yields the logit

$$g(\mathbf{x}_{ij}, \boldsymbol{\beta}_{PA}) = \mathbf{x}'_{ij}\boldsymbol{\beta}_{PA} \ . \tag{8.26}$$

Probabilities based on the logit in equation (8.26) reflect the proportion of subjects in the population with outcome present among subjects with covariates $\mathbf{x}_{ij}$. We note that we have not specified the statistical distribution of the random effects, only that the marginal or population proportions have logit function given by equation (8.26). The lack of distributional assumptions presents problems when trying to estimate $\boldsymbol{\beta}_{PA}$ that we discuss shortly. The population average model does not make use of the fact that we may have covariates whose values could be different at different occasions measuring the same subject. For example the interpretation of a coefficient equal to ln(2) for having had a cold during the previous two months is that the odds of a severe asthma attack among those who had a cold is twice the odds among those who did not have a cold. Thus the coefficient describes the effect of the covariate in broad groups of subjects rather than in individual subjects. If the coefficient for race is ln(2) then the log odds of a severe asthma attack among non-whites is twice that of whites. Since a characteristic like race cannot change over multiple measurements on the same subject the population average model is best suited for this covariate and for others that describe broad groups of subjects.

Both the cluster-specific and population average model may be fit to data containing subject-specific and cluster-level covariates. The choice of which model to use should consider what types of inferences the fitted model is intended to provide. As described via the two covariates "having had a cold" and "race", the cluster-specific model is most useful when the goal is to provide inferences incorporating individual subject covariate values. Alternatively, the population average model is likely to be more useful in addressing epidemiologic type assessment of exposure effects through outcome experience in larger groups of subjects.

A third, slightly more specialized, model is the transitional model. This is a cluster-specific model where one or more of the previously observed values of the outcome or other covariates is used. For example, we might include the observation of whether or not a severe asthma attack had occurred in the first two months when modeling the event in the third and fourth months. We illustrate this model with an example after considering the cluster-specific and population average models in more detail.

As we alluded to above, estimation in correlated data models is not as straightforward or easily described as in the uncorrelated data setting where a likelihood function can be derived from the binomial distribution. We begin with the population average model. The approach used is the method of generalized estimating equations, usually abbreviated as GEE. Liang and Zeger (1986) and Zeger, Liang and Albert (1988) first used GEE with the population average model.

The GEE approach uses a set of equations that look like weighted versions of the likelihood equations shown in Chapters 1 and 2. The weights involve an approximation of the underlying covariance matrix of the correlated within-cluster observations. This requires making an assumption about the nature of the correlation. The default assumption used by most packages is called *exchangeable correlation* and assumes that the correlation between pairs of responses is constant, $\text{Cor}(Y_{ij}, Y_{il}) = \rho$ for $j \neq l$. Three other possible correlation structures that can be specified in most packages are *independent*, *auto-regressive* and *unstructured*. Under the independent model $\text{Cor}(Y_{ij}, Y_{il}) = 0$ for $j \neq l$ and the GEE equations simplify to the likelihood equations obtained from the binomial likelihood in Chapter 2. We do not consider this correlation structure further. The auto-regressive structure is appropriate when there is a time or order component associated with the observations. The correlation among responses depends on the lag between the observations and is assumed to be constant for equally lagged observations. Since settings where there is an explicit time component are a bit specialized we do not consider this type of correlation further in this text. Under unstructured correlation one assumes that the correlation of the possible pairs of responses is different, $\text{Cor}(Y_{ij}, Y_{il}) = \rho_{jl}$ for $j \neq l$. At first glance this might seem to be the model of choice. However, it may be used only after estimating a large number of parameters that are, for the most part, of secondary importance. In most applications we are only interested in estimating the regression coefficients and need to account for correlation in responses to obtain correct estimates of the standard errors of the estimated coefficients. For this reason Liang and Zeger (1986) refer to the choice of correlation structure to use in the GEE as the *working correlation*. The idea is that one chooses a correlation structure for estimation that is plausible for the setting and this structure is then adjusted in the estimator of the variance. It turns out that, in a wide variety of settings, assuming "exchangeable correlation" gives good results. Thus we develop the GEE method in some detail using exchangeable correlation as the working correlation.

We need some additional notation to fully describe the application of GEE to the population average model. We denote the logistic probability obtained from the logit in equation (8.26) as

$$\pi_{PA}(\mathbf{x}_{ij}) = \frac{e^{\mathbf{x}'_{ij}\boldsymbol{\beta}_{PA}}}{1 + e^{\mathbf{x}'_{ij}\boldsymbol{\beta}_{PA}}}. \tag{8.27}$$

We use two matrices to describe the within-cluster covariance of the correlated observations of the outcome variable. The first is a $n_i \times n_i$ diagonal matrix containing the variances under the model in equation (8.27) denoted

$$\mathbf{A}_i = \text{diag}\left[\pi_{PA}(\mathbf{x}_{ij}) \times \left(1 - \pi_{PA}(\mathbf{x}_{ij})\right)\right] \tag{8.28}$$

and the second is the $n_i \times n_i$ *exchangeable correlation matrix* denoted

$$\mathbf{R}_i(\rho) = \begin{bmatrix} 1 & \rho & \cdots & \rho \\ \rho & 1 & & \rho \\ \vdots & & \ddots & \vdots \\ \rho & \rho & \cdots & 1 \end{bmatrix}. \tag{8.29}$$

Using the fact that the correlation is defined as the covariance divided by the product of the standard deviations it follows that the covariance matrix in the ith cluster is

$$\mathbf{V}_i = \mathbf{A}_i^{0.5}\mathbf{R}_i(\rho)\mathbf{A}_i^{0.5} \tag{8.30}$$

where $\mathbf{A}_i^{0.5}$ is the diagonal matrix whose elements are the square roots of the elements in the matrix in equation (8.28). The contribution to the estimating equations for the ith cluster is

$$\mathbf{D}_i'\mathbf{V}_i^{-1}\mathbf{S}_i$$

where $\mathbf{D}_i' = \mathbf{X}_i'\mathbf{A}_i$, $\mathbf{X}_i$ is the $n_i \times (p+1)$ matrix of covariate values and $\mathbf{S}_i$ is the vector with jth element the residual $s_{ij} = \left(y_{ij} - \pi_{PA}(\mathbf{x}_{ij})\right)$. The full set of estimating equations is

$$\sum_{i=1}^{m} \mathbf{D}_i'\mathbf{V}_i^{-1}\mathbf{S}_i = \mathbf{0}. \tag{8.31}$$

We denote the solution to the GEE in equation (8.31) as $\hat{\boldsymbol{\beta}}_{PA}$. Implicit in the solution of these equations is an estimator of the correlation parameter, ρ. Typically this is based on the average correlation among within-cluster empirical residuals and as such it is also adjusted with each iterative change in the solution for $\hat{\boldsymbol{\beta}}_{PA}$.

A useful exercise is to show that under the assumption of no correlation, $\rho = 0$, the GEE in equation (8.31) simplify to the likelihood equations for the multiple logistic regression model shown in Chapters 2 and 3.

Liang and Zeger (1986) show that the estimator, $\hat{\boldsymbol{\beta}}_{PA}$, is asymptotically normally distributed with mean $\boldsymbol{\beta}_{PA}$. They derive, as an estimator of the covariance matrix, the estimator that is often referred to as the *information sandwich estimator*. The "bread" of the sandwich is based on the observed information matrix under the assumption of exchangeable correlation. The bread for the ith cluster is

$$\begin{aligned}\mathbf{B}_i &= \mathbf{D}_i'\mathbf{V}_i^{-1}\mathbf{D}_i \\ &= \mathbf{X}_i'\mathbf{A}_i\left(\mathbf{A}_i^{0.5}\mathbf{R}_i(\rho)\mathbf{A}_i^{0.5}\right)^{-1}\mathbf{A}_i\mathbf{X}_i'.\end{aligned}$$

The "meat" of the sandwich is an information matrix that uses empirical residuals to estimate the within-cluster covariance matrix. The meat for the ith cluster is

$$\begin{aligned}\mathbf{M}_i &= \mathbf{D}_i'\mathbf{V}_i^{-1}\mathbf{C}_i\mathbf{V}_i^{-1}\mathbf{D}_i \\ &= \mathbf{X}_i'\mathbf{A}_i\left(\mathbf{A}_i^{0.5}\mathbf{R}_i(\rho)\mathbf{A}_i^{0.5}\right)^{-1}\mathbf{C}_i\left(\mathbf{A}_i^{0.5}\mathbf{R}_i(\rho)\mathbf{A}_i^{0.5}\right)^{-1}\mathbf{A}_i\mathbf{X}_i',\end{aligned}$$

where $\mathbf{C}_i$ is the outer product of the empirical residuals. Specifically, the jkth element of this $n_i \times n_i$ matrix is

$$c_{jk} = \left(y_{ij} - \pi_{PA}(\mathbf{x}_{ij})\right) \times \left(y_{ik} - \pi_{PA}(\mathbf{x}_{ik})\right).$$

The equation for the estimator is obtained by evaluating all expressions at the estimator $\hat{\boldsymbol{\beta}}_{PA}$ and the respective values of the covariates, namely

$$\widehat{\mathrm{Cov}}\left(\hat{\boldsymbol{\beta}}_{PA}\right) = \left(\sum_{i=1}^{m}\hat{\mathbf{B}}_i\right)^{-1} \times \left(\sum_{i=1}^{m}\hat{\mathbf{M}}_i\right) \times \left(\sum_{i=1}^{m}\hat{\mathbf{B}}_i\right)^{-1}. \tag{8.32}$$

We note that some packages may offer the user the choice of using the information sandwich estimator, also called the *robust estimator*, in equation (8.32) or one based on the observed information matrix for the specified correlation structure, the bread $\mathbf{B}_i$. We think that unless there is strong evidence from other studies or clinical considerations that the working correlation structure is correct, one should use the estimator in equation (8.32).

One can use the estimated coefficients and estimated standard errors to estimate odds ratios and to perform tests for individual coefficients. Joint hypotheses must be tested using multivariable Wald tests since the GEE approach is not based on likelihood theory. This does make model building a bit more cumbersome since in most packages it is more complicated to perform multivariable Wald tests than likelihood ratio tests.

It is possible to formulate a likelihood function for the cluster-specific model described in equation (8.25). If we assume that the random effects follow a normal distribution with mean zero and constant variance, $\alpha_i \sim N(0, \sigma_\alpha^2)$, then the contribution of the ith cluster to the likelihood function is

$$P(\boldsymbol{\beta}_s)_i = \int_{-\infty}^{\infty} \left[\prod_{j=1}^{n_i} \frac{e^{y_{ij} \times (\alpha_i + \mathbf{x}'_{ij} \boldsymbol{\beta}_s)}}{1 + e^{\alpha_i + \mathbf{x}'_{ij} \boldsymbol{\beta}_s}} \right] \frac{1}{\sqrt{2\pi}} \frac{1}{\sigma_\alpha} \exp\left(-\frac{\alpha_i^2}{2\sigma_\alpha^2} \right) d\alpha_i \qquad (8.33)$$

and the full log likelihood is

$$L(\boldsymbol{\beta}_s) = \sum_{i=1}^{m} \ln\left[P(\boldsymbol{\beta}_s)_i \right]. \qquad (8.34)$$

The problem is that complicated numerical methods are needed to evaluate the log likelihood, obtain the likelihood equations and then solve them. These methods are well beyond the mathematical level of this text and thus we do not consider them further. We note that in addition to the typical output about coefficients most packages include an estimate of the variance of the random effects and an estimate of its standard error. Newer versions of some major software packages have the capability to fit models based on the log likelihood in equation (8.34) (e.g., the NLMIXED procedure in SAS version 8.0 and the xtlogit command in STATA version 6.0).

Neuhaus, Kalbfleisch and Hauk (1991) and Neuhaus (1992) present summary results that compare the magnitude of the coefficients from the cluster-specific model and population average model. These authors show, for coefficients whose value is near zero, that

$$\beta_{PA} \approx \beta_s[1-\rho(0)] \tag{8.35}$$

where $\rho(0)$ is the intracluster correlation among the observations of the binary outcome. This result demonstrates that we expect the estimates from fitted population average models to be closer to the null value, zero, than estimates from the fitted cluster-specific model. The shrinkage to the null in equation (8.35) can also be obtained from results examining the effect of failing to include an important covariate in the model, see Neuhaus and Jewell (1993) and Chao, Palta and Young (1997).

We fit models to computer-generated data to illustrate the effect of the intracluster correlation on the difference between the cluster-specific and population average coefficients. In each case, the fitted model contained a single continuous covariate distributed normal with mean zero, standard deviation 3 and true cluster-specific coefficient $\beta_s = 1$. The random effects were generated from a normal distribution with mean zero and standard deviation $\sigma_\alpha = 0, 0.5, 1.0, 1.5, \ldots, 10.0$. As we noted earlier in this section, the intracluster correlation increases with increasing σ_α. In these examples the resulting intracluster correlations, $\rho(0)$, range from 0 to about 0.84. For each set of parameter values we generated data for 200 clusters of size four. Hence the equation of the logit is $g(x_{ij}, \beta_s) = \alpha_i + x_{ij}$ with $i = 1, 2, \ldots, 200$, $j = 1, 2, 3, 4$, $x_{ij} \sim N(0, 9)$ and $\alpha_i \sim \mathrm{N}(0, \sigma_\alpha^2)$. We fit cluster-specific and population average models containing the covariate x. The values of the respective estimated coefficients are plotted versus the intracluster correlation in Figure 8.5. In addition we plot an approximate population average coefficient obtained using equation (8.35), i.e., $\tilde{\beta}_{PA} \approx \hat{\beta}_s[1-\rho(0)]$.

The results shown in Figure 8.5 demonstrate that the attenuation to the null described in equation (8.35) holds in this example. We note that the estimate of the cluster-specific coefficient tends to fluctuate about the true value of 1.0 with increased variability for large values of the intracluster correlation. We observed this same general pattern for varying numbers of clusters and observations per cluster.

Neuhaus (1992) shows that the variability in the estimates of the coefficients depends on the total sample size and intracluster correlation. In practice, the variability in the estimates of the population average coefficient depends to a greater extent on the number of clusters while that of the cluster-specific coefficient depends more on the total sample size and the intracluster correlation. The results in Neuhaus

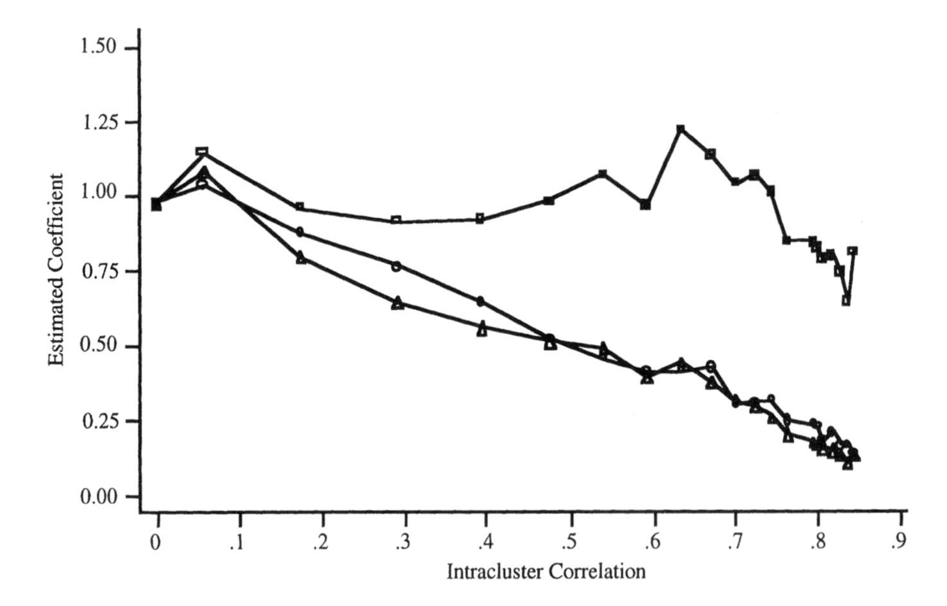

Figure 8.5 Plot of the estimated cluster specific coefficient (□), estimated population average coefficient (○) and approximate estimated population average coefficient (△) versus the intracluster correlation obtained from fitting models with 200 clusters of size 4.

(1992) also show that the Wald statistics for population average coefficients under exchangeable correlation and the cluster-specific model should be approximately the same. This result also follows from the approximation shown in equation (8.25).

As an example, we created a hypothetical data set based on the low birth weight data described in Section 1.6.2. We excluded the woman of age 45 years, leaving 188 women. We do not go into the details of how the data were constructed, as they are not essential to the discussion of fitting models. A hypothetical additional number of births was generated for each woman and varied between 1 and 3 yielding an average number of 2.6 births per woman. We did not retain all the covariates in the low birth weight data. Instead we retained as cluster-specific covariates, the age of the mother and the weight of the mother at the last menstrual period associated with each birth. As examples of cluster-level covariates we kept the race of the mother and smoking status, which was assumed not to change. The total data set has information on 488 births. One may obtain the data from either of the two web sites listed

Table 8.24 Listing of the Data for Three Women in the Longitudinal Low Birth Weight Data Set

ID	OBS	SMOKE	RACE	AGE	LWT	BWT	LOW
1	1	1	3	28	120	2865	0
1	2	1	3	33	141	2609	0
2	1	0	1	29	130	2613	0
2	2	0	1	34	151	3125	0
2	3	0	1	37	144	2481	1
43	1	1	2	24	105	2679	0
43	2	1	2	30	131	2240	1
43	3	1	2	35	121	2172	1
43	4	1	2	41	141	1853	1

in the preface in a file named CLSLOWBWT.DAT. Data for three mothers are shown in Table 8.24. We note that the identification codes are not the same as those in the original low birth weight data. We wish to remind the reader that these data are hypothetical and reflect the statistical method used to generate them and thus do not reflect actual birth histories. As we show when we fit models, the data do provide results typical of cluster-specific and population average models.

In Table 8.24 we see that the woman with ID = 1 had two births. She smoked during both pregnancies and has RACE = 3 (Other). She was 28 years old at the first birth and 33 at her second birth. Prior to her first pregnancy she weighed 120 pounds and prior to her second pregnancy she weighed 141 pounds. Her first baby weighed 2865 grams and thus has LOW = 0. Her second baby weighed 2609 grams and also has LOW = 0. The woman with ID = 2 had three births, the third of which weighed less than 2500 grams and has LOW = 1. The woman with ID = 43 had four births. She smoked during all four pregnancies and has RACE = 2 (Black). She was 24 years old at her first birth and 41 years old at her fourth birth. Prior to her first pregnancy she weighed 105 pounds and 141 pounds prior to her fourth pregnancy. Each of her last three babies weighed less than 2500 grams and thus were of low weight. Data for the other 185 women record similar information on their respective birth histories.

The results of fitting a population average model containing AGE, LWT and SMOKE (using STATA's xtlogit command with option pa) are shown in Table 8.25. Estimates of the standard errors are based on

Table 8.25 Estimated Coefficients, Robust Standard Errors, Wald Statistics, Two-tailed *p*-Values and 95 Percent Confidence Intervals for a Population Average Model with Exchangeable Correlation

Variable	Coef.	Robust Std. Err.	z	P>|z|	95% CI
AGE	0.058	0.0195	2.99	0.003	0.0201, 0.0967
LWT	−0.009	0.0041	−2.23	0.026	−0.0172, −0.0011
SMOKE	0.702	0.2829	2.48	0.013	0.1472, 1.2562
Constant	−1.342	0.5895	−2.28	0.023	−2.4975, −0.1866

the robust estimator in equation (8.32). We see that all three covariates are significant at the five percent level. We defer discussion of estimated odds ratios from this model until after we present the results from the cluster-specific model.

The results of fitting the cluster-specific model containing AGE, LWT and SMOKE (using STATA's xtlogit command with option re) are shown in Table 8.26. The table contains two panels of output. The top panel contains the usual results describing the estimates of the coefficients. The bottom panel contains results describing the estimate of the variance of the random effect. For numerical stability reasons STATA chooses to estimate the log of the variance described in the row labeled Ln_Sig2 in Table 8.26. The resulting estimate of the standard deviation, displayed in the row labeled Sigma, is obtained as the square root of exponentiation of the estimate of the log variance, e.g., $4.006 = \sqrt{\exp(2.776)}$. The results in the row labeled "Rho" are not the estimated intracluster correlations, $\hat{\rho}(0)$. Instead these values describe the proportion of the total variance accounted for by the random effect. Specifically, $0.941 = 4.006^2 / (1 + 4.006^2)$. The estimate of the intracluster correlation (obtained using STATA's loneway command with response variable LOW and grouping variable ID) is $\hat{\rho}(0) = 0.606$. Other packages provide similar or equivalent output on the fitted model.

We note that the Wald statistics for the respective coefficients in Table 8.25 and Table 8.26 are of similar magnitude but are not as close to each other as theory behind the development of equation (8.35) would suggest.

Table 8.26 Estimated Coefficients, Standard Errors, Wald Statistics, Two-tailed p-Values and 95 Percent Confidence Intervals for a Cluster-specific Model

Variable	Coef.	Std. Err.	z	P>|z|	95% CI
AGE	0.141	0.0493	2.86	0.004	0.044, 0.238
LWT	−0.015	0.0082	−1.85	0.064	−0.031, 0.001
SMOKE	1.861	0.6392	2.91	0.004	0.608, 3.114
Constant	−4.642	1.6778	−2.77	0.006	−7.931, −1.354
Ln_Sig2	2.776	0.3675			2.055, 3.496
Sigma	4.006	0.7361			2.795, 5.743
Rho	0.941	0.0203			0.887, 0.971

Log Likelihood = −232.9881

The likelihood ratio test of the model in Table 8.26 versus the usual logistic regression model is used to test $H_0: \sigma_\alpha = 0$. In this example, twice the difference in the log likelihoods is

$$G = -2[(-288.7622) - (-232.9881)] = 111.548$$

and $P[\chi^2(1) > 111.548] < 0.001$. The problem with the likelihood ratio statistic is that the null value, zero, lies on the boundary of the parameter space, which violates one of the assumptions of the test. Nevertheless, the statistic may be used to give an indication of the magnitude of the variance of the random effect. In this example the point estimate, its corresponding Wald test and the likelihood ratio statistic all suggest significant random effect variation.

In order to describe the effect of the intracluster correlation we calculate the approximate estimate from equation (8.35) and present all three estimates in Table 8.27. The results show that the shrinkage to the null is well described by the approximation formula. This is particularly interesting in this example as the result in equation (8.35) was derived assuming the true cluster-specific coefficient is near zero. Yet we see that equation (8.35) provides a good approximation for even the coefficient for SMOKE. In additional calculations, similar to those whose results are shown in Figure 8.5, we found that the shrinkage to the null is well described by equation (8.35) for coefficients as large as 5.0. In general, the approximation improved as the intracluster correlation became larger.

Table 8.27 Estimated Coefficients From the Cluster-Specific Model, Population Average Model and the Approximation to the Population Average Model from Equation (8.35)

Variable	Cluster-Specific Coef., $\hat{\beta}_s$	Population Average Coef., $\hat{\beta}_{PA}$	Approximate Pop. Average $\hat{\beta}_s(1-\hat{\rho}(0))$
AGE	0.141	0.058	0.055
LWT	−0.015	−0.009	−0.006
SMOKE	1.861	0.702	0.733

We calculate estimated odds ratios in the usual manner for both models. They are presented in Table 8.28. One calculates the end-points for a confidence interval estimate for an odds ratio in the same manner as used for all other fitted logistic regression models. We have not included confidence intervals in Table 8.28 since our purpose here is to compare and contrast the odds ratios from the two models. In practice, one would use the model that best addresses the study objectives. Results would then include appropriately calculated confidence intervals.

The interpretation of the estimated odds ratios from a population average model is a bit easier than for a cluster-specific model. They compare odds computed from proportions of subjects in the population at the different levels of the comparison covariate holding all other co-

Table 8.28 Estimated Odds Ratios from the Cluster-Specific Model and Population Average Model

Variable	Cluster-Specific Odds Ratio	Population Average Odds Ratio
AGE+	2.024	1.336
LWT*	0.861	0.914
SMOKE	6.430	2.017

+: Odds ratio for a 5 year increase in age

*: Odds ratio for a 10 pound increase in weight

variates fixed. For example, the estimated population average odds ratio for smoking during pregnancy (SMOKE) is 2.02. The interpretation is that the odds of low birth weight computed from the proportion of women who smoke is twice that based on the proportion of women who do not smoke, holding age at delivery and weight at the last menstrual period constant. The population average odds ratio for a five year difference in age at delivery is 1.34. The interpretation is that the odds of low birth weight computed from the proportion of women who are five years older than some reference level for age is 1.34 times higher than that based on the proportion of women who are at the reference age, holding weight at the last menstrual period and smoking status constant. The fact that age is linear in the logit implies this odds ratio holds for a five year difference at any age. The population average odds ratio for a 10 pound increase in weight at the last menstrual period is 0.91. The interpretation is that the odds of low birth weight computed from the proportion of women who are 10 pounds heavier than some reference level is 9 percent less than the odds of low birth weight based on the proportion of women who are at the reference weight level, holding age at pregnancy and smoking status constant. The fact that LWT is linear in the logit implies this odds ratio holds for a 10 pound difference at any weight. These estimated odds ratios describe risk of the event via proportions in the population and as such are more analogous to odds ratios from the logistic regression models described in Chapter 3. A fact noted earlier in this section is that the population average model with $\rho = 0$ is the usual logistic regression model. Hence, the population average model is likely to be the best model when the objectives of the study are to describe in broad terms the effects of covariates. This broad interpretation comes at the cost of not using information available in repeated measurements of a covariate on study subjects.

The interpretation of odds ratios from a fitted cluster-specific model applies to subjects with a common but unobserved value of the random effect α_i. This could be a single subject or a group of subjects. For example the estimate of the cluster-specific odds ratio for smoking during pregnancy is 6.43. The interpretation is that by smoking during pregnancy the woman has increased her odds of a low weight baby by 6.43 times the odds if she did not smoke holding age and weight constant. In this case the odds ratio makes sense since smoking during pregnancy is a modifiable risk factor. However the odds ratio for a non-modifiable factor such as race is more difficult to interpret. One would have to resort to comparisons of hypothetical groups of subjects with the same random effect who differ in their race holding other covariates constant. We leave the details as an exercise.

The estimate of the cluster-specific odds ratio for a five year increase in age is 2.02. The interpretation is that the odds of a woman having a low weight baby in five years is 2.02 times the odds at the current age holding weight and smoking status constant. This estimate takes full advantage of the fact that we have observed women over at least a portion of their reproductive years and thus expresses the influence of chronological aging at the individual level. The same would be true of any similarly observed longitudinal factor.

The cluster-specific odds ratio for a 10 pound increase in weight is 0.86. The interpretation is that the odds of a woman having a low weight baby if she gained 10 pounds is 14 percent less than the odds at her current weight, holding age at pregnancy and smoking status constant. This odds ratio suggests, likely incorrectly, that by simply gaining weight a woman can substantially reduce the risk of a low weight baby. In rough terms, the estimated coefficients in Table 8.26 suggest, incorrectly, that a 10 pound increase in weight could counteract a one year increase in age. In this case, the odds ratio is an artifact of weight gain of the women in the study over time. What is needed in the model is a more objective measure of size of the woman such as height or body mass index. However if the weight gain over time is similar for both small and large women then the odds ratio correctly estimates the effect of body size on baby weight among women with the same random effect and holding all other covariates constant. We leave fitting a cluster-specific model using weight of the mother at last menstrual period of the first pregnancy as a cluster-level covariate as an exercise.

The covariates age, weight and smoking status provide good examples of the strengths and weaknesses of population average and cluster-specific models. In a sense, the odds ratios for covariates like smoking status and weight are easier to interpret from population average models since they describe effects in broad groups of subjects in the population. The clear weakness of the population average model is that it cannot address effects such age. The cluster-specific model is best suited for this covariate as one does not have to argue that the inferences apply to some hypothetical and unobservable group of subjects with the same random effect. However, one must pay close attention to determine whether covariates measured repeatedly are true longitudinal covariates or repeated imprecise measures of a non-longitudinal covariate. An example of this would be body weight when basic body size is actually the covariate of interest. Both models address important clinical questions and have their place in an analysis of clustered binary data.

There are a number of important practical issues that we have not discussed. Model building issues have been ignored yet they are as vital in developing models for correlated data as they were with uncorrelated data. The modeling paradigm presented in detail in Chapter 4 may be applied with the models discussed in this section. Statistical variable selection methods such as stepwise and best subsets are not currently available for fitting correlated data models in software packages. Thus one must use some form of purposeful selection using Wald tests with the population average model and Wald or likelihood ratio tests with the cluster-specific model. Checking the scale of continuous covariates is just as important as with non-correlated data models. One can always use the method of design variables since computer intensive methods such as fractional polynomials have not as yet been implemented for use with correlated data models. An alternative approach would be to assume the observations are in fact not correlated and use the usual logistic regression model with fractional polynomials to identify a potential non-linear transformation. One would then try this transformation when fitting the appropriate correlated data model. Interactions should be specified and checked for inclusion in the same manner as described in Chapter 4. Diagnostic statistics, such as those described in Chapter 5, have not as yet been extended for use in model checking with correlated data models. However, one could approximate the analysis by assuming the observations are not correlated and using the methods in Chapter 5. Although not specifically developed for this situation, this analysis is better than not doing any model checking.

One must be careful when fitting cluster-specific models. The numerical methods are sensitive to the number of clusters and cluster size. The numerical methods seem to work best when the cluster size is not too large. STATA suggests that cluster sizes should be less than 10 unless the intracluster correlation is small. In addition one must be aware that if the intracluster correlation is quite small then the software may either fail to converge to a solution or output an estimate of σ_α^2 that is effectively zero. For example in these settings STATA typically stops and reports an estimate of the log variance of -14.0. In this case one should abandon the cluster-specific model in favor of the usual logistic regression model since the two models are equivalent when $\sigma_\alpha = 0$. Another potential problem with the cluster-specific model is the log likelihood in equation (8.34) is based on the assumption that the ran-

Table 8.29 Estimated Coefficients and Standard Errors Obtained from STATA, SAS and EGRET

	STATA		SAS		EGRET	
Variable	Coef.	Std. Err.	Coef.	Std. Err.	Coef.	Std. Err.
AGE	0.141	0.0493	0.146	0.0530	0.166	0.0579
LWT	−0.015	0.0082	−0.022	0.0113	−0.028	0.0112
SMOKE	1.861	0.6392	1.816	0.7520	1.668	0.6879
Constant	−4.642	1.6778	−3.561	1.5872	−3.133	1.4900
Sigma	4.006	0.7361	14.618*	5.0170	3.749	0.6252
	LogL = −232.988		LogL= −233.326		LogL = −233.35	

*: SAS reports the estimate of the variance

dom effects are distributed normally. This assumption cannot be checked since we have no observations of the random effect. Thus evidence of lack of fit of a model could be due to a poorly specified systematic component and/or non-normal random effects. Current software does not permit other distributions. Thus, if casewise diagnostic statistics (assuming independence) provide no indication that the systematic component of the model is inadequate, then one should probably abandon the cluster-specific model in favor of the population average model.

An additional difficulty one may encounter when fitting cluster-specific models is that different software packages may give different parameter estimates. The reason is that the solution to the likelihood equations depends on the particular numerical method used to evaluate the log likelihood in equation (8.34). In addition, the rules used by the package to stop the iteration process are not universally applied. For example, SAS's NLMIX procedure has numerous options and criteria that the user can specify. However only expert users should consider using anything but the default settings. STATA's xtlogit command uses the same basic method to evaluate equation (8.34) as SAS but has far fewer optimization options. Again we think that modifying these options should be left to experienced users. A third package, EGRET (EGRET for windows (1999)), uses an entirely different method to evaluate equation (8.34). In order to compare these three packages we show in Table 8.29 the results of fitting the same cluster-specific model using each program's default settings. We note that none of the estimates has exactly the same value. All the estimates of the coefficients are within 10 percent of each other, except for LWT. For this covariate the estimate from STATA is 46 percent smaller than the estimates from SAS and EGRET. However, all three estimates are within two estimated standard errors of each other. Thus, estimates of the odds ratios for AGE and SMOKE, while not identical, would certainly convey the same message. The estimated odds ratios for a 10 pound increase in weight from STATA, SAS and EGRET are: 0.86, 0.80 and 0.76 respectively. The estimated reduction in risk ranges from 14 to 24 percent; but their respective confidence intervals (not shown) have considerable overlap.

Based on the results shown in Table 8.29 and the general sensitivity of the numerical calculations, we think it is best to proceed cautiously when fitting cluster-specific models.

It is possible to perform overall tests of fit for correlated data models. Recent unpublished work by Evans (1998) examined the performance of the Hosmer-Lemeshow test and extensions of the Pearson chi square and other tests described in Section 5.2.2 to the correlated data setting. His results indicate that the usual Hosmer-Lemeshow test may be used in some settings to assess fit of population average models. In general, one must avoid using the test when there are many tied or nearly tied values in the estimates of the probabilities. This is likely to occur under one or more of the following conditions: the model contains only a few cluster-level covariates, the intracluster correlation among the responses is large, and settings with a few clusters and few observations per cluster. If, in a particular setting, none of these conditions holds, then the test can be used. For example, the fitted population average model in Table 8.25 is based on 188 clusters, each containing between two and four observations and only moderate intracluster correlation $\hat{\rho}(0)=0.606$. This represents a setting where the test may be used. In this case the Hosmer-Lemeshow test and its p-value can be easily obtained in STATA by using the lfit command with the beta option. This yields a value of the Hosmer-Lemeshow test of $\hat{C}=11.67$ which, with eight degrees-of-freedom, yields $p=0.167$ that supports model fit. We do not present the two by ten table of observed and predicted frequencies as there was close agreement in all ten deciles of risk.

Evans' (1998) extension of the normal approximation to the Pearson chi square test requires calculations that are much more complicated than those required in the usual logistic regression model described in Section 5.2.2. As such we do not discuss it further in this text. However, Evans' simulation results show that the test is effective for assessing fit. Perhaps it will be added to future releases of software packages.

The Hosmer-Lemeshow test may be used with the cluster-specific model. However its application requires that one use fitted values that include an estimate of the random effect term, α_i, as well as the regression coefficients, i.e.,

$$\hat{\pi}(\mathbf{x}_{ij}) = \frac{e^{\hat{\alpha}_i + \mathbf{x}'_{ij}\hat{\boldsymbol{\beta}}_s}}{1 + e^{\hat{\alpha}_i + \mathbf{x}'_{ij}\hat{\boldsymbol{\beta}}_s}} .$$

These fitted values may be obtained from SAS's version 8.0 NLMIXED procedure. However the Hosmer-Lemeshow test itself is not available in SAS. We used this test in STATA to check the fit of SAS's version of the fitted model, shown in Table 8.29. The calculations involve obtaining the estimates of the random effects using the predict option from within NLMIXED and then calculating the test "by hand". The value of the test is $\hat{C} = 24.70$ which, with eight degrees-of-freedom, yields $p = 0.002$. Hence we conclude that there is significant evidence of lack of fit. Examination of the table (not shown) of observed and expected numbers of low and normal weight births in the 10 deciles of risk showed that the model underestimated the number of normal weight births in the top two deciles of risk, $\hat{\pi} > 0.88$. This result suggests that the model might be improved. We leave as an exercise examination of the inclusion of interactions and the affect they have on the fit of the model.

We note that one should always avoid using the test with $\alpha_i = 0$ for all clusters. For example, calculation of the test using $\alpha_i = 0$ with coefficients from the fitted model from STATA in Table 8.26 yields $\hat{C} = 413.36$ with $p < 0.001$.

Evans' (1998) extension of the normal approximation to the Pearson chi square test for the cluster-specific model is considerably more complicated than the extension to the population average model. As such we do not discuss it further in this text. Evans' simulations show the test performs effectively. Hopefully it will be included in future releases of the software packages.

In some settings we may want to include a previously observed value(s) of the outcome as a covariate(s) for future observations. These models are typically referred to as *transitional models*. The logit in this setting is of the form

$$g(y_{i-1}, \theta, \mathbf{x}_{ij}, \boldsymbol{\beta}_T) = \mathbf{x}'_{ij}\boldsymbol{\beta}_T + y_{i-1}\theta . \tag{8.36}$$

In this model the coefficient θ is the log odds of the ith observation of the outcome conditional on the value of the previous outcome.

The longitudinal low birth weight data provide the possibility for several versions of such a covariate. For example, we could model low birth weight in the second and subsequent births by including a covariate indicating whether the previous birth was of low weight (the model shown in equation (8.36)). Another possible covariate is whether the woman ever had a low birth weight baby, (i.e., include $z_i = \max(y_{i-1}, y_{i-2}, \ldots, y_1)$ as a covariate in the model). A third version is to include information on the status of all previous births (i.e., include $y_{i-1}, y_{i-2}, \ldots, y_1$ in the model for the ith birth). In these models one assumes that including any previously observed value of the outcome accounts for any intracluster correlation and we use the usual logistic regression model to fit the data.

As an example we show in Table 8.30 the results of fitting a model that includes the value of LOW from the previous birth. We leave fitting models containing the other two versions of the covariate as exercises. When we fit this model the sample size is reduced from 488 to 300 since we exclude the first birth of each woman. There is no previous birth history for the first birth.

The estimate of the odds ratio for previous birth of a low weight baby is $\widehat{\text{OR}} = 30.4$. The interpretation is that the odds of a low weight birth among women whose previous pregnancy resulted in a low weight birth is 30 times the odds for women whose previous pregnancy did not result in a low weight baby, controlling for age, weight and smoking status. The coefficient estimates for AGE and LWT are about the same order of magnitude as the estimates from the population average model

Table 8.30 Estimated Coefficients, Standard Errors, Wald Statistics, Two-tailed p-Values and 95 Percent Confidence Intervals for a Fitted Logistic Regression Model Containing the Covariate Previous Birth of Low Weight, (n = 300)

Variable	Coef.	Std. Err.	z	P>|z|	95% CI
AGE	0.080	0.0338	2.38	0.017	0.014, 0.146
LWT	−0.017	0.0066	−2.54	0.011	−0.030, −0.004
SMOKE	1.687	0.3613	4.67	<0.001	0.979, 2.395
PREV_LOW	3.415	0.3892	8.77	<0.001	2.652, 4.177
Constant	−2.491	1.2596	−1.98	0.048	−4.960, −0.022

in Table 8.25 while the estimate of the coefficient for SMOKE is closer to that of the cluster-specific model in Table 8.26. The odds ratio for previous birth of low weight may seem unrealistically large but in any case it is the dominant factor in the model.

In practice we think that transitional models should only be used in settings where there is an explicit time ordering in the repeated observations of the outcome.

In summary, the cluster-specific and population average models provide useful and powerful modeling tools when observations of the outcome variable are correlated. The correlation must be due to recognizable factors in the design of the study that allow one to explicitly identify clusters, or sets of observations, that are correlated and those that are uncorrelated. The cluster-specific model is likely to be most useful for describing the effect of covariates that are repeatedly measured on the same subject. The population average model is best suited to describe the effect of covariates that are constant within clusters. However, both models may be fit with both types of covariates. One must pay particular attention to signs of numerical problems when fitting cluster-specific models. These include failure of the program to converge to a solution and a "zero" estimate of the variance of the random effect.

Logistic regression models for correlated binary data is an area of active statistical research with new developments appearing on a regular basis. As these developments become accepted in the statistical community as being sound and worthwhile modeling tools, developers of software packages can be expected to add them to their routines.

8.4 EXACT METHODS FOR LOGISTIC REGRESSION MODELS

The methods used for testing and inference up to this point in the text have assumed, in addition to other mathematical assumptions, that the sample size is sufficiently large for parameter estimates to be normally distributed and for the likelihood ratio and Wald tests to follow chi square and normal distributions, respectively. There may be occasions where one would like to fit a logistic regression model but the sample size is such that these large sample assumptions are clearly not justified. Recent advances in computational methods now make it possible to fit models in such settings.

The problem of fitting a logistic regression model and then making inferences and tests about the parameters when the sample size is small is a complicated version of Fisher's exact test for a 2×2 contingency table. Cox and Snell (1989) note that the extension of the theory of Fisher's exact test to logistic regression models has been known since the 1970's. However the computations required are extremely complex and were considered impractical until efficient algorithms were developed by Tritchler (1984), Hirji, Mehta, and Patel (1987 and 1988) and Hirji (1992). Mehta and Patel (1995) review the theory and provide a number of insightful examples. The exact methods have been incorporated into the statistical software package, LogXact 4 for Windows (2000). We use this package to fit the models in this section.

The central idea behind the theory of exact methods for logistic regression is to construct a statistical distribution that can, with efficient algorithms, be completely enumerated. The starting point in this process is to construct a conditional likelihood similar to that used in Chapter 7 for matched studies. Assume we have n independent observations of a binary outcome and a vector of $p+1$ covariates (i.e., $(y_i, \mathbf{x}_i)$, $i = 1, 2, \ldots, n$). We assume that the functional form of the logit is $g(\mathbf{x}, \boldsymbol{\beta}) = \sum_{j=0}^{p} x_j \beta_j$ with $x_0 = 1$. In settings where we are primarily interested in the slope coefficients we consider the intercept, β_0, as the nuisance parameter and condition on its sufficient statistic, $n_1 = \sum_{i=1}^{n} y_i$. As shown in Mehta and Patel (1995) the resulting conditional likelihood is

$$P\left(Y_1 = y_1, Y_2 = y_2, \ldots, Y_n = y_n \middle| n_1\right) = \frac{\exp\left(\sum_{i=1}^{n} y_i \times \sum_{j=1}^{p} x_{ij} \beta_j\right)}{\sum_{l \in R} \exp\left(\sum_{l=1}^{n} y_l \times \sum_{j=1}^{p} x_{lj} \beta_j\right)} \tag{8.37}$$

where R denotes the collection of

$$\binom{n}{n_1} = \frac{n!}{[n_1!] \times [(n - n_1)!]}$$

possible allocations of 0 and 1 to $(y_1, y_2, \ldots, y_n)$ such that $n_1 = \sum_{i=1}^{n} y_i$. The form of the likelihood in equation (8.37) suggests that the sufficient statistic for β_j is

$$t_j = \sum_{i=1}^{n} y_i x_{ij} \,. \qquad (8.38)$$

(Cox and Hinkley (1974) present a discussion of sufficient statistics and their role in conditional inference.) Let the vector of sufficient statistics for the slope coefficients be denoted by $\mathbf{t}' = (t_1, t_2, \ldots, t_p)$. The exact distribution of the collection of p sufficient statistics is given by the equation

$$\mathrm{P}(T_1 = t_1, T_2 = t_2, \ldots, T_p = t_p) = \frac{c(\mathbf{t})\exp\left(\sum_{j=1}^{p} t_j \beta_j\right)}{\sum_{\mathbf{u} \in S} c(\mathbf{u})\exp\left(\sum_{l=1}^{p} u_l \beta_l\right)}, \qquad (8.39)$$

where $c(\mathbf{t})$ denotes the number of possible allocations of 0 and 1 to $(y_1, y_2, \ldots, y_n)$ such that $t_j = \sum_{i=1}^{n} y_i x_{ij}$ and S denotes the set of allocations of 0 and 1 to $(y_1, y_2, \ldots, y_n)$ such that $n_1 = \sum_{i=1}^{n} y_i$ and $u_j = \sum_{l=1}^{n} y_i x_{lj}$ denotes the resulting value of the jth sufficient statistic for the lth allocation. The distribution in equation (8.39) is used to obtain point and confidence interval estimates of the regression coefficients as well as tests of hypotheses that coefficients are equal to zero. The calculations required for the multivariable problem are quite complex. Thus we illustrate the exact methods with a model containing a single dichotomous covariate.

As an example, suppose we wish to model risk factors for having a low birth weight baby among women 30 years or older in the low birth weight study described in Section 1.6.2. There are 27 such women and 4 had a low birth weight baby. It is clear that, with only 27 observations and four $LOW = 1$ outcomes, we should not use methods requiring large sample sizes for their validity. Consider the covariate recording the number of previous pre-term deliveries dichotomized into none (0) or at least one (1) and denoted PTD. The cross-classification of LOW by PTD is shown in Table 8.31.

The results in Table 8.31 show that the observed value of the sufficient statistic for the intercept term is $t_0 = 4$ and for the coefficient for PTD it is $t_1 = 2$. The later result follows from the fact that only two

Table 8.31 Cross-Classification of Low Birth Weight (LOW) by History of Pre-term Delivery (PTD) Among Women 30 Years of Age or Older

	PTD		
LOW	0	1	Total
0	19	4	23
1	2	2	4
Total	21	6	27

subjects had $LOW = 1$ and $PTD = 1$ (i.e., $2 = \sum_{i=1}^{27} LOW_i \times PTD_i$). It follows from equation (8.39) that the exact probability is

$$P(T_1 = t_1) = \frac{c(t_1)\exp(t_1\beta_1)}{\sum_{u \in S} c(u)\exp(u\beta_1)}. \tag{8.40}$$

The possible values of the sufficient statistic are $t_1 = 0,1,2,3,4$. Thus the term $c(t_1)$ describes the number of possible allocations of 23 values of zero and 4 values of one to 27 subjects with the resulting value of $t_1 = \sum_{i=1}^{27} LOW_i \times PTD_i$. For example, with the help of LogXact 4, we obtain these and they are in the column labeled "Count" in Table 8.32. There we see that are 5985 sequences of 23 zeros and 4 ones, where $0 = \sum_{i=1}^{27} LOW_i \times PTD_i$. The simplest exact inferential question is a test of the hypothesis that $\beta_1 = 0$. The values of equation (8.40) under the null hypothesis are given in the last column in Table 8.32. These probabilities are calculated using the fact that $1 = \exp(t_1 \times 0)$ and S contains 17,550 sequences. Thus the first probability is

$$P(t_1 = 0) = \frac{5985}{17550} = 0.34103,$$

and the others are calculated in a similar manner. We calculate the two tailed p-value by summing the probabilities in Table 8.32 over values of the sufficient statistic that are as likely, or less likely, to have a smaller probability than the observed value of $t = 2$. Thus we obtain

Table 8.32 Enumeration of the Exact Probability Distribution of the Sufficient Statistic for the Coefficient of PTD

t_1	Count: $c(t)$	Probability Under $H_o: \beta_1 = 0$
0	5985	0.34103
1	7980	0.45469
2	3150	0.17949
3	420	0.02393
4	15	0.00086
Total	17550	1.0

$$p = 0.17949 + 0.02393 + 0.00086 = 0.20428.$$

We note that the value is identical to the two sided p-value for Fisher's exact test computed from Table 8.31. In this case we cannot conclude that having a history of pre-term delivery is a significant risk factor for having a low weight birth among women who are 30 years of age or older.

The exact conditional maximum likelihood point estimate of the coefficient is the value that maximizes the probability given in equation (8.40) which, given the counts in Table 8.32, is

$$P(T_1 = 2) = \frac{c(2)\exp(2\beta_1)}{\sum_{u \in S} c(u)\exp(u\beta_1)}$$

$$= \frac{3150\exp(2\beta_1)}{5985\exp(0\beta_1) + 7980\exp(1\beta_1) + 3150\exp(2\beta_1) + 420\exp(3\beta_1) + 15\exp(4\beta_1)}.$$

Even in this rather simple example the computations require a package like LogXact 4. For comparative purposes we show the results from fitting the conditional exact maximum likelihood estimate (CMLE) as well as those from fitting the usual logistic regression model in Table 8.33. In this example, as shown in Chapter 3, the usual MLE is simply the log of the odds ratio from Table 8.31.

Table 8.33 Results of Fitting the Usual Logistic Model (MLE) and the Exact Conditional Model (CMLE) to the Data in Table 8.31

	Method	Coeff.	Std.Err.	95% CI
PTD	MLE	1.558	1.1413	−0.679, 3.795
	CMLE	1.482	1.1059	−1.383, 4.370
Constant	MLE	−2.251	0.7434	−0.794, 0.409
	CMLE	*	*	*

*: Not computed using CMLE in this case

Both the point estimate of the coefficient for PTD and the estimate of the associated standard error are slightly smaller when the exact conditional model is used. The endpoints of the confidence internal for the MLE are obtained in the usual manner as $\hat{\beta}_1 \pm 1.96\widehat{\text{SE}}(\hat{\beta}_1)$. The endpoints of the CMLE are obtained from the following procedure.

Assume that the possible range of the sufficient statistic, given the observed value of t_0, is $t_{\min} \le t_1 \le t_{\max}$. In our example the range is $0 \le t_1 \le 4$. The lower endpoint of a $100(1-\alpha)\%$ confidence interval is the value of β_1 such that

$$\alpha/2 = \sum_{k=t_{1obs}}^{t_{\max}} \mathrm{P}(T_1 = k), \tag{8.41}$$

where t_{1obs} denotes the observed value of t_1, 2 in our example, and $\mathrm{P}(T_1 = k)$ is given in equation (8.40). If $t_{1obs} = t_{\min}$ then the lower limit is set to $-\infty$. The upper endpoint of a $100(1-\alpha)\%$ confidence interval is the value of β_1 such that

$$\alpha/2 = \sum_{k=t_{\min}}^{t_{1obs}} \mathrm{P}(T_1 = k). \tag{8.42}$$

If $t_{1obs} = t_{\max}$ then the lower limit is set to $+\infty$. The solutions to equations (8.41) and (8.42) for a 95 percent confidence interval in our example are shown in Table 8.33. We note that the CMLE interval is considerably wider than the MLE interval, reflecting the increased uncertainty in our estimate due to the small sample size.

Table 8.34 Cross-Classification of Low Birth Weight (LOW) by Smoking Status of the Mother during Pregnancy (SMOKE) Among Women 30 Years of Age or Older

	SMOKE		
LOW	0	1	Total
0	17	6	23
1	0	4	4
Total	21	6	27

As a second example consider the cross classification of smoking status during pregnancy versus low birth weight among women 30 years of age or older shown in Table 8.34. We note that the table contains a cell with zero frequency. As shown in Chapter 4, Section 5 conventional logistic regression software cannot be used in this case. However we are able to obtain a two-tailed p-value, point and confidence interval estimate using exact methods.

The exact probability distribution under the hypothesis of no effect due to smoking during pregnancy, $\beta_1 = 0$, is shown in Table 8.35. The p-value in this case is 0.01197 since no other value had as small or smaller probability than the observed value of 4. Since the observed value of the sufficient statistic is $4 = t_{\max}$ the upper limit of the 95 percent confidence interval is $+\infty$ and the solution to equation (8.41) is 0.308. In settings where $t_{1obs} = t_{\min}$ or $t_{1obs} = t_{\max}$ the CMLE does not

Table 8.35 Enumeration of the Exact Probability Distribution of the Sufficient Statistic for the Coefficient of SMOKE

t_1	Count: $c(t)$	Probability Under $H_0: \beta_1 = 0$
0	2380	0.13561
1	6800	0.38746
2	6120	0.34872
3	2040	0.11624
4	210	0.01197
Total	17550	1.0

have a finite solution and Hirji, Tsiatis and Mehta (1989) suggest using the median unbiased estimator (MUE). This estimator is defined as the average of the endpoints of a 50 percent confidence interval estimator. In settings where $t_{1obs} = t_{\min}$ and the lower limit is $-\infty$ the MUE is set equal to the upper limit of the 50 percent interval. In settings where $t_{1obs} = t_{\max}$ and the upper limit is $+\infty$ the MUE is set equal to the lower limit of the 50 percent interval. In our example we have $t_{1obs} = t_{\max}$ and the lower limit of the 50 percent interval is $\hat{\beta}_{1MUE} = 2.510$. That is the solution to equation (8.41) using $\alpha = 0.5$ is 2.510. Thus use of exact methods yields point and interval estimates as well as a test of significance when none are computable using conventional approaches to MLE.

We obtain point and interval estimators of odds ratios in the usual manner by exponentiating the respective estimators for the coefficient. The odds ratio for smoking during pregnancy obtained from the MUE is $\widehat{OR} = 12.3$ and the endpoints of the 95 percent confidence interval are $(1.36, +\infty)$. The interpretation is that the odds of a low weight baby among women 30 years or older who smoke during pregnancy is 12.3 times the odds of women 30 years or older who do not smoke during pregnancy and it could be as little as 1.36 times with 95 percent confidence.

One can use exact methods to fit multivariable logistic regression models and perform tests of subsets of parameters. Thus it is theoretically possible to use exact methods with the modeling paradigm described in detail in Chapter 4. However the required computations are extensive and can be quite time consuming, even on a fast computer. Thus we recommend that one restrict use of exact analyses to those settings where the sample sizes are small enough to question the use of the large sample assumption. The exception to this recommendation might be a setting where one has a zero frequency cell in an important, for clinical reasons, dichotomous covariate or a polychotomous covariate whose categories should not be combined to eliminate the zero frequency.

The exact methods as described above focus on exact CMLE of the slope coefficients. It is possible to extend the approach to estimation of all coefficients. The basic idea is the same but one estimates each parameter conditioning on the sufficient statistic for all other parameters. The result is a fitted model similar to ones discussed in detail in Chapter 4. As an example we present in Table 8.36 the results of fitting both the usual and exact logistic models using women 25 years or older in the

Table 8.36 Results of Fitting the Usual Logistic Model (MLE) and the Exact Conditional Model (CMLE) and in the Low Birth Weight Study to Women 25 Years or Older

	Method	Coeff.	Std.Err.	95% CI
LWT	MLE	−0.019	0.0117	−0.042, 0.004
	CMLE	−0.018	0.0113	−0.043, 0.002
SMOKE	MLE	0.249	0.6087	−0.944, 1.442
	CMLE	0.256	0.5933	−1.111, 1.567
PTD	MLE	1.393	0.6687	0.082, 2.703
	CMLE	1.310	0.6440	−0.014, 2.798
Constant	MLE	1.097	1.5599	−1.961, 4.154
	CMLE	0.331	0.7381	−1.331. 2.105

low birth weight study. The covariates in the model are weight at the last menstrual period (LWT), smoking status during pregnancy (SMOKE) and history of prior pre-term delivery (PTD). There are 69 women in this subgroup and 19 low weight births. We note that LogXact 4 took one hour and thirty minutes to perform the necessary computations on a 400MHz computer and the optimal amount of memory needed was 18MB. This shows that even with a small problem fitting exact logistic regression models requires a fast computer with a lot of memory and plenty of patience.

The estimates of the slope coefficients in Table 8.36 are similar and would result in effectively equivalent estimates of their respective odds ratios. However, the exact confidence intervals are much wider, reflecting the increased variability due to the small sample size. The two estimates of the intercept coefficients appear to be quite different but, in fact, there is considerable overlap in the two confidence intervals. We leave as an exercise determining the effect this difference has on the estimated probabilities.

In addition to fitting all parameters it is possible to evaluate fit and compute diagnostic statistics as described in Chapter 5 and Chapter 7. The LogXact 4 package has the capability to compute the Hosmer-Lemeshow goodness-of-fit test and the casewise diagnostic statistics. However we do not recommend using the p-value for the Hosmer-Lemeshow test based on a chi square distribution with eight degrees-of-freedom as it is based on the large sample assumption that one is trying to avoid by using exact methods. Instead we suggest that one visually check the agreement between the observed and expected frequencies in the two by ten table. One should examine the diagnostic statistics using the plots discussed in Chapter 5. Models can then be refit deleting suspect cases. We leave assessing the fit of the model in Table 8.36 as an exercise.

In summary we feel that exact methods for logistic regression should be considered when one is fitting models with a small sample size or unbalanced data that result in zero frequency cells.

8.5 SAMPLE SIZE ISSUES WHEN FITTING LOGISTIC REGRESSION MODELS

In our experience there are two sample size questions, prospective and retrospective. The prospective question is: How many subjects do I need to observe to have specified power to detect that the new treatment is significantly better than the old or placebo treatment? The retrospective question is: Do I have enough data to fit this model? There has been surprisingly little work on sample size for logistic regression. The available methods to address sample size selection have been implemented in just a few specialty software packages. The key element in assessing if one has adequate data to fit a particular model involves the number of events per covariate. Recent research by Peduzzi, Concato, Kemper, Holford and Feinstein (1996) provides some guidance. In this Section we consider methods for choosing a sample size first and then discuss the importance of having an adequate number of events per covariate.

The basic sample size question is as follows: What sample size does one need to test the null hypothesis that a particular slope coefficient is equal to zero (without loss of generality we assume it is the first of p covariates in the model) versus the alternative that it is equal to some specified value, i.e., $\text{H}_0:\beta_1 = 0$ versus $\text{H}_a:\beta_1 = \beta_1^*$. If the logistic regression model is to contain only this single dichotomous covariate, then one may use conventional sample size methods to test for the equality of two proportions (see Fleiss (1981) or Lemeshow, Hosmer, Klar and Lwanga (1990)). Alternatively one may use results in Whitemore (1981) for a logistic regression model containing a single dichotomous covariate. The difference in the two approaches is that the former is based on the sampling distribution of the difference in two proportions and the later on the sampling distribution of the log of the odds ratio.

We illustrate the two approaches using the data from the UMARU IMPACT Study (UIS) described in Section 1.6.4. Suppose that we consider these data as being either pilot data or data from an earlier study. We use it to help us determine what sample size we would need in a new study to test for a 50 percent increase in the odds of remaining drug free for one year for the longer versus the shorter treatment. In terms of the logistic regression model the null and alternative hypotheses are $H_o: \beta_1 = \ln(1) = 0$ versus $H_a: \beta_1 = \ln(1.5)$. To determine the sample size with either approach we need an estimate of the response probability under the shorter treatment, $P_0 = P(Y = 1 \mid x = 0)$. Cross classifying the outcome variable (DFREE) by the treatment covariate (TREAT) results in the observation that 21.4 percent of those on the shorter duration treatment remained drug free for 12 months. We round this down to 20 percent and use this as our response probability. The response probability yielding an odds ratio of 1.5 is

$$P_1 = P(Y = 1 \mid x = 1)\frac{1.5 \times 0.2}{(1 - 0.2) + 1.5 \times 0.2} = 0.2728 .$$

Thus, stated in terms of proportions the null and alternative hypotheses are $H_o: P_0 = P_1 = 0.2$ and $H_a: P_0 = 0.2, P_1 = 0.2728$.

The sample size one needs in each of two groups for a one sided test at the α level of significance of $H_o: P_0 = P_1$ and power $1 - \theta$ for the alternative $H_a: P_0 < P_1$ is given by the equation

$$n = \frac{\left(z_{1-\alpha}\sqrt{2\overline{P}(1-\overline{P})} + z_{1-\theta}\sqrt{P_0(1-P_0) + P_1(1-P_1)}\right)^2}{(P_1 - P_0)^2} \qquad (8.43)$$

where $\overline{P} = (P_0 + P_1)/2$ and $z_{1-\alpha}$ and $z_{1-\theta}$ denote the upper α and θ percent points respectively of the standard normal distribution. We use a one sided test here for better comparability with the results in Whitemore (1981). For a two sided test one would use $z_{1-\alpha/2}$ in place of $z_{1-\alpha}$ in equation (8.43).

Thus the number we would need in our two treatment groups for a 5 percent level test to have power 80 percent is

$$n = \frac{\left(1.645\sqrt{2 \times 0.2364 \times 0.7636} + 0.842\sqrt{0.2 \times 0.8 + 0.2728 \times 0.7272}\right)^2}{(0.2728 - 0.2)^2}$$
$$= 420.29,$$

or 421 subjects in each group for a total sample size of approximately 842 subjects.

Whitemore (1981) approaches the sample size problem via the sampling distribution of the Wald statistic for the estimate of the logistic regression coefficient. For a univariable logistic regression model containing a single dichotomous covariate, x, coded 0 or 1 the total sample size needed to test $H_o: \beta_1 = 0$ versus $H_a: \beta_1 = \beta_1^*$ is

$$n = (1 + 2P_0) \times \frac{\left(z_{1-\alpha}\sqrt{\frac{1}{1-\pi} + \frac{1}{\pi}} + z_{1-\theta}\sqrt{\frac{1}{1-\pi} + \frac{1}{\pi e^{\beta_1^*}}}\right)^2}{P_0 \beta_1^{*2}} . \tag{8.44}$$

where $\pi = P(X = 0)$ denotes the fraction of subjects in the study expected to have $x = 0$. In our example we want the sample size for an odds ratio of 1.5 or $\beta_1^* = \ln(1.5)$ and we plan to use equal numbers of subjects in the two treatment groups. Thus the value of equation (8.44) with $\pi = 0.5$ is

$$n = (1 + 2 \times 0.2) \times \frac{\left(1.645\sqrt{\frac{1}{0.5} + \frac{1}{0.5}} + 0.842\sqrt{\frac{1}{0.5} + \frac{1}{0.5e^{[\ln(1.5)]}}}\right)^2}{0.2 \times [\ln(1.5)]^2}$$
$$= 1.4 \times \frac{\left(1.645 \times 2 + 0.842\sqrt{2 + 2 \times 0.6666}\right)^2}{0.2 \times 0.1644}$$
$$= 992.19 .$$

This suggests that, rounding up to be divisible by 2, we would need approximately 994 subjects or 497 in each group. This is 76 more subjects per group than the sample size given by equation (8.43). The difference in the two sample sizes stems from a number of assumptions made by Whitemore to obtain equation (8.44). This equation is derived under the assumption that the logistic probabilities are small. The lead term in equation (8.44) is proposed as a way to adjust the sample size when this is not the case. To our knowledge no research has been published that compares the results from equations (8.43) and (8.44) in a systematic manner. Our recommendation for univariable models is that one should use equation (8.43) as it relies on fewer assumptions than equation (8.44).

If the single covariate we plan to include in the model is continuous, then we may use results for this setting derived by Whitemore (1981) and Hsieh (1989). We must assume that the covariate is standardized to have mean 0 and standard deviation 1.0. Thus the logistic regression coefficient is the effect of a one standard deviation increment in the unstandardized covariate. The sample size needed for a one sided test, at the α level of significance and power $1-\theta$, of $\text{H}_0: \beta_1 = 0$ versus $\text{H}_a: \beta_1 = \beta_1^*$ is given by the equation

$$n = (1 + 2P_0\delta) \times \frac{\left(z_{1-\alpha} + z_{1-\theta} e^{-0.25\beta_1^{*2}}\right)^2}{P_0 \beta_1^{*2}}, \tag{8.45}$$

where

$$\delta = \frac{1 + \left(1 + \beta_1^{*2}\right) e^{1.25\beta_1^{*2}}}{1 + e^{-0.25\beta_1^{*2}}}. \tag{8.46}$$

and P_0 is the value of the logistic probability evaluated at the mean of the standardized covariates, i.e.,

$$P_0 = \frac{e^{\beta_0}}{1 + e^{\beta_0}}. \tag{8.47}$$

As an example, suppose we consider the covariate age in the UIS study and ignore all of the other covariates. In these data the mean age of the subjects is approximately 32 years with a standard deviation of 6 years. We would like to determine what sample size we would need in order to be able to detect that the effect of a one standard deviation increase in age is a 50 percent increase in the odds of remaining drug free (i.e., $\beta_1^* = \ln(1.5)$). To obtain an estimate of P_0 in (8.47) we fit a univariable logistic regression model containing the standardized covariate $AGES = (AGE - 32)/6$. The estimate of the intercept term is $\hat{\beta}_0 =$ -1.079 (results not shown) and equation (8.47) becomes

$$P_0 = \frac{e^{-1.079}}{1+e^{-1.079}} = 0.25 \ .$$

The value of equation (8.46) in this example is

$$\delta = \frac{1+\left(1+\left[\ln(1.5)\right]^2\right)e^{1.25[\ln(1.5)]^2}}{1+e^{-0.25[\ln(1.5)]^2}} = 1.24$$

and the sample size from equation (8.45)

$$n = (1+2\times 0.25\times 1.24)\times \frac{\left(1.645+0.842e^{-0.25[\ln(1.5)]^2}\right)^2}{0.25\left[\ln(1.5)\right]^2}$$

$$= 237.19 \, .$$

This result suggests that if the true effect of age is to increase by 50 percent the odds of remaining drug free for every increment of 6 years, then we need a total 238 subjects in our study. This same result may also be obtained from Table II in Hsieh (1989). In addition it may also be obtained from the PASS 6.0 (1996) software package.

However it is extremely rare in practice to have final inferences based on a univariable logistic regression model. The major problem is that the only sample size results currently available are for multivariable models containing continuous covariates that are assumed to be distributed normal, exponential or Poisson. For example Hsieh's multivariable adaptation of equation (8.45) suggests using the sample size given by the equation

$$n = \frac{\left(1+2P_0\delta\right)}{\left(1-\rho^2\right)} \times \frac{\left(z_{1-\alpha}+z_{1-\theta}e^{-0.25\beta_1^{*2}}\right)^2}{P_0\beta_1^{*2}}, \qquad (8.48)$$

where ρ^2 is the squared multiple correlation of the covariate of interest, x_1, and the remaining $p-1$ covariates in the model. This is the equation used by the PASS 6.0 package.

In spite of these rather stringent assumptions we think that one can use a modification of equation (8.45) or equation (8.48) as a first step in obtaining an approximate sample size in practical settings (i.e., ones

Table 8.37 Results of Fitting a Logistic Regression Model to the UIS Data (n = 575)

Variable	Coef.	Std. Err.	z	P>z
AGES	0.306	0.1038	2.94	0.003
NDRGTXS	−0.316	0.1283	−2.46	0.014
IVHX_2	−0.593	0.2864	−2.07	0.038
IVHX_3	−0.760	0.2490	−3.05	0.002
RACE	0.208	0.2215	0.94	0.347
TREAT	0.439	0.1991	2.20	0.028
Constant	−1.041	0.2097	−4.96	<0.001

where models are fit containing a mix of continuous and discrete covariates).

As a first multivariable example we consider the sample size needed to test for an age effect of ln(1.5) per 6 year increment where we also include the number of previous drug treatments, history of IV drug use, race and treatment in the model. In this example we consider treatment to be just another potential confounder of the age effect. The results of fitting a logistic regression model to the UIS data with age standardized, $AGES = (AGE - 32)/6$, and number of previous drug treatments standardized, $NDRGTXS = (NDRGTX - 5)/5$ are shown in Table 8.37.

Based on the results in Table 8.37 the estimated probability of remaining drug free with all covariates equal to zero is

$$P_0 = \frac{e^{-1.041}}{1 + e^{-1.041}} = 0.261.$$

In this case a subject with all covariates equal to zero corresponds to a subject who is 32 years old, has had 5 previous drug treatments, no previous history of IV drug use, is white and on the shorter treatment.

Suppose we perform our test at the $\alpha = 0.05$ level and would like power $1 - \theta = 0.8$. Use of a multiple linear regression package with AGES as the dependent variable and the remaining variables as covariates yields $R^2 = 0.1473$. The value of equation (8.46) is the same as that determined for the univariable model, $\delta = 1.24$, and the sample size from equation (8.48) is

$$n=\frac{(1+2\times0.261\times1.24)}{(1-0.1473)}\times\frac{\left(1.645+0.842\times e^{-0.25[\ln(1.5)]^2}\right)^2}{0.261\times[\ln(1.5)]^2}=270.92.$$

Thus application the Hsieh modification of the Whitemore formula suggests that only about 271 subjects are needed to have 80 percent power to test for the stated effect of age. We note that if the average fitted logistic probability is approximately equal to $P_0=0.261$ then we would expect to have 71 "events" or subjects who remain drug free for 12 months. We comment on the importance of this number shortly.

As a second multivariable example we consider sample size for a study where treatment is the main covariate of interest. What sample size is necessary to have 80 percent power to detect a treatment coefficient ln(1.5) when we adjust for age, the number of previous drug treatments, history of IV drug use and race? Application of Hsieh's correction factor for multiple covariates to equation (8.44) yields sample size

$$n=\frac{(1+2P_0)}{1-\rho^2}\times\frac{\left(z_{1-\alpha}\sqrt{\frac{1}{1-\pi}+\frac{1}{\pi}}+z_{1-\theta}\sqrt{\frac{1}{1-\pi}+\frac{1}{\pi e^{\beta_1^*}}}\right)^2}{P_0\beta_1^{*2}}. \tag{8.49}$$

In this case, since the covariate of interest is dichotomous, we suggest using one of the R^2 measures discussed in Chapter 5. One possibility is the squared correlation between the values of the dichotomous covariate and fitted values from a logistic regression of this covariate on all other variables in the model (i.e., the value of equation (5.6)). In our example this yields $\rho^2=(0.1123)^2=0.0126$. Thus the multivariable adjusted sample size from equation (8.49) is

$$\begin{aligned}n&=\frac{(1+2\times0.261)}{(1-0.0126)}\times\frac{\left(1.645\sqrt{\frac{1}{0.5}+\frac{1}{0.5}}+0.842\sqrt{\frac{1}{0.5}+\frac{1}{0.5e^{[\ln(1.5)]}}}\right)^2}{0.261\times[\ln(1.5)]^2}\\&=1.541\times\frac{\left(1.645\times2+0.842\sqrt{2+2\times0.6666}\right)^2}{0.261\times0.1644}\\&=836.86.\end{aligned}$$

This suggests that, rounding up to be divisible by 2, a total sample size of about 838, or 419 per treatment group would be required.

There are a number of potential problems with the sample size formula in equation (8.49). One is the ad-hoc use of the Hsieh's correction factor to account for multiple covariates. A second problem involves the earlier noted discrepancy in sample sizes suggested by equation (8.43) and equation (8.44). We think that the sample size suggested by equation (8.49) may be unnecessarily large but could be the starting point for a more in depth sample size analysis using pilot data do some model fitting. For example, one way to assess the precision obtained from modeling with a sample of 838 subjects is to construct a pseudo-study by combining the original 575 subjects with a random sample of 263 of the 575 subjects. We would then fit the proposed multivariable logistic regression model and examine the estimated coefficient for treatment, its estimated standard error, Wald statistic and p-value. These results can be used to provide guidance as to how significant the results may be expected to be in the new, larger study. Ideally we would repeat this process a number of times to obtain an approximate sampling distributions of the estimated quantities. If in the end we think the estimated standard error is too small and confidence intervals are too narrow then we would repeat the process using a smaller sample size. This could be repeated until we had empirical evidence that the sample size provides about the desired precision in the multivariable model.

A second consideration, and one relevant to any model being fit, is the issue of events per covariate. Peduzzi, Concato, Kemper, Holford and Feinstein (1996) examine the issue of how many events per covariate are needed to obtain reliable estimates of regression coefficients when fitting a logistic regression model. Peduzzi et. al. consider single term main effects models. In order to extend their ideas to more complex models that may have multiple terms for a number of covariates, we prefer to use the terminology *events per parameter*. In general the relevant quantity is the frequency of the least frequent outcome, $m=\min(n_1,n_0)$. In our experience this is usually the number of subjects with the event present, $y=1$ but it could just as well be the number with the event absent, $y=0$. Peduzzi et. al. show that a minimum of 10 events per parameter are needed to avoid problems of over estimated and under estimated variances and thus poor coverage of Wald-based confidence intervals and Wald tests of coefficients. Thus the simplest answer to the "do I have enough data" question is to suggest that the model contain no more than $p+1\le\min(n_1,n_0)/10$ parameters. For example in the UIS study we have $147 = \min(147, 428)$ events. The rule of 10 suggests that models should contain no more than 14 parameters. The model fit in Chapter 4 and evaluated in Chapter 5, see Table 5.10, using the UIS data contains 11 parameters. Note that with the sample size of 271 that we obtain when the goal is to test the coefficient age we expect about 71 events. In this case the rule of 10 suggests that models should contain no more than 7 parameters.

As is the case with any overly simple solution to a complex problem, the rule of 10 should only be used as a guideline and a final determination must consider the context of the total problem. This includes the actual number of events, the total sample size and most importantly the mix of discrete, continuous and interaction terms in the model. Peduzzi et. al. considered only discrete covariates and provide no information about the bivariate distributions of outcome by covariates. We think that the ten events per parameter rule may work well for continuous covariates and discrete covariates with a balanced distribution. However, we are less certain about its applicability in settings where the distribution of discrete covariates is weighted heavily to one value. Here one may require that the minimum observed frequency be, say 10, in the contingency table of outcome by covariate. Research is needed to determine if 10 is too stringent a requirement.

In summary, having an adequate sample size is just as important when fitting logistic regression models as any other regression model. However, the performance of model-based estimates may be determined more by the number of events rather than the total sample size.

EXERCISES

1. Data from the mammography experience study are described in Section 8.1. Use a subset of these data and fit a multinomial logistic regression model. For example, you may choose to use only the first 200 subjects. The purpose of the exercise is to obtain practice when there are more than two categories of outcome. Hence, any alternative strategy for identifying a subset of subjects is acceptable.

2. The data for the low birth weight study are described in Section 1.6.2. These data are used in Section 8.2 to illustrate ordinal logistic regression models via the four category outcome BWT4,

$$BWT4 = \begin{cases} 0 \text{ if } BWT > 3500 \\ 1 \text{ if } 3000 < BWT \leq 3500 \\ 2 \text{ if } 2500 < BWT \leq 3000 \\ 3 \text{ if } BWT \leq 2500. \end{cases}$$

Use the outcome variable *BWT*4 and fit the multinomial or baseline logistic regression model.

In each of the above problems the steps in fitting the model should include: (1) a complete univariate analysis, (2) an appropriate selection of variables for a multivariate model (this should include scale identification for continuous covariates and assessment of the need for interactions), (3) an assessment of fit of the multivariate model, (4) preparation and presentation of a table containing the results of the final model (this table should contain point and interval estimates for all relevant odds ratios), and (5) conclusions from the analysis.

3. Using the final models identified in problems 1 and 2 compare the estimates of the coefficients obtained from fitting the multinomial logistic regression model to those obtained from fitting the adjacent-category, continuation-ratio and proportional odds ordinal logistic regression models. For the mammography experience data recode the outcome variable, ME, to 0 = Never, 1 = Over one year ago and 2 = Within one year in order that its codes increase with frequency of use.

4. The following exercise is designed to enhance the idea expressed in Figure 8.2 and Figure 8.3 that one way to obtain the proportional odds model is via categorization of a continuous variable.
   (a) Form the scatter plot of BWT versus LWT.
   (b) Fit the linear regression of BWT on LWT and add the estimated regression line to the scatterplot in 4(a). Let $\hat{\lambda}_0$ denote the estimate of the intercept, $\hat{\lambda}_1$ the estimate of the slope and s the root mean squared error from the linear regression.
   (c) It follows from results for the logistic distribution that the relationship between the root mean squared error in the normal errors linear regression and the scale parameter for logistic errors

linear regression is approximately $\hat{\sigma} = s\sqrt{3}/\pi$. Use the results from the linear regression in 4(b) and obtain $\hat{\sigma}$.

(d) Use the results from 4(b) and 4(c) and show that the estimates presented in Table 8.20 are approximately $\hat{\tau}_1 = \left(2500 - \hat{\lambda}_0\right)/\hat{\sigma}$, $\hat{\tau}_2 = \left(3000 - \hat{\lambda}_0\right)/\hat{\sigma}$, $\hat{\tau}_3 = \left(3500 - \hat{\lambda}_0\right)/\hat{\sigma}$ and $\hat{\beta}_1 = \hat{\lambda}_1/\hat{\sigma}$.

(e) By hand draw a facsimile of the density function shown in Figure 8.3 with the three vertical lines at the values 2500, 3000 and 3500. Using the results in equation (8.20), equation (8.21) and the estimates in Table 8.20 compute the value of the four areas under the hand-drawn curve. Using these specific areas demonstrate that the relationship shown in equation (8.22) holds at each cutpoint.

(f) Repeat problem 4.5 for $LWT = 135$ and show by direct calculation using areas under the two curves that the relationship in equation (8.24) holds at each cutpoint.

5. Using the data from the longitudinal low birth weight study and considering all the covariates (be sure to consider the possibility of interactions among the covariates):
   (a) Find the best cluster-specific and population average models.
   (b) Evaluate the fit of the two models obtained in problem 5(a).
   (c) Prepare separate table for each model obtained in 5(a) containing estimates of the odds ratios with 95 percent confidence intervals.
   (d) Compare the interpretation of the point estimates of the odds ratios from the cluster-specific model and population average model.

6. Using the cluster-specific and population average models obtained in problem 5(a) explore alternative ways of including the weight of the mother at the last menstrual period. For example, one alternative is to use the weight at the first birth as a cluster level covariate. Others representations are possible. For each alternative fit the cluster-specific and population average model, estimate an odds ratio for weight and compare their interpretation.

7. Using the covariates in the population average model obtained in problem 5(a) explore alternative ways of including history of a low birth weight baby. For reach model compute and interpret the esti-

mate of the odds ratio and a 95 percent confidence interval for the previous low birth weight covariate. Recall that we fit these models using the usual logistic regression model.

(a) Fit the model that adds the outcome of the previous birth to the model. In this problem explore the use of two versions of this covariate: one that assigns a missing value for the first birth and one that assigns a value of zero to the first birth.

(b) Fit the model that includes a dichotomous covariate indicating if any previous birth was of low weight. In this problem explore the use of two versions of this covariate: one that assigns a missing value for the first birth and one that assigns a value of zero to the first birth.

8. Using the data from the ICU Study described in Chapter 1, Section 1.6.1, attempt to fit the usual logistic regression model containing type of admission (TYP) using subjects 25 years of age or younger. Why does the usual MLE have problems in this example? Fit the exact logistic regression model. Compute the point and 95 percent confidence interval estimates of the odds ratio.

9. Repeat problem 8 fitting models containing systolic blood pressure (SYS).

10. Evaluate the fit of the usual and exact logistic regression models shown in Table 8.36.

11. Consider the low birth weight study. What sample size would be needed in a new study to be able to detect that the odds of a low birth weight baby among women who smoke during pregnancy is 2.5 times that of women who do not smoke, using a 5 percent type I error probability and 80 percent power?

12. Consider the low birth weight study. What sample size would be needed in a new study to be able to detect that the odds of a low birth weight baby decrease at a rate of 10 percent per 10 pound increase in weight at the last menstrual period, using a 5 percent type I error probability and 80 percent power?

13. Repeat problem 11 assuming that you plan to use a model that contains age, weight of the mother at the last menstrual period and race.

14. Repeat problem 12 assuming that you plan to use a model that contains age, smoking status during pregnancy and race.

15. If the sample size obtained in problems 13 or 14 is larger than the original study size of 189 then use the suggested method for obtaining a larger study to explore the effect the larger study size has on estimated coefficients, standard errors and confidence intervals.

ADDENDUM
Applied Logistic Regression, Second Edition, by David W. Hosmer and Stanley Lemeshow. © 2000 by John Wiley & Sons, Inc.
ISBN 0-471-35632-8.

Tables 4.6, 4.8. 4.10

Table 4.6 Summary of the Use of the Method of Fractional Polynomials for NDRGTX

	df	Deviance	*G* for Model vs. Linear	Approx. *p*-Value	Powers
Not in model	0	626.176			
Linear	1	619.248	0.000	0.008*	1
$J=1$	2	618.818	0.430	0.512+	0.5
$J=2$	4	613.451	5.797	0.068#	−1, −1

* Compares linear model to model without NDRGTX.
+ Compares the $J=1$ model to the linear model
# Compares the $J=2$ model to the $J=1$ model

Table 4.8 Log-likelihood, Likelihood Ratio Test Statistic (*G*), Degrees of Freedom (df), and *p*-Value for Interactions of Interest When Added to the Main Effects Model in Table 4.7.

Interaction	Log-Likelihood	*G*	df	*p*-value
Main Effects Model	−306.7256			
AGE×NDRGTX*	−302.8314	7.79	2	0.020
AGE×IVHX	−306.3559	0.74	2	0.691
AGE×RACE	−306.6269	0.20	1	0.657
AGE×TREAT	−305.3410	2.76	1	0.096
AGE×SITE	−305.9265	1.60	1	0.206
NDRGTX*×IVHX	−304.0092	5.43	4	0.246
NDRGTX*×RACE	−304.6541	4.14	2	0.126
NDRGTX*×TREAT	−305.2580	2.94	2	0.231
NDRGTX*×SITE	−306.7239	0.01	2	0.998
IVHX×RACE	−305.8361	1.78	2	0.411
IVHX×TREAT	−306.7051	0.04	2	0.980
IVHX×SITE	−306.2910	0.87	2	0.648
RACE×TREAT	−306.2541	0.94	1	0.332
RACE×SITE	−302.4533	8.54	1	0.004
TREAT×SITE	−306.7087	0.03	1	0.854

*: All Interactions involving NDRGTX are formed using NDRGFP1 and NDRGFP2

Table 4.10 Results of Applying Stepwise Variable Selection Using the Maximum Likelihood Method to the UIS Data Presented at Each Step in Terms of the p-Values to Enter, Below the Horizontal Line, and the p-Value to Remove, Above the Horizontal Line in Each Column. The Asterisk Denotes the Maximum p-Value to Remove at Each Step.

Variable/Step	0	1	2	3	4
NDRGTX	0.0006	0.0006	0.0006	0.0164	0.0062
TREAT	0.0229	0.0249	0.0249*	0.0312	0.0224
IVHX	0.0013	0.0273	0.0332	0.0332*	0.0027
AGE	0.2371	0.0458	0.0356	0.0021	0.0021*
RACE	0.0315	0.0663	0.0914	0.2107	0.1350
SITE	0.1968	0.3644	0.3382	0.4987	0.5688
BECK	0.4250	0.5582	0.5436	0.7744	0.9938

REFERENCES

Agresti, A. (1990). *Categorical Data Analysis*, Wiley, Inc., New York.

Agresti, A., Booth, J. G., Hobert, J. P., and Caffo, B. (2000). Random effects modeling of categorical response data. *Sociological Methodology*, To appear.

Albert, A., and Anderson, J. A. (1984). On the existence of maximum likelihood estimates in logistic models. *Biometrika*, **71**, 1–10.

Ambler, G., and Royston, P. (1999). Fractional polynomial model selection procedures: Investigation of type I error rate, Submitted for publication.

Ananth, C. V., and Kleinbaum, D. G. (1997). Regression models for ordinal data: A review of methods and application. *International Journal of Epidemiology*, **26**, 1323-1333.

Anderson, T. W. (1984). *An Introduction to Multivariate Statistical Analysis*, Second Edition, Wiley, Inc., New York.

Ashby, M., Neuhaus, J. M., Hauck, W. W., Bacchetti, P., Heilbron, D. C., Jewell, N. P., Segal, M. R., and Fusaro, R. E. (1992). An annotated bibliography of methods for analyzing corrrelated categorical data. *Statistics in Medicine*, **11**, 67-99.

Bachand, A. M., and Hosmer, D. W. (1999). Defining confounding when the outcome is dichotomous, Submitted for Publication.

Bedrick, E. J., and Hill, J. R. (1996). Assessing the fit of logistic regression models to individual matched sets of case–control data. *Biometrics*, **52**, 1–9.

Begg, C. B., and Gray, R. (1984). Calculation of polychotomous logistic regression parameters using individualized regressions. *Biometrika*, **71**, 11–18.

Belsley, D. A., Kuh, E., and Welsch, R. E. (1980). *Regression Diagnostics: Identifying Influential Data and Sources of Collinearity*. Wiley, Inc., New York.

Bendel, R. B., and Afifi, A. A. (1977). Comparison of stopping rules in forward regression. *Journal of the American Statistical Association*, **72**, 46–53.

Bishop, Y. M. M., Feinberg, S. E., and Holland, P. (1975). *Discrete Multivariate Analysis: Theory and Practice*. MIT Press, Boston.

BMDP Statistical Software (1992). BMDP Classic 7.0 for DOS. SPSS, Inc., Chicago.

Box, G. E. P., and Tidwell, P. W. (1962). Transformation of the independent variables. *Technometrics*, **4**, 531–550.

Breslow, N. E. (1996). Statistics in epidemiology: The case-control study. *Journal of the American Statistical Association*, **91**, 14–28.

Breslow, N. E., and Cain, K. C. (1988). Logistic regression for two-stage case-control data. *Biometrika*, **75**, 11–20.

Breslow, N. E., and Clayton, D. G. (1993). Approximate inference in generalized linear mixed models. *Journal of the American Statistical Association*, **88**, 9-25.

Breslow, N. E., and Day, N. E. (1980). *Statistical Methods in Cancer Research.* Vol. 1 : The analysis of case-control studies. International Agency on Cancer, Lyon, France.

Breslow, N. E., and Zaho, L. P. (1988). Logistic regression for stratified case–control studies. *Biometrics*, **44**, 891–899.

Brown, C. C. (1982). On a goodness-of-fit test for the logistic model based on score statistics. *Communications in Statistics*, **11**, 1087–1105.

Bryson, M. C., and Johnson, M. E. (1981). The incidence of monotone likelihood in the Cox model. *Technometrics*, **23**, 381–384.

Chambless, L. E., and Boyle, K. E. (1985). Maximum likelihood methods for complex sample data: Logistic regression and discrete proportional hazards models. *Communications in Statistics: Theory and Methods*, **14**, 1377–1392.

Chao, W-H., Palta, M., and Young, T. (1997). Effect of omitted confounders in the analysis of correlated binary data. *Biometrics*, **53**, 678-689.

Cleveland, W. S. (1993). *Visualizing Data,* Hobart Press, Summit, NJ.

Collett, D. (1991) *Modelling Binary Data.* Chapman & Hall, London.

Cook, R. D. (1977). Detection of influential observations in linear regression. *Technometrics*, **19**, 15–18.

Cook, R. D. (1979). Influential observations in linear regression. *Journal of the American Statistical Association*, **74**, 169–174.

Cook, R. D., and Weisberg, S. (1982). *Residuals and Influence in Regression.* Chapman & Hall, New York.

Copas, J. B. (1983). Plotting p against x. *Applied Statistics*, **32**, 25–31.

Cornfield, J. (1951). A method of estimating comparative rates from clinical data; Applications to cancer of the lung, breast and cervix. *Journal of the National Cancer Institute*, **11**, 1269–1275.

Cornfield, J. (1962). Joint dependence of the risk of coronary heart disease on serum cholesterol and systolic blood pressure: A discriminant function analysis. *Federation Proceedings*, **21**, 58–61.

Costanza, M. C., and Afifi, A. A. (1979). Comparison of stopping rules in forward stepwise discriminant analysis. *Journal of the American Statistical Association*, **74**, 777–785.

Costanza, M. E., Stoddard, A. M., Gaw, V., and Zapka, J. G. (1992). The risk factors of age and family history and their relationship to screening mammography utilization. *Journal of the American Geriatrics Society*, **40**, 774–778.

Coull, B. A., and Agresti, A. (2000). Random effects modeling of multiple binomial responses using the multivariate binomial logit-normal distribution. *Biometrics*, **56**, 73-80.

Cox, D. R. (1970). *The Analysis of Binary Data.* Methuen, London.

Cox, D. R., and Hinkley, D. V. (1974). *Theoretical Statistics.* Chapman & Hall, London.

Cox, D. R., and Snell, E. J. (1989). *Analysis of Binary Data,* Second Edition. Chapman & Hall, London.

Day, N. E., and Byar, D. P. (1979). Testing hypotheses in case-control studies - equivalence of Mantel-Haenszel statistics and logit score tests. *Biometrics*, **35**, 623–630.

Diggle, P. J., Liang, K-Y., Zeger, S. L. (1994). *Analysis of Longitudinal Data.* Oxford University Press, New York.

Dobson, A. (1990). *An Introduction to Generalized Linear Models.* Chapman & Hall, London.

Efron, B. (1975). The efficiency of logistic regression compared to normal discriminant function analysis. *Journal of the American Statistical Association,* **70**, 892–898.

Efron, B. (1986). How biased is the apparent error rate? *Journal of the American Statistical Association,* **81**, 461–470.

EGRET for Windows (1999). Software for the Analysis of Biomedical and Epidemiological Studies, Cytel Software, Cambridge, MA.

Evans, M., Hastings, N., and Peacock, B. (1993). *Statistical Distributions*: Second Edition, Wiley, Inc., New York.

Evans, S. R. (1998). Goodness-of-fit in two models for clustered binary data. Unpublished doctoral dissertation. University of Massachusetts, Amherst, MA.

Farewell, V. T. (1979). Some results on the estimation of logistic models based on retrospective data. *Biometrika*, **66**, 27–32.

Fears, T. R., and Brown, C. C. (1986). Logistic regression methods for retrospective case-control studies using complex sampling procedures. *Biometrics*, **42**, 955–960.

Flack, V. F., and Chang, P. C. (1987). Frequency of selecting noise variables in subset regression analysis: A simulation study. *American Statistician*, **41**, 84–86.

Fleiss, J. (1979). Confidence intervals for the odds ratio in case-control studies: State of the art. *Journal of Chronic Diseases*, **32**, 69–77.

Fleiss, J. (1981). *Statistical Methods for Rates and Proportions*. Wiley, Inc., New York.

Fleiss, J. (1986). *The Design and Analysis of Clinical Experiments*. Wiley, Inc., New York.

Fowlkes, E. B. (1987). Some diagnostics for binary regression via smoothing. *Biometrika*, **74**, 503–515.

Freedman, D. A. (1983). A note on screening regression equations. *American Statistician,* **37**, 152–155.

Furnival, G. M., and Wilson, R. W. (1974). Regression by leaps and bounds. *Technometrics*, **16**, 499–511.

Gart, J. J., and Thomas, D. G. (1972). Numerical results on approximate confidence limits for the odds ratio. *Journal of the Royal Statistical Society, Series B*, **34**, 441–447.

Glynn, R. J., and Rosner, B. (1994). Comparison of alternative regression models for paired binary data. *Statistics in Medicine*, **13**, 1023-1036.

Greenland, S. (1989). Modelling variable selection in epidemiologic analysis. *American Journal of Public Health*, **79**, 340–349.

Griffiths, W. E., and Pope, P. J. (1987). Small sample properties of probit models. *Journal of the American Statistical Association*, **82**, 929–937.

Grizzle, J., Starmer, F., and Koch, G. (1969). Analysis of categorical data by linear models. *Biometrics*, **25**, 489–504.

Guerro, V. M., and Johnson, R. A. (1982). Use of the Box–Cox transformation with binary response models. *Biometrika*, **69**, 309–314.

Halpern, M., Blackwelder, W. C., and Verter, J. I. (1971). Estimation of the multivariate logistic risk function: A comparison of the discriminant function and maximum likelihood approaches. *Journal of Chronic Disease*, **24**, 125–158.

Harrell, F. E., Lee, K. L., and Mark, D. B. (1996). Tutorial in biostatistics: Multivariable prognostic models: Issues in developing models, evaluating assumptions and measuring and reducing errors. *Statistics in Medicine*, **15**, 361–387.

Haseman, J. K., and Hogan, M. D. (1975). Selection of the experimental unit in teratology studies. *Teratology*, **12**, 165–172.

Haseman, J. J., and Kupper, L. L. (1979). Analysis of dichotomous response data from certain toxicological experiments. *Biometrics*, **35**, 281–293.

Hastie, T., and Tibshirani, R. (1986). Generalized additive models. *Statistical Science*, **3**, 297–318.

Hastie, T., and Tibshirani, R. (1987). Generalized additive models: Some applications. *Journal of the American Statistical Association*, **82**, 371–386.

Hastie, T., and Tibshirani, R. (1990). *Generalized Additive Models*, Chapman & Hall, London.

Hauck, W. W., and Donner, A. (1977). Wald's test as applied to hypotheses in logit analysis. *Journal of the American Statistical Association,* **72**, 851–853.

Hirji, K. E., Mehta, C. R., and Patel, N. R. (1987). Computing distributions for exact logistic regression. *Journal of the American Statstical Association*, **82**, 1110-1117.

Hirji, K. E., Mehta, C. R., and Patel, N. R. (1988). Exact inference for matched case-control studies. *Biometrics*, **44**, 803-814.

Hirji, K. F., Tsiatis, A. A., and Mehta, C. R. (1989). Median unbiased estimation for binary data. *The American Statistician,* **43**, 7-11

Hirji, K. E. (1992). Exact distributions for polytomous data. *Journal of the American Statistical Association,* **87**, 487-492.

Hjort, N. L. (1988). Estimating the logistic regression equation when the model is incorrect. Technical Report, Norwegian Computing Center, Oslo, Norway.

Hjort, N. L. (1999). Estimation in moderately misspecified models. Statistical Research Report, Department of Mathematics, University of Oslo, Oslo.

Hosmer, D. W., Jovanovic, B., and Lemeshow, S. (1989). Best subsets logistic regression. *Biometrics*, **45**, 1265-1270.

Hosmer, D. W., and Lemeshow, S. (1980). A goodness-of-fit test for the multiple logistic regression model. *Communications in Statistics*, **A10**, 1043–1069.

Hosmer, D. W., Lemeshow, S., and Klar, J. (1988). Goodness-of-fit testing for multiple logistic regression analysis when the estimated probabilities are small. *Biometrical Journal*, **30**, 911–924.

Hosmer, T., Hosmer, D. W., and Fisher, L. L. (1983). A comparison of the maximum likelihood and discriminant function estimators of the coefficients of the logistic regression model for mixed continuous and discrete variables. *Communications in Statistics*, **B12**, 577–593.

Hosmer, D. W., Hosmer, T., Le Cessie, S., and Lemeshow, S. (1997). A comparison of goodness-of-fit tests for the logistic regression model, *Statistics in Medicine*, **16**, 965–980.

Hosmer, D. W., and Lemeshow, S. (1999). *Applied Survival Analysis: Regression Modeling of Time to Event Data.* Wiley, Inc., New York.

Hsieh, F. Y. (1989). Sample size tables for logistic regression. *Statistics in Medicne*, **8**, 795-802.

Jennings, D. E. (1986a). Judging inference adequacy in logistic regression. *Journal of the American Statistical Association*, **81**, 471–476.

Jennings, D. E. (1986b). Outliers and residual distributions in logistic regression. *Journal of the American Statistical Association*, **81**, 987–990.

Kay, R., and Little, S. (1986). Assessing the fit of the logistic model: A case study of children with haemalytic uraemic syndrome. *Applied Statistics*, **35**, 16–30.

Kay, R., and Little, S. (1987). Transformation of the explanatory variables in the logistic regression model for binary data. *Biometrika*, **74**, 495–501.

Kelsey, J. L., Thompson, W. D., and Evans, A. S. (1986). *Methods in Observational Epidemiology.* Oxford University Press, New York.

Kleinbaum, D. G. (1994). *Logistic Regression: A Self-Learning Text.* Springer-Verlag, New York.

Kleinbaum, D. G., Kupper, L. L., Muller, K. E., and Nazim, A. (1998). *Applied Regression Analysis and Other Multivariable Methods, 3rd Edition.* Duxbury, Belmont, CA.

Korn, E. L., and Graubard, B. I. (1990). Simultaneous testing of regression coefficients with complex survey data: Use of Bonferroni t statistics, *American Statistician*, **44**, 270–276.

Lachenbruch, P. A. (1975). *Discriminant Analysis.* Hafner, New York.

Landwehr, J. M., Pregibon, D., and Shoemaker, A. C. (1984). Graphical methods for assessing logistic regression models. *Journal of the American Statistical Association*, **79**, 61–71.

Lawless, J. F., and Singhal, K. (1978). Efficient screening of non–normal regression models. *Biometrics*, **34**, 318–327.

Lawless, J. F., and Singhal, K. (1987a). ISMOD: An all subsets regression program for generalized linear models.I. Statistical and computational background. *Computer Methods and Programs in Biomedicine*, **24**, 117–124.

Lawless, J. F., and Singhal, K. (1987b). ISMOD: An all subsets regression program for generalized linear models II. Program guide and examples. *Computer Methods and Programs in Biomedicine*, **24**, 117–124.

Le Cessie, S., and van Houwelingen, J. C. (1991). A goodness-of-fit test for binary data based on smoothing residuals. *Biometrics*, **47**, 1267–1282.

Le Cessie, S., and van Houwelingen, J. C. (1995). Testing the fit of a regression model via score tests in random effects models. *Biometrics*, **51**, 600–614.

Lee, K., and Koval, J. J. (1997). Determination of the best significance level in forward stepwise logistic regression. *Communication in Statistics B*, **26**, 559–575.

Legler, J. M., and Ryan, L. M. (1997). Latent variable models for teratogenesis using multiple binary outcomes. *Journal of the American Statistical Association*, **92**, 13–20.

Lemeshow, S., and Hosmer, D. W. (1982). The use of goodness-of-fit statistics in the development of logistic regression models. *American Journal of Epidemiology*, **115**, 92–106.

Lemeshow, S., and Hosmer, D. W. (1983). Estimation of odds ratios with categorical scaled covariates in multiple logistic regression analysis. *American Journal of Epidemiology*, **119**, 147–151.

Lemeshow, S., Hosmer, D. W., Klar, J., and Lwanga, S.K. (1990). *Adequacy of Sample Size in Health Studies*. Wiley, Chichester.

Lemeshow, S., Teres, D., Avrunin, J. S., and Pastides, H. (1988). Predicting the outcome of intensive care unit patients. *Journal of the American Statistical Association*, **83**, 348–356.

Lesaffre, E. (1986). Logistic discriminant analysis with applications in electrocardiography. Unpublished D.Sc. Thesis, University of Leuven, Belgium.

Lesaffre, E., and Albert, A. (1989). Multiple-group logistic regression diagnostics. *Applied Statistics*, **38**, 425-440.

Liang, K. Y., and Zeger, S. L. (1986). Longitudinal data analysis using generalized linear models. *Biometika*, **73**, 13-22.

Lin, D. Y., Pstay, B. M., and Kronmal, R. A. (1998). Assessing the sensitivity of regression results to unmeasured confounders in observational studies, *Biometrics*, **54**, 948–963.

LogXact 4 for Windows (2000). *Logistic Regression Software Featuring Exact Methods*. Cytel Software, Cambridge, MA.

Lynn, H. S., and McCulloch, C. E. (1992). When does it pay to break the matches for an analysis of a matched-pairs design. *Biometrics*, **48**, 397–409.

Maldonado, G., and Greenland, S. (1993). Interpreting model coefficients when the true model form is unknown. *Epidemiology*, **4**, 310–318.

Mallows, C. L. (1973). Some comments on Cp. *Technometrics*, **15**, 661–676.

Math Type 3.6: Mathematical Equation Editor (1998). Design Sciences, Inc. Long Beach, CA.

McCullagh, P. (1985a). On the asymptotic distribution of Pearson's statistics in linear exponential family models. *International Statistical Review*, **53**, 61–67.

McCullagh, P. (1985b). Sparse data and conditional tests. *Proceedings of the* 45th *Session of the ISI* (Amsterdam), Invited Paper 28, **3**, 1–10.

McCullagh, P. (1986). The conditional distribution of goodness-of-fit statistics for discrete data. *Journal of the American Statistical Association,* **81**, 104–107.

McCullagh, P., and Nelder, J. A. (1989). *Generalized Linear Models,* Second Edition. Chapman & Hall, London.

McCusker, J., Vickers-Lahti, M., Stoddard, A. M., Hindin, R., Bigelow, C., Garfield, F., Frost, R., Love, C., and Lewis, B. F. (1995). The effectiveness of alternative planned durations of residential drug abuse treatment. *American Journal of Public Health,* **85**, 1426–1429.

McCusker, J., Bigelow, C., Frost, R., Garfield, F., Hindin, R., Vickers-Lahti, M., and Lewis, B. F. (1997). The effects of planned duration of residential drug abuse treatment on recovery and HIV risk behavior. *American Journal of Public Health,* **87**, 1637–1644.

McCusker, J., Bigelow, C., Vickers-Lahti, M., Spotts, D., Garfield, F., and Frost, R. (1997). Planned duration of residential drug abuse treatment: efficacy versus treatment. *Addiction*, **92**, 1467–1478.

McFadden, D. (1974). Conditional logit analysis of qualitative choice behavior, pp105–142 in *Frontiers in Econometrics*, Edited by P. Zarembka, Academic Press, New York.

Mehta, C. R., and Patel, N. R. (1995). Exact logistic regression: Theory and examples. *Statistics in Medicine*, **14**, 2143–2160.

Mickey, J., and Greenland, S. (1989). A study of the impact of confounder-selection criteria on effect estimation. *American Journal of Epidemiology*, **129**, 125–137.

Microsoft Corporation (1998). Microsoft® Word 98, Word Processing Program for the Apple® Macintosh™, Microsoft Corporation, Bellvue.

Miettinen, O. S. (1976). Stratification by multivariate confounder score. *American Journal of Epidemiology*, **104**, 609–620.

Mittlböck, M., and Schemper, M. (1996). Explained variation for logistic regression. *Statistics in Medicine*, **15**, 1987–1997.

Mittlböck, M., and Schemper, M. (1999). Computing measures of explained variation for logistic regression models. *Computer Methods and Programs in Biomedicine*, **58**, 17–24.

Moolgavkar, S., Lustbader, E., and Venzon, D. J. (1985). Assessing the adequacy of the logistic regression model for matched case–control studies. *Statistics in Medicine*, **4**, 425–435.

Moore, D. S. (1971). A chi-square test with random cell boundaries. *Annals of Mathematical Statistics*, **42**, 147–156.

Moore, D. S., and Spruill, M. C. (1975). Unified large–sample theory of general chi–square statistics for tests of fit. *Annals of Statistics*, **3**, 599–516.

Neuhaus, J. M., Kalbfleisch, J. D., and Hauck, W. W. (1991). A comparison of cluster-specific and population-average approaches for analyzing correlated binary data. *International Statistical Review*, **59**, 25-35.

Neuhaus, J. M. (1992). Statistical methods for longitudinal and clustered designs with binary data. *Statistical Methods in Medical Research*, **1**, 249-273.

Neuhaus, J. M., and Segal, M. R. (1993). Design effects for binary regression models fitted to dependent data. *Statistics in Medicine*, **12**, 1259–1268.

Neuhaus, J. M., and Jewell, N. P. (1993). A geometric approach to assess bias due to omitted covariates in generalized linear models. *Biometrika*, **80**, 807-815.

NHANES III Reference Manuals and Reports (1996). CD ROM issued by the National Center for Health Statistics/Centers for Disease Control and Prevention, 6–0178, October 1996.

Osius, G, and Rojek, D. (1992). Normal goodness-of-fit tests for multinomial models with large degrees-of-freedom. *Journal of the American Statistical Association,* **87**, 1145–1152.

PASS Users Guide (1996). PASS 6.0: Power and Sample Size for Windows. Number Cruncher Statistical Software, Kaysville, UT.

Pastides, H., Kelsey, J. L., Holford, T. R., and LiVolsi, V. A. (1985). The epidemiology of fibrocystic breast disease. *American Journal of Epidemiology*, **121**, 440–447.

Pastides, H., Kelsey, J. L., LiVolsi, V. A., Holford, T., Fischer, D., and Goldberg, I. (1983). Oral contraceptive use and fibrocystic breast disease with special reference to its histopathology. *Journal of the National Cancer Institute*, **71**, 5–9.

Peduzzi, P. N., Hardy, R. J., and Holford, T. R. (1980). A stepwise selection procedure for nonlinear regression models. *Biometrics*, **36**, 511–516.

Peduzzi, P. N., Concato, J., Kemper, E., Holford, T. R., and Feinstein, A. (1996). A simulation study of the number of events per variable in logistic regression analysis. *Journal of Clinical Epidemiology*, **99**, 1373–1379.

Pendergast, J. F., Gange, S. J., Newton, M. A., Lindstrom, M. J., Palta, M., and Fisher, M. R. (1996). A survey of methods for analyzing clustered binary response data. *International Statistical Review* **64**, 89-118.

Pierce, D. A., and Sands, B. R. (1975). Extra-Bernoulli variation in binary data. *Technical Report No. 46.* Department of Statistics, Oregon State University.

Pregibon, D. (1980). Goodness-of-link tests for generalized linear models. *Applied Statistics*, **29**, 15–24.

Pregibon, D. (1981). Logistic regression diagnostics. *Annals of Statistics*, **9**, 705–724.

Pregibon, D. (1984). Data analytic methods for matched case-control studies. *Biometrics*, **40**, 639–651.

Prentice, R. L. (1976). A generalization of the probit and logit methods for dose response curves. *Biometrics*, **32**, 761–768.

Prentice, R. L. (1986). A case-cohort design for epidemiologic cohort studies and disease prevention trials. *Biometrika*, **73**, 1–11.

Prentice, R. L., and Pyke, R. (1979). Logistic disease incidence models and case-control studies. *Biometrika*, **66**, 403–411.

Rao, C. R. (1973). *Linear Statistical Inference and Its Application*, Second Edition. Wiley, Inc., New York.

Roberts, G., Rao, J. N. K., and Kumar, S. (1987). Logistic regression analysis of sample survey data. *Biometrika*, **74**, 1–12.

Rosner, B. (1984). Multivariate methods in opthalmology with application to other paired-data situations. *Biometrics*, **40**, 1025-1035.

Rothman, K. J., and Greenland, S. (1998). *Modern Epidemiology*, Third edition. Lippincott-Raven, Philadelphia.

Royston, P., and Altman, D. G. (1994). Regression using fractional polynomials of continuous covariates: Parsimonious parametric modelling (with discussion). *Applied Statistics*, **43**, 429–467.

Royston, P., and Ambler, G. (1998). Fitting generalized additive models in Stata. *StataTechnical Bulletin,* **STB–42,** 38–43.

Royston, P., and Ambler, G. (1998). Multivariable fractional polynomials. *Stata Technical Bulletin,* **STB–43**, 24–32.

Royston, P., and Ambler, G. (1999). Mulivariable fractional poynomials: update. *Stata Technical Bulletin,* **STB–49,** 17–22.

Ryan, L. M. (1992). Quantitative risk assessment for developmental toxicity, *Biometrics*, **48**, 163–174.

Ryan, T. (1997). *Modern Regression Methods.* Wiley, Inc., New York.

Santner, T. J., and Duffy, D. E. (1986). A note on A. Albert's and J. A. Anderson's conditions for the existence of maximum likelihood estimates in logistic regression models. *Biometrika*, **73**, 755–758.

SAS Institute Inc. (1999). *SAS Guide for Personal Computers, Version 6.12.* SAS Institute Inc., Cary, NC.

SAS Institute Inc. (2000). *SAS Guide for Personal Computers, Version 8.0.* SAS Institute Inc., Cary, NC.

Sauerbrei, W., and Royston, P. (1999). Building multivariable prognostic and diagnostic models: Transformations of the predictors using fractional polynomials. *Journal of the Royal Statistical Society, Series* A, **162**, 71–94.

Schaefer, R. L. (1986). Alternative estimators in logistic regression when the data are collinear. *Journal of Statistical Computation and Simulation*, **25**, 75–91.

Schlesselman, J. J. (1982). *Case-Control Studies.* Oxford University Press, New York.

Scott, A. J., and Wild, C. J. (1986). Fitting models under case-control or choice based sampling. *Journal of the Royal Statistical Association, Series B*, **48**, 170–182.

Shah, B. V., Barnwell, B. G., and Bieler, G. S. (1996). SUDAAN User's Manual, Release 7.0. Research Triangle Park, NC: Research Triangle Institute.

Skinner, C. J., Holt, D., and Smith, T. M. F. (1989). *Analysis of Complex Surveys.* Wiley, Inc., New York.

SPSS Statistical Software (1998). Version 9.0 for Windows. SPSS Inc. Chicago.

StataCorp (1999). Stata Statistical Software: Release 6.0. Stata Corporation, College Station, TX.

Stukel, T. A. (1988). Generalized logistic models. *Journal of the American Statistical Association*, **83**, 426–431.

Su, J. Q., and Wei, L. J. (1991). A lack-of-fit test for the mean function in a generalized linear model. *Journal of the American Statistical Association*, **86**, 420–426.

Surgeon General (1964). *Smoking and Health. Report on the Advisory Committee to the Surgeon General of the Public Heath Service,* U.S. Department of Health, Education and Welfare, Washington, D.C.

Tarone, R. E. (1985). On heterogeneity tests based on efficient scores. *Biometrika*, **72**, 91–95.

Thomas, D. R., and Rao, J. N. K. (1987). Small-sample comparisons of level and power for simple goodness-of-fit statistics under cluster sampling. *Journal of the American Statistical Association*, **82**, 630–636.

Tritchler, D. (1984). An algorithm for exact logistic regression. *Journal of the American Statistical Association*, **79**, 709–711.

Truett, J., Cornfield, J., and Kannel, W. (1967). A multivariate analysis of the risk of coronary heart disease in Framingham. *Journal of Chronic Diseases*, **20**, 511–524.

Tsiatis, A. A. (1980). A note on a goodness-of-fit test for the logistic regression model. *Biometrika*, **67**, 250–251.

Weesie, J. (1998). Windmeijer's goodness-of-fit test for logistic regression. *Stata Technical Bulletin,* **STB–44,** 22–27.

White, H. (1982). Maximum likelihood estimation of misspecified models. *Econometrika*, **50**, 1–25.

White, H. (1989). *Estimation, Inference and Specification Analysis.* Cambridge University Press, New York.

Whitemore, A. S. (1981). Sample size for logistic regression with small response probability. *Journal of the American Statistical Association*, **76**, 27-32.

Williams, D. (1975). The analysis of binary responses from toxicological experiments involving reproduction and teratogenicity. *Biometrics*, **31**, 949–952.

Williams, D. A. (1982). Extra-binomial variation in linear logistic models. *Applied Statistics*, **31**, 144–148.

Windmeijer, F. A. G. (1990). The asymptotic distribution of the sum of weighted squared residuals in binary choice models. *Statistica Neerlandica*, **44**, 69–78.

Wolfe, R. (1998). Continuation-ratio models for ordinal response data. *Stata Technical Bulletin*, **STB-44**, 18–21.

Xu, H. (1996). Extensions of the Hosmer-Lemeshow goodness-of-fit test. Unpublished Masters Thesis, School of Public Health and Health Sciences, University of Massachusetts, Amherst, MA.

Zapka, J. G., Stoddard, A. M., Maul, L., and Costanza, M. E. (1991). Interval adherence to breast cancer screening by mammography in women over 50. *Medical Care*, **29**, 697–707.

Zapka, J. G., Hosmer, D. W., Harris, D. E., Costanza, M. E., and Stoddard, A. M. (1992). Change in mammography use: economic, need and services factors. *American Journal of Public Health,* **82**, 1345–1351.

Zeger, S. L., Liang, K-Y., and Albert, P. A. (1988). Models for longitudinal data: A generalized estimating equation approach. *Biometrics*, **44**, 1049-1060.

Zhang, B. (1999). A chi-squared goodness-of-fit for logistic regression models based on case-control data. *Biometrics*, **86**, 531–539.

Zhang, J., and Yu, K. F. (1998). What's the relative risk? *Journal of the American Medical Association*, **280**, 1690–1691.

Index

WILEY SERIES IN PROBABILITY AND STATISTICS
ESTABLISHED BY WALTER A. SHEWHART AND SAMUEL S. WILKS

Probability and Statistics Section

*ANDERSON · The Statistical Analysis of Time Series
ARNOLD, BALAKRISHNAN, and NAGARAJA · A First Course in Order Statistics
ARNOLD, BALAKRISHNAN, and NAGARAJA · Records
BACCELLI, COHEN, OLSDER, and QUADRAT · Synchronization and Linearity: An Algebra for Discrete Event Systems
BARNETT · Comparative Statistical Inference, *Third Edition*
BASILEVSKY · Statistical Factor Analysis and Related Methods: Theory and Applications
BERNARDO and SMITH · Bayesian Statistical Concepts and Theory
BILLINGSLEY · Convergence of Probability Measures, *Second Edition*
BOROVKOV · Asymptotic Methods in Queuing Theory
BOROVKOV · Ergodicity and Stability of Stochastic Processes
BRANDT, FRANKEN, and LISEK · Stationary Stochastic Models
CAINES · Linear Stochastic Systems
CAIROLI and DALANG · Sequential Stochastic Optimization
CONSTANTINE · Combinatorial Theory and Statistical Design
COOK · Regression Graphics
COVER and THOMAS · Elements of Information Theory
CSÖRGŐ and HORVÁTH · Weighted Approximations in Probability Statistics
CSÖRGŐ and HORVÁTH · Limit Theorems in Change Point Analysis
*DANIEL · Fitting Equations to Data: Computer Analysis of Multifactor Data, *Second Edition*
DETTE and STUDDEN · The Theory of Canonical Moments with Applications in Statistics, Probability, and Analysis
DEY and MUKERJEE · Fractional Factorial Plans
*DOOB · Stochastic Processes
DRYDEN and MARDIA · Statistical Shape Analysis
DUPUIS and ELLIS · A Weak Convergence Approach to the Theory of Large Deviations
ETHIER and KURTZ · Markov Processes: Characterization and Convergence
FELLER · An Introduction to Probability Theory and Its Applications, Volume I, *Third Edition,* Revised; Volume II, *Second Edition*
FULLER · Introduction to Statistical Time Series, *Second Edition*
FULLER · Measurement Error Models
GHOSH, MUKHOPADHYAY, and SEN · Sequential Estimation
GIFI · Nonlinear Multivariate Analysis
GUTTORP · Statistical Inference for Branching Processes
HALL · Introduction to the Theory of Coverage Processes
HAMPEL · Robust Statistics: The Approach Based on Influence Functions
HANNAN and DEISTLER · The Statistical Theory of Linear Systems
HUBER · Robust Statistics

*Now available in a lower priced paperback edition in the Wiley Classics Library.

Probability and Statistics (Continued)

HUSKOVA, BERAN, and DUPAC · Collected Works of Jaroslav Hajek—with Commentary
IMAN and CONOVER · A Modern Approach to Statistics
JUREK and MASON · Operator-Limit Distributions in Probability Theory
KASS and VOS · Geometrical Foundations of Asymptotic Inference
KAUFMAN and ROUSSEEUW · Finding Groups in Data: An Introduction to Cluster Analysis
KELLY · Probability, Statistics, and Optimization
KENDALL, BARDEN, CARNE, and LE · Shape and Shape Theory
LINDVALL · Lectures on the Coupling Method
McFADDEN · Management of Data in Clinical Trials
MANTON, WOODBURY, and TOLLEY · Statistical Applications Using Fuzzy Sets
MORGENTHALER and TUKEY · Configural Polysampling: A Route to Practical Robustness
MUIRHEAD · Aspects of Multivariate Statistical Theory
OLIVER and SMITH · Influence Diagrams, Belief Nets and Decision Analysis
*PARZEN · Modern Probability Theory and Its Applications
PRESS · Bayesian Statistics: Principles, Models, and Applications
PUKELSHEIM · Optimal Experimental Design
RAO · Asymptotic Theory of Statistical Inference
RAO · Linear Statistical Inference and Its Applications, *Second Edition*
RAO and SHANBHAG · Choquet-Deny Type Functional Equations with Applications to Stochastic Models
ROBERTSON, WRIGHT, and DYKSTRA · Order Restricted Statistical Inference
ROGERS and WILLIAMS · Diffusions, Markov Processes, and Martingales, Volume I: Foundations, *Second Edition;* Volume II: Îto Calculus
RUBINSTEIN and SHAPIRO · Discrete Event Systems: Sensitivity Analysis and Stochastic Optimization by the Score Function Method
RUZSA and SZEKELY · Algebraic Probability Theory
SCHEFFE · The Analysis of Variance
SEBER · Linear Regression Analysis
SEBER · Multivariate Observations
SEBER and WILD · Nonlinear Regression
SERFLING · Approximation Theorems of Mathematical Statistics
SHORACK and WELLNER · Empirical Processes with Applications to Statistics
SMALL and McLEISH · Hilbert Space Methods in Probability and Statistical Inference
STAPLETON · Linear Statistical Models
STAUDTE and SHEATHER · Robust Estimation and Testing
STOYANOV · Counterexamples in Probability
TANAKA · Time Series Analysis: Nonstationary and Noninvertible Distribution Theory
THOMPSON and SEBER · Adaptive Sampling
WELSH · Aspects of Statistical Inference
WHITTAKER · Graphical Models in Applied Multivariate Statistics
YANG · The Construction Theory of Denumerable Markov Processes

Applied Probability and Statistics Section

ABRAHAM and LEDOLTER · Statistical Methods for Forecasting
AGRESTI · Analysis of Ordinal Categorical Data
AGRESTI · Categorical Data Analysis

*Now available in a lower priced paperback edition in the Wiley Classics Library.

Applied Probability and Statistics (Continued)

ANDERSON, AUQUIER, HAUCK, OAKES, VANDAELE, and WEISBERG · Statistical Methods for Comparative Studies
ARMITAGE and DAVID (editors) · Advances in Biometry
*ARTHANARI and DODGE · Mathematical Programming in Statistics
ASMUSSEN · Applied Probability and Queues
*BAILEY · The Elements of Stochastic Processes with Applications to the Natural Sciences
BARNETT and LEWIS · Outliers in Statistical Data, *Third Edition*
BARTHOLOMEW, FORBES, and McLEAN · Statistical Techniques for Manpower Planning, *Second Edition*
BASU and RIGDON · Statistical Methods for the Reliability of Repairable Systems
BATES and WATTS · Nonlinear Regression Analysis and Its Applications
BECHHOFER, SANTNER, and GOLDSMAN · Design and Analysis of Experiments for Statistical Selection, Screening, and Multiple Comparisons
BELSLEY · Conditioning Diagnostics: Collinearity and Weak Data in Regression
BELSLEY, KUH, and WELSCH · Regression Diagnostics: Identifying Influential Data and Sources of Collinearity
BHAT · Elements of Applied Stochastic Processes, *Second Edition*
BHATTACHARYA and WAYMIRE · Stochastic Processes with Applications
BIRKES and DODGE · Alternative Methods of Regression
BLISCHKE AND MURTHY · Reliability: Modeling, Prediction, and Optimization
BLOOMFIELD · Fourier Analysis of Time Series: An Introduction, *Second Edition*
BOLLEN · Structural Equations with Latent Variables
BOULEAU · Numerical Methods for Stochastic Processes
BOX · Bayesian Inference in Statistical Analysis
BOX and DRAPER · Empirical Model-Building and Response Surfaces
*BOX and DRAPER · Evolutionary Operation: A Statistical Method for Process Improvement
BUCKLEW · Large Deviation Techniques in Decision, Simulation, and Estimation
BUNKE and BUNKE · Nonlinear Regression, Functional Relations and Robust Methods: Statistical Methods of Model Building
CHATTERJEE and HADI · Sensitivity Analysis in Linear Regression
CHERNICK · Bootstrap Methods: A Practitioner's Guide
CHILÈS and DELFINER · Geostatistics: Modeling Spatial Uncertainty
CHOW and LIU · Design and Analysis of Clinical Trials: Concepts and Methodologies
CLARKE and DISNEY · Probability and Random Processes: A First Course with Applications, *Second Edition*
*COCHRAN and COX · Experimental Designs, *Second Edition*
CONOVER · Practical Nonparametric Statistics, *Second Edition*
CORNELL · Experiments with Mixtures, Designs, Models, and the Analysis of Mixture Data, *Second Edition*
*COX · Planning of Experiments
CRESSIE · Statistics for Spatial Data, *Revised Edition*
DANIEL · Applications of Statistics to Industrial Experimentation
DANIEL · Biostatistics: A Foundation for Analysis in the Health Sciences, *Sixth Edition*
DAVID · Order Statistics, *Second Edition*
*DEGROOT, FIENBERG, and KADANE · Statistics and the Law
DODGE · Alternative Methods of Regression
DOWDY and WEARDEN · Statistics for Research, *Second Edition*
DUNN and CLARK · Applied Statistics: Analysis of Variance and Regression, *Second Edition*
*ELANDT-JOHNSON and JOHNSON · Survival Models and Data Analysis
*FLEISS · The Design and Analysis of Clinical Experiments

*Now available in a lower priced paperback edition in the Wiley Classics Library.

Applied Probability and Statistics (Continued)

FLEISS · Statistical Methods for Rates and Proportions, *Second Edition*
FLEMING and HARRINGTON · Counting Processes and Survival Analysis
GALLANT · Nonlinear Statistical Models
GLASSERMAN and YAO · Monotone Structure in Discrete-Event Systems
GNANADESIKAN · Methods for Statistical Data Analysis of Multivariate Observations, *Second Edition*
GOLDSTEIN and LEWIS · Assessment: Problems, Development, and Statistical Issues
GREENWOOD and NIKULIN · A Guide to Chi-Squared Testing
*HAHN · Statistical Models in Engineering
HAHN and MEEKER · Statistical Intervals: A Guide for Practitioners
HAND · Construction and Assessment of Classification Rules
HAND · Discrimination and Classification
HEIBERGER · Computation for the Analysis of Designed Experiments
HEDAYAT and SINHA · Design and Inference in Finite Population Sampling
HINKELMAN and KEMPTHORNE: · Design and Analysis of Experiments, Volume 1: Introduction to Experimental Design
HOAGLIN, MOSTELLER, and TUKEY · Exploratory Approach to Analysis of Variance
HOAGLIN, MOSTELLER, and TUKEY · Exploring Data Tables, Trends and Shapes
HOAGLIN, MOSTELLER, and TUKEY · Understanding Robust and Exploratory Data Analysis
HOCHBERG and TAMHANE · Multiple Comparison Procedures
HOCKING · Methods and Applications of Linear Models: Regression and the Analysis of Variables
HOGG and KLUGMAN · Loss Distributions
HOSMER and LEMESHOW · Applied Logistic Regression, *Second Edition*
HØYLAND and RAUSAND · System Reliability Theory: Models and Statistical Methods
HUBERTY · Applied Discriminant Analysis
HUNT and KENNEDY · Financial Derivatives in Theory and Practice
JACKSON · A User's Guide to Principle Components
JOHN · Statistical Methods in Engineering and Quality Assurance
JOHNSON · Multivariate Statistical Simulation
JOHNSON and KOTZ · Distributions in Statistics
JOHNSON, KOTZ, and BALAKRISHNAN · Continuous Univariate Distributions, Volume 1, *Second Edition*
JOHNSON, KOTZ, and BALAKRISHNAN · Continuous Univariate Distributions, Volume 2, *Second Edition*
JOHNSON, KOTZ, and BALAKRISHNAN · Discrete Multivariate Distributions
JOHNSON, KOTZ, and KEMP · Univariate Discrete Distributions, *Second Edition*
JUREČKOVÁ and SEN · Robust Statistical Procedures: Aymptotics and Interrelations
KADANE · Bayesian Methods and Ethics in a Clinical Trial Design
KADANE AND SCHUM · A Probabilistic Analysis of the Sacco and Vanzetti Evidence
KALBFLEISCH and PRENTICE · The Statistical Analysis of Failure Time Data
KELLY · Reversability and Stochastic Networks
KHURI, MATHEW, and SINHA · Statistical Tests for Mixed Linear Models
KLUGMAN, PANJER, and WILLMOT · Loss Models: From Data to Decisions
KLUGMAN, PANJER, and WILLMOT · Solutions Manual to Accompany Loss Models: From Data to Decisions
KOTZ, BALAKRISHNAN, and JOHNSON · Continuous Multivariate Distributions, Volume 1, *Second Edition*
KOVALENKO, KUZNETZOV, and PEGG · Mathematical Theory of Reliability of Time-Dependent Systems with Practical Applications
LACHIN · Biostatistical Methods: The Assessment of Relative Risks

*Now available in a lower priced paperback edition in the Wiley Classics Library.

Applied Probability and Statistics (Continued)

LAD · Operational Subjective Statistical Methods: A Mathematical, Philosophical, and Historical Introduction
LANGE, RYAN, BILLARD, BRILLINGER, CONQUEST, and GREENHOUSE · Case Studies in Biometry
LAWLESS · Statistical Models and Methods for Lifetime Data
LEE · Statistical Methods for Survival Data Analysis, *Second Edition*
LEPAGE and BILLARD · Exploring the Limits of Bootstrap
LINHART and ZUCCHINI · Model Selection
LITTLE and RUBIN · Statistical Analysis with Missing Data
LLOYD · The Statistical Analysis of Categorical Data
MAGNUS and NEUDECKER · Matrix Differential Calculus with Applications in Statistics and Econometrics, *Revised Edition*
MALLER and ZHOU · Survival Analysis with Long Term Survivors
MANN, SCHAFER, and SINGPURWALLA · Methods for Statistical Analysis of Reliability and Life Data
McLACHLAN and KRISHNAN · The EM Algorithm and Extensions
McLACHLAN · Discriminant Analysis and Statistical Pattern Recognition
McNEIL · Epidemiological Research Methods
MEEKER and ESCOBAR · Statistical Methods for Reliability Data
*MILLER · Survival Analysis, *Second Edition*
MONTGOMERY and PECK · Introduction to Linear Regression Analysis, *Second Edition*
MYERS and MONTGOMERY · Response Surface Methodology: Process and Product in Optimization Using Designed Experiments
NELSON · Accelerated Testing, Statistical Models, Test Plans, and Data Analyses
NELSON · Applied Life Data Analysis
OCHI · Applied Probability and Stochastic Processes in Engineering and Physical Sciences
OKABE, BOOTS, and SUGIHARA · Spatial Tesselations: Concepts and Applications of Voronoi Diagrams
PANKRATZ · Forecasting with Dynamic Regression Models
PANKRATZ · Forecasting with Univariate Box-Jenkins Models: Concepts and Cases
PIANTADOSI · Clinical Trials: A Methodologic Perspective
PORT · Theoretical Probability for Applications
PUTERMAN · Markov Decision Processes: Discrete Stochastic Dynamic Programming
RACHEV · Probability Metrics and the Stability of Stochastic Models
RÉNYI · A Diary on Information Theory
RIPLEY · Spatial Statistics
RIPLEY · Stochastic Simulation
ROLSKI, SCHMIDLI, SCHMIDT, and TEUGELS · Stochastic Processes for Insurance and Finance
ROUSSEEUW and LEROY · Robust Regression and Outlier Detection
RUBIN · Multiple Imputation for Nonresponse in Surveys
RUBINSTEIN · Simulation and the Monte Carlo Method
RUBINSTEIN and MELAMED · Modern Simulation and Modeling
RYAN · Statistical Methods for Quality Improvement, *Second Edition*
SCHIMEK · Smoothing and Regression: Approaches, Computation, and Application
SCHUSS · Theory and Applications of Stochastic Differential Equations
SCOTT · Multivariate Density Estimation: Theory, Practice, and Visualization
*SEARLE · Linear Models
SEARLE · Linear Models for Unbalanced Data
SEARLE, CASELLA, and McCULLOCH · Variance Components
SENNOTT · Stochastic Dynamic Programming and the Control of Queueing Systems
STOYAN, KENDALL, and MECKE · Stochastic Geometry and Its Applications, *Second Edition*

*Now available in a lower priced paperback edition in the Wiley Classics Library.

Applied Probability and Statistics (Continued)

STOYAN and STOYAN · Fractals, Random Shapes and Point Fields: Methods of Geometrical Statistics
THOMPSON · Empirical Model Building
THOMPSON · Sampling
THOMPSON · Simulation: A Modeler's Approach
TIJMS · Stochastic Modeling and Analysis: A Computational Approach
TIJMS · Stochastic Models: An Algorithmic Approach
TITTERINGTON, SMITH, and MAKOV · Statistical Analysis of Finite Mixture Distributions
UPTON and FINGLETON · Spatial Data Analysis by Example, Volume 1: Point Pattern and Quantitative Data
UPTON and FINGLETON · Spatial Data Analysis by Example, Volume II: Categorical and Directional Data
VAN RIJCKEVORSEL and DE LEEUW · Component and Correspondence Analysis
VIDAKOVIC · Statistical Modeling by Wavelets
WEISBERG · Applied Linear Regression, *Second Edition*
WESTFALL and YOUNG · Resampling-Based Multiple Testing: Examples and Methods for *p*-Value Adjustment
WHITTLE · Systems in Stochastic Equilibrium
WOODING · Planning Pharmaceutical Clinical Trials: Basic Statistical Principles
WOOLSON · Statistical Methods for the Analysis of Biomedical Data
*ZELLNER · An Introduction to Bayesian Inference in Econometrics

Texts and References Section

AGRESTI · An Introduction to Categorical Data Analysis
ANDERSON · An Introduction to Multivariate Statistical Analysis, *Second Edition*
ANDERSON and LOYNES · The Teaching of Practical Statistics
ARMITAGE and COLTON · Encyclopedia of Biostatistics: Volumes 1 to 6 with Index
BARTOSZYNSKI and NIEWIADOMSKA-BUGAJ · Probability and Statistical Inference
BENDAT and PIERSOL · Random Data: Analysis and Measurement Procedures, *Third Edition*
BERRY, CHALONER, and GEWEKE · Bayesian Analysis in Statistics and Econometrics: Essays in Honor of Arnold Zellner
BHATTACHARYA and JOHNSON · Statistical Concepts and Methods
BILLINGSLEY · Probability and Measure, *Second Edition*
BOX · R. A. Fisher, the Life of a Scientist
BOX, HUNTER, and HUNTER · Statistics for Experimenters: An Introduction to Design, Data Analysis, and Model Building
BOX and LUCEÑO · Statistical Control by Monitoring and Feedback Adjustment
BROWN and HOLLANDER · Statistics: A Biomedical Introduction
CHATTERJEE and PRICE · Regression Analysis by Example, *Third Edition*
COOK and WEISBERG · Applied Regression Including Computing and Graphics
COOK and WEISBERG · An Introduction to Regression Graphics
COX · A Handbook of Introductory Statistical Methods
DILLON and GOLDSTEIN · Multivariate Analysis: Methods and Applications
*DODGE and ROMIG · Sampling Inspection Tables, *Second Edition*
DRAPER and SMITH · Applied Regression Analysis, *Third Edition*
DUDEWICZ and MISHRA · Modern Mathematical Statistics
DUNN · Basic Statistics: A Primer for the Biomedical Sciences, *Second Edition*
EVANS, HASTINGS, and PEACOCK · Statistical Distributions, *Third Edition*
FISHER and VAN BELLE · Biostatistics: A Methodology for the Health Sciences

*Now available in a lower priced paperback edition in the Wiley Classics Library.

Texts and References (Continued)

FREEMAN and SMITH · Aspects of Uncertainty: A Tribute to D. V. Lindley
GROSS and HARRIS · Fundamentals of Queueing Theory, *Third Edition*
HALD · A History of Probability and Statistics and their Applications Before 1750
HALD · A History of Mathematical Statistics from 1750 to 1930
HELLER · MACSYMA for Statisticians
HOEL · Introduction to Mathematical Statistics, *Fifth Edition*
HOLLANDER and WOLFE · Nonparametric Statistical Methods, *Second Edition*
HOSMER and LEMESHOW · Applied Logistic Regression, *Second Edition*
HOSMER and LEMESHOW · Applied Survival Analysis: Regression Modeling of Time to Event Data
JOHNSON and BALAKRISHNAN · Advances in the Theory and Practice of Statistics: A Volume in Honor of Samuel Kotz
JOHNSON and KOTZ (editors) · Leading Personalities in Statistical Sciences: From the Seventeenth Century to the Present
JUDGE, GRIFFITHS, HILL, LÜTKEPOHL, and LEE · The Theory and Practice of Econometrics, *Second Edition*
KHURI · Advanced Calculus with Applications in Statistics
KOTZ and JOHNSON (editors) · Encyclopedia of Statistical Sciences: Volumes 1 to 9 with Index
KOTZ and JOHNSON (editors) · Encyclopedia of Statistical Sciences: Supplement Volume
KOTZ, REED, and BANKS (editors) · Encyclopedia of Statistical Sciences: Update Volume 1
KOTZ, REED, and BANKS (editors) · Encyclopedia of Statistical Sciences: Update Volume 2
LAMPERTI · Probability: A Survey of the Mathematical Theory, *Second Edition*
LARSON · Introduction to Probability Theory and Statistical Inference, *Third Edition*
LE · Applied Categorical Data Analysis
LE · Applied Survival Analysis
MALLOWS · Design, Data, and Analysis by Some Friends of Cuthbert Daniel
MARDIA · The Art of Statistical Science: A Tribute to G. S. Watson
MASON, GUNST, and HESS · Statistical Design and Analysis of Experiments with Applications to Engineering and Science
MURRAY · X-STAT 2.0 Statistical Experimentation, Design Data Analysis, and Nonlinear Optimization
PURI, VILAPLANA, and WERTZ · New Perspectives in Theoretical and Applied Statistics
RENCHER · Linear Models in Statistics
RENCHER · Methods of Multivariate Analysis
RENCHER · Multivariate Statistical Inference with Applications
ROSS · Introduction to Probability and Statistics for Engineers and Scientists
ROHATGI · An Introduction to Probability Theory and Mathematical Statistics
RYAN · Modern Regression Methods
SCHOTT · Matrix Analysis for Statistics
SEARLE · Matrix Algebra Useful for Statistics
STYAN · The Collected Papers of T. W. Anderson: 1943–1985
TIERNEY · LISP-STAT: An Object-Oriented Environment for Statistical Computing and Dynamic Graphics
WONNACOTT and WONNACOTT · Econometrics, *Second Edition*
WU and HAMADA · Experiments: Planning, Analysis, and Parameter Design Optimization

*Now available in a lower priced paperback edition in the Wiley Classics Library.

WILEY SERIES IN PROBABILITY AND STATISTICS
ESTABLISHED BY WALTER A. SHEWHART AND SAMUEL S. WILKS

Survey Methodology Section

BIEMER, GROVES, LYBERG, MATHIOWETZ, and SUDMAN · Measurement Errors in Surveys
COCHRAN · Sampling Techniques, *Third Edition*
COUPER, BAKER, BETHLEHEM, CLARK, MARTIN, NICHOLLS, and O'REILLY (editors) · Computer Assisted Survey Information Collection
COX, BINDER, CHINNAPPA, CHRISTIANSON, COLLEDGE, and KOTT (editors) · Business Survey Methods
*DEMING · Sample Design in Business Research
DILLMAN · Mail and Telephone Surveys: The Total Design Method, *Second Edition*
GROVES and COUPER · Nonresponse in Household Interview Surveys
GROVES · Survey Errors and Survey Costs
GROVES, BIEMER, LYBERG, MASSEY, NICHOLLS, and WAKSBERG · Telephone Survey Methodology
*HANSEN, HURWITZ, and MADOW · Sample Survey Methods and Theory, Volume 1: Methods and Applications
*HANSEN, HURWITZ, and MADOW · Sample Survey Methods and Theory, Volume II: Theory
KISH · Statistical Design for Research
*KISH · Survey Sampling
KORN and GRAUBARD · Analysis of Health Surveys
LESSLER and KALSBEEK · Nonsampling Error in Surveys
LEVY and LEMESHOW · Sampling of Populations: Methods and Applications, *Third Edition*
LYBERG, BIEMER, COLLINS, de LEEUW, DIPPO, SCHWARZ, TREWIN (editors) · Survey Measurement and Process Quality
SIRKEN, HERRMANN, SCHECHTER, SCHWARZ, TANUR, and TOURANGEAU (editors) · Cognition and Survey Research

*Now available in a lower priced paperback edition in the Wiley Classics Library.